T0339496

HEAT TRANSPORT AND ENERGETICS OF THE EARTH AND ROCKY PLANETS

HEAT TRANSPORT AND ENERGETICS OF THE EARTH AND ROCKY PLANETS

ANNE M. HOFMEISTER

Department of Earth and Planetary Science, Washington University,
St. Louis, MO, United States

ELSEVIER

Elsevier
Radarweg 29, PO Box 211, 1000 AE Amsterdam, Netherlands
The Boulevard, Langford Lane, Kidlington, Oxford OX5 1GB, United Kingdom
50 Hampshire Street, 5th Floor, Cambridge, MA 02139, United States

Notices
Knowledge and best practice in this field are constantly changing. As new research and experience broaden our understanding, changes in research methods, professional practices, or medical treatment may become necessary.

Practitioners and researchers must always rely on their own experience and knowledge in evaluating and using any information, methods, compounds, or experiments described herein. In using such information or methods they should be mindful of their own safety and the safety of others, including parties for whom they have a professional responsibility.

To the fullest extent of the law, neither the Publisher nor the authors, contributors, or editors, assume any liability for any injury and/or damage to persons or property as a matter of products liability, negligence or otherwise, or from any use or operation of any methods, products, instructions, or ideas contained in the material herein.

British Library Cataloguing-in-Publication Data
A catalogue record for this book is available from the British Library

Library of Congress Cataloging-in-Publication Data
A catalog record for this book is available from the Library of Congress

ISBN: 978-0-12-818430-1

For Information on all Elsevier publications
visit our website at https://www.elsevier.com/books-and-journals

Publisher: Candice Janco
Acquisition Editor: Marissa LeFleur and Amy Shapiro
Editorial Project Manager: Kelsey Connors and Theresa Yannetty
Production Project Manager: Debasish Ghosh
Cover Designer: Greg Harris

Typeset by MPS Limited, Chennai, India

Working together
to grow libraries in
developing countries

www.elsevier.com • www.bookaid.org

Contents

II

The thermal state and evolution of Earth

6. Thermal models of the continental lithosphere

ROBERT E. CRISS

7. Thermal models of the oceanic lithosphere and upper mantle

8. Thermal structure of the lower mantle and core

III

Thermal evolution of Earth and other rocky bodies

9. Thermo-chemical evolution of the Earth

10. Thermal history of the terrestrial planets

ROBERT E. CRISS AND ANNE M. HOFMEISTER

11. A sedimentary origin for chondritic meteorites

ANNE M. HOFMEISTER AND ROBERT E. CRISS

12. Conclusions and future work

Preface

The chemical and thermal evolution of the Earth is actively debated and is of great interest to natural scientists. Heat flow in large bodies being linked to material properties, energy inventories, and mass flow is central to this problem, and is a major focus of this book. Because heat flow depends on time and space, the initial conditions are very important: The gravitational processes that formed the Solar System remain relevant today. Because much is unknown, comparing Earth to other rocky planets with ancient surfaces is very useful. Moreover, such comparisons highlight Earth's unique behavior. Critical analysis of old observations and considering recent data on heat transfer and melting of planetary materials led us to revise the geotherm, propose a new mechanism for plate tectonics, and postulate a novel origin of chondritic meteorites. This book also provides new calculations of the thermal structure of Earth's interior that are based on real material properties.

The theoretical basis for this treatment is presented in our recently published book "Measurements, Mechanisms, and Models of Heat Transport," which covers heat flow on microscopic scales and in the laboratory. Motion of heat in large bodies involves additional, macroscopic mechanisms. Body size matters for reasons other than cooling time frames. Distance and mass govern gravitational forces, which generally oppose thermal forces such as buoyancy, and created the axial spin of planets. The effect of self-gravitation on the thermal state of the Earth has been misunderstood because gravity is not considered in classical thermodynamics. Neither has the effect of length-scales on heat transport received sufficient attention, yet the physical property of thermal diffusivity and Fourier's second equation relate temperature to the lumped combination of length and time. Lastly, today's view of the Earth as a whole is mostly garnered from numerical models. Important observations have been short-shifted, despite these being numerous and constraining. In addition, some historical relics impede progress: for example, Kevin's disproven concept that contraction produces starlight still exists in the faulty notion that planetary core formation releases enormous amounts of heat.

To better understand the interior of the Earth (which is a moderately hot, evolving, and self-gravitating body) requires addressing the above issues and others. Although much relevant material exists, it is scattered among the literature of diverse fields, including materials science, geoscience, planetary science, and astronomy. This disorganization has caused key links to be overlooked, which has fostered misunderstandings and precluded the development of a coherent picture. Moreover, recent discoveries regarding microscopic behavior, which are the focus of "Measurements, Mechanisms, and Models of Heat Transport," have not been applied to geophysical problems.

The present book develops new concepts that reconcile available data with physical principles using analytical mathematics, while recognizing that the interiors of Earth and other rocky bodies are mostly hidden from view. In reviewing the literature for the book and over the past several years, we were quite surprised at the large number of inconsistencies between observations and many current models in geophysics, planetary science, and related fields. These inconsistencies appear to be stem from overspecialization, overextension, and placing too much trust on ideas developed hundreds of years ago when data were few and uncertain. Concepts in the book are based on what is truly known rather than on what is inferred or believed.

The book is organized in 3 parts. Part I presents relevant observations on the Earth which is the best constrained planet and the focus of the book. Processes that move heat on small and large scales are covered, and the large energy reservoirs generated by radionuclides, axial spin, and externally imposed gravitational forces are evaluated. Mathematical formulations useful for analyzing bodies of diverse size are included. We explain how the core formed first, due to magnetic interactions, friction welding, and cold welding. We propose that motions of the tectonic plates are driven by the 3-dimensional, time-dependent, gravitational configuration of the Moon-Earth-Sun system. We address why plate tectonics is unique to the Earth. Our proposal is based on the reality that gravitational forces, not heat, are responsible for practically all large-scale motion in the universe. Although the focus is on the behavior of Earth's interior, much of the discussion in Part I is general.

Part II delves into the current thermal state of the Earth by combining models and observations. It critically reviews existing paradigms, many of which are based on problematic mathematics. This part is organized from Earth's surface down, because the outermost layers are the best constrained. Whereas the continental crust can be treated as 1-dimensional, the oceanic crust is a 2-dimensional feature, and the mantle and core must be treated in spherical coordinates. Hence, these regions are treated in separate chapters.

Part III first discusses the thermal history of Earth, and continues in a separate chapter that compares Earth to the other rocky bodies of the Solar System, including the Moon. This chapter completes the picture of heat flow in large rocky bodies while exploring examples of what the Earth is not. The smallest bodies, chondrules, are covered separately to constrain the initial conditions, which set the stage for thermal evolution. The final chapter summarizes.

The quality and clarity of the book were substantially improved by review. Most chapters were critically reviewed by Bob Criss (Washington University, St. Louis), who is co-author on several. Both Bob and Everett Criss (Panasonic Avionics, Inc., Irvine, CA) contributed many ideas and crucial discussion. Everett co-authored one of the most important chapters in the book, and alerted me to cold welding. Reinhardt Criss (Manufacturing Technology, Inc. South Bend, IN) suggested the likelihood of friction welding. Bob's efforts on thermally modeling the continents and in comparing the terrestrial planets were essential to completing the book and providing its coherency.

Preparation of this book was supported by my active NSF grant EAR-1524495. However, the content of the book, the findings and opinions, are those of the authors, but not necessarily of NSF. Preparation was aided by the hard-working librarians Ryan Wallace and Clara McLeod at Washington U. Many websites are referenced per the suggestion of Genevieve Criss, chemistry teacher at Young Women's College Preparatory High School, Rochester, NY. This work is based on two decades of work in heat transfer, and my longer career in spectroscopy, with a recent focus on gravitation and applications of physical principles to the Earth and planets, in collaboration with Bob and Everett Criss.

Much appreciation is also due to the staff at Elsevier, for guidance and patience. I thank Marisa LaFleur, the acquisition editor, for inviting me to contribute and for obtaining helpful reviews. Acquisitions editor Amy Shapiro helped in the subsequent stages. Special thanks are due to the developmental editors, Theresa Yannetty and Kelsey Connors, for encouragement and substantial efforts in bringing the book to the finish line. Narmatha Mohan helped with permissions. I thank the production team headed by Debasish Ghosh for the impressive appearance of the final product.

Lastly, the book provides alternative hypotheses to explain observations in the geologic sciences and related fields, especially where controversy exists. These ideas are all related to the flow of heat. Last but not least, I thank Gillian Foulger (University Durham, UK) and Warren Hamilton (Colorado School of Mines) for stimulating discussions and encouragement. We hope that the reader will be stimulated to think critically, and beyond what is proposed here.

Anne M. Hofmeister
St. Louis, MO, United States
July 23, 2019

PART I

Data and basic theory

Observational constraints on the thermal and compositional structure of the earth

Heat Transport and Energetics of the Earth and Rocky Planets
DOI: https://doi.org/10.1016/B978-0-12-818430-1.00001-X

3

"All we know are the facts, ma'am." *Sargent Joe Friday (Dragnet TV series, 1951−1959)*

To solve problems in heat transfer, information is needed on system temperature, heat flux, heat sources, and material properties, specifically the heat capacity, density, and either the thermal conductivity or thermal diffusivity. The size of the body and geometry are also important, as are the presence of internal layers, regional inhomogeneity, and phase transitions. This information defines the boundary conditions that describe the body, which are used to infer interior temperatures and their change over time. Whether a system is currently steady-state or evolving is important. For the latter case, initial conditions are germane to deciphering thermal evolution. Complications arise when internal masses are in motion, as well as heat. A microscopic picture of heat flow that is based on laboratory measurements is covered in Hofmeister (2019). Discussion and equations are only repeated here if essential.

Most bodies in the Solar System have ancient, heavily cratered, inactive surfaces with many commonalities, although little is known about their inaccessible interiors. Earth is currently very active and is unlike any other large rocky body. The complexity of Earth's volcanic and tectonic activity, which are central to its heat and mass transfer processes, has fueled considerable debate about its internal workings. Therefore, much effort has been directed to constrain various properties of the interior and of the surface. Interpreting the copious, yet incomplete, information needed to construct a thermal model is daunting, especially insofar as Earth's thermal, chemical, and physical evolution are linked. This is the case because the generation and ascent of magma have created compositional layering and fostered the upward concentration of the heat-emitting isotopes, which location enhances cooling (e.g., Hofmeister and Criss, 2013). To make significant inroads into this complicated problem requires deciphering what information provides the most useful and accurate constraints, and focusing on that information.

Historically, the focus has been on the surface heat flux (e.g., MacDonald, 1959) and its possible connection with internal processes. Surface heat flux, along with surface temperature, material properties, and a model (e.g., Fourier's equation) has been used to ascertain interior temperatures. However, temperatures at a few interior locations have been established by means that are independent of thermal models. Important constraints imposed by isothermal phase boundaries have been overlooked. Much evidence has been subjectively interpreted in view of the current paradigm for plate motion, which involves convection of the entire, solid, and presumably chemically homogenous mantle. Yet, many first order problems persist, as even acknowledged by proponents of whole mantle convection (Bercovici, 2015).

Accordingly, this chapter discusses information relevant to heat flow and assumptions made during interpretation. Information is needed on the materials in Earth's internal layers, because a layered configuration influences conduction and impedes convection. Averages are relevant to Earth's behavior as a whole, even though the uppermost layers are dichotomous, and data from many research areas in Earth science pertain, so we have relied on books and review volumes (e.g., Anderson, 2007; Turekian and Holland, 2014).

Section 1.1 discusses constraints on the internal structure of the Earth provided by seismology. Section 1.2 describes constraints on the outermost layers provided by the rock record, and what can be inferred about the interior from meteorites and comets. Section 1.3 describes how isothermal layers are created and how they can buffer temperatures, using the core as an example. Section 1.4 describes data on surface gradients and flux, and uses this and data in Sections 1.1–1.3 to construct a preliminary model of Earth's thermal structure. Section 1.5 concerns the largest inventory of energy in the Earth (its spin), evidence for spin loss during differential rotation, and the amount of heat generated. As summarized here, many observations support spin, rather than mantle convection, as driving plate tectonics. Section 1.6 estimates the contribution of radionuclides to surface heat flux using recent findings. The chapters to follow further develop the ideas presented in this chapter, which summarize previous work.

1.1 Locations and descriptions of Earth's internal layers from seismology

1.1.1 Basic information

Earthquakes release large amounts of energy inside the Earth, which propagate from their source regions as acoustic waves, and are subsequently detected by distant receivers on the surface. The elapsed time (t) is the integral of the inverse velocity over the fastest path length for the various types of waves:

$$t = \int_{source}^{reciever} \frac{dl}{u(l)} \tag{1.1}$$

Because the path is unknown, models are required to provide u. Furthermore, Earth is inhomogeneous and layered, so seismology uses complicated models of the refraction and reflection of these essentially elastic waves to extract velocities. For the basics, see Stein and Wysession (2003) and Anderson (2007). Energy is lost during the propagation of seismic waves because without losses, after each event the Earth would ring like a bell for a long, long time. However, these losses (attenuation) are extremely small compared to losses during heat propagating in solids, which involves a different mechanism than acoustic waves.

Many different types of seismic waves exist, but the most important are the longitudinal compression waves and the transverse shear waves, whose respective speeds u_p and u_s can be measured in laboratory experiments. Importantly, liquids cannot transmit shear waves, because fluids flow under any stress. Seismic velocities are controlled largely by density, as long known (e.g., Shankland, 1972), and so are affected by mineral structure and orientation. Because Earth is self-gravitating and large, its interior is highly

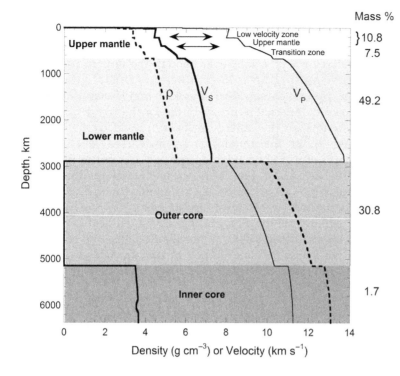

FIGURE 1.1 Layers in the earth from a global average of seismic data. PREM velocities and densities from tables in Anderson (2007). White and gray rectangles are the major regions. "Upper mantle" usually refers to the region between ~220 and 410 km, but sometimes is used to include everything above 670 km. Because an x−y plot distorts the relative sizes of the layers in a sphere, their relative masses are shown on the right. anisotropy above ~220 km is not shown in this figure.

compressed. Stability requires that density increases downwards. Fortunately, the Earth's mass is known from orbits of the Moon and artificial satellites, and information also exists on its moment of inertia and oblate shape. These constraints are important to developing accurate models of Earth's interior.

Although seismic models are being continually improved and refined, these are only approximations to a very complex reality (Stein and Wysession, 2003). Also important is that acoustic velocities (3–14 km s^{-1}) are $10^{12}\times$ faster than plate motions (~few cm year^{-1}), so seismic models provide a present-day snapshot of Earth's interior. This includes the 3-d tomographic images of the mantle. As shown in Fig. 1.1, resolution of Earth's internal character decreases with depth.

1.1.2 Compositional layers in the earth

Fig. 1.1 shows the general increases in density and the acoustic velocities with depth, which are consistent with gravitational stratification. The model shown is a global average known as PREM, the preliminary earth reference model of Dziewonski and Anderson (1981). The major features of PREM and other models were deduced in the pioneering, velocity only, model of Jeffreys and Bullen from 1940 (see Stein and Wysession, 2003). A key discovery of Jeffreys and Bullen was the existence of sharp changes in velocities at only a few depths: these seismic discontinuities define the major regions of Earth's interior (Fig. 1.1).

1.1.2.1 The two cores

The most extreme change in seismic velocity occurs at 2990 km depth, where u_s becomes null, signifying melting, as discovered by Jeffreys. From laboratory data on density and speeds, and the common occurrence of iron meteorites, the core is considered to be an iron-nickel alloy. No samples are available and tradeoffs exist in modeling the seismic velocities so the detailed composition of the core is debated, particularly as regards its contents of light elements (Li and Fei, 2014).

Further in, shear waves reappear, defining the inner core as a solid. This region is sufficiently compressed to stabilize the denser, solid phase.

1.1.2.2 The two crusts

Rocks of the topmost layer (the crust) are summarized here and detailed in Section 1.2. The crust has two distinct lateral components: oceanic and continental. This large heterogeneity and anisotropy is more than skin deep.

Oceanic crust covers 60% of the globe but only composes only $\sim 0.1\%$ of Earth's total mass. This crust typically lies 4 km below sea level, as it is composed of dense, low-silica mafic rocks. A very well-defined seismic discontinuity (the Moho) occurs 5–10 km below the ocean floor, at an average depth of 7.5 km. This large change in seismic velocity defines the transition from the crust to the upper mantle (Fig. 1.2), constituted of denser ultramafic rocks with lower silica and alumina, and higher magnesium (e.g., Anderson, 2007).

The continental crust covers 40% of the globe, when the offshore parts of the continental shelves are included. Continental crust represents a only a small fraction, $\sim 0.37\%$, of Earth's total mass, which is a bit larger than the mass of the quite thin oceanic crust. Because continental rocks have a rather low density associated with their generally high SiO_2 content (felsic minerals), the average elevation of the continents above sea level is 840 km, which excludes the submerged shelves. The rocks are varied and so is the thickness of the continental crust. Rather than a flat termination similar to the Moho, many seismic studies have suggested that the base of the continental crust lies at depths ranging from 30 to 70 km (Fig. 1.2). In contrast, 3-dimensional tomographic studies suggest that continental roots are even thicker, in places reaching the 410 km discontinuity. This depth has been revised considerably upwards by recent models of a certain type of shear wave by Tharimena et al. (2017), which indicate a 7–9% velocity drop at depths of 130–190 km beneath old continental interiors, called shields or cratons. Gradual decreases in velocities in earlier models (Fig. 1.2) provide support for the ~ 190 km limit. Thus, the continents rest on a layer of upper mantle material that is relatively thin below the centers of ancient cratons. The thick and deep roots under the cratons appear to be compositionally distinct from the upper continental crust (above ~ 30 km) that is directly probed (Section 1.2).

1.1.2.3 Top of the upper mantle

Although the profiles of u_p differ under the various tectonic regions, below a depth of 350 km, the profiles for different provinces are all similar; the same applies to profiles of u_s (Fig. 1.2). Thus, below this depth in the upper mantle (UM), all of Earth's layers appear

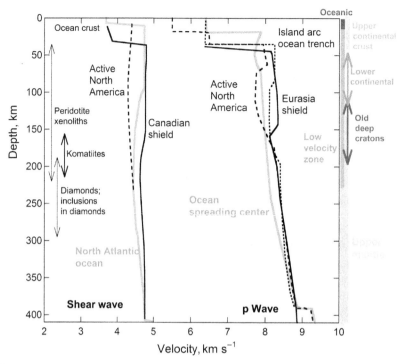

FIGURE 1.2 Uppermost layers in the earth from regional studies. Compressional velocities from Walck (1984), who modeled the ocean spreading center in the gulf of California (GCA). Dotted Line = the arc near Japan. Shear velocities from Grand and Helmberger (1984) who modeled Mexico and north, and the Atlantic in their earlier studies. Gray = oceans. Solid Black = shields. Dashes = tectonically active N. America. Double arrows to the left summarize approximate depths of origin for mantle samples. Bars to the right show layer Thicknesses, see text.

to be laterally uniform. This appears to be a global condition. Before discussing interpretations of these deep layers, a few more features of the upper layers need mentioning.

Seismic velocities in the uppermost mantle beneath oceanic crust gradually decrease with depth, but at and below about 220 km, velocities resume their normal increase with depth. The anomalous region above 220 km called the low velocity zone (LVZ). One interpretation is that the LVZ contains ~1% partial melt (Anderson, 2007), which is inferred from the increase of temperature with depth, and observations that oceanic crust forms from melt. Thus, the LVZ appears to be a weak, partially molten, but not chemically distinct part of the upper mantle.

Somewhat analogous effects occur within a narrow range of depths beneath several continents, where the shear velocity decreases with depth, contrary to its expected increase (Fig. 1.2). This effect, observed independently by Tharimena et al. (2017), appears to be analogous to the LVZ under the oceans. Upper mantle underlies both oceanic and continental crust, such that its weak, top section varies in thickness due to the thickness of the overlying layers. This finding is consistent with differential rotation indicated by plate motions (Section 1.5).

1.1.3 Interpreting velocities using laboratory data

Acoustic velocities depend on temperature, pressure, and the structure and chemical composition of the various phases in any given zone. Because the number of variables far exceeds the constraints from seismology, ascertaining the composition of Earth's interior involves tradeoffs, as is obvious from matrix algebra.

The deep interior is particularly a problem because close-packed, dense phases are expected. Oxides and silicates essentially consist of a sublattice of anions, whose structure defines the "holes" that the cations occupy. This description holds quite well at mantle pressures (Prewitt and Downs, 1998). Consequently, the pressure (P) derivative for the bulk modulus (B) vary little, where

$$\alpha = \frac{1}{V}\frac{\partial V}{\partial T}\Big|_P; \quad \beta = -\frac{1}{V}\frac{\partial V}{\partial P}\Big|_T = \frac{1}{B_T}, \tag{1.2}$$

T is temperature, V is volume, α is thermal expansivity, and β is compressibility. Nominally, $\partial B_T/\partial P = 4$. Bulk moduli are linked to density (ρ), and largely control u_p, via:

$$\rho u_p^2 = B_S + \frac{4}{3}G \quad \text{and} \quad \rho u_s^2 = G, \tag{1.3}$$

where B_S is the isentropic bulk modulus and G is the shear modulus. The bulk moduli are related:

$$B_S = B_T(1 + \alpha\gamma_{th}T); \qquad c_P = c_V(1 + \alpha\gamma_{th}T) \tag{1.4}$$

where the thermal Grüneisen parameter is defined by:

$$\gamma_{th} = \frac{\alpha B_T}{\rho c_V} = \frac{\alpha B_S}{\rho c_P}. \tag{1.5}$$

Commonly, B_S is referred to as the adiabatic bulk modulus under the incorrect belief that reversible adiabatic processes are isentropic (S is constant). This belief stems from considering the ideal gas (Fegley, 2015). Instead, from thermodynamic identities, adiabats are isentropics if pressure is externally applied to the system (Hofmeister, 2019, Equations 1.16−1.21), which describes laboratory measurements of elastic waves, but certainly not the self-compressed Earth.

The shear velocity u_S and G vary more with pressure, but $\partial G/\partial P = 1$ to 2 describes diverse phases. Bass (1995) and Hofmeister and Mao (2003) provide tabulated data, whereas the figures of Duffy and Anderson (1988) show parallel trends of measured acoustic velocities for the various candidate minerals with depth.

Consequently, compositions of the mantle below where rocks are sampled are equivocal, and temperatures are assumed in interpreting seismic results. The adiabatic approximation is invalid as well as the isentropic because Earth includes heat sources that vary with time and position.

Discontinuities in the seismic velocities at 410 and 670 km are interpreted as phase transitions, but such assignments assume that the chemical compositions of the upper and lower mantles are identical (Agee, 1998). In contrast, isotopic evidence proves that the UM + TZ (transition zone), from whence the deepest geological samples originate, is chemically heterogeneous. Recent studies point to chaotic distribution (Armienti and Gasperini, 2010).

1.1.4 The lithosphere

The lithosphere is not a chemically distinct layer, but rather is defined as the outer portion of the globe that is rigid. Rigidity is a mechanical definition. Anderson (2007) noted that elasticity studies suggest that its thickness is ~ 50 km.

In contrast, thermal models are more typically used to define lithosphere thickness. These do not use a mechanical definition. In fact, plate models assume a constant basal temperature and fix the plate thickness, most commonly at 100 or 120 km (e.g., Stein and Stein, 1992). Clearly, these thermal models do not determine the thickness of the lithosphere, but instead assume it. Thus, observations have become blurred with assumptions. Conductive cooling models for spherical bodies are discussed in Chapter 2 and Chapter 7.

So how thick is the lithosphere? Section 1.2 shows that the LVZ is weak and global. The overlying crust is cold and rigid, but brittle, as evidenced by earthquakes, plate boundaries and other fractures. The mantle beneath the LVZ is increasingly compressed and densified, regardless of its particular temperature, because the seismic velocities increase with depth. Below 350 km, the material is strong mainly due to compression, and resists deformation. Extreme strength is indicated by models of mantle viscosity as 4×10^{20} Pa s from glacial rebound studies (Lambeck et al., 1998).

Thus, the LVZ is the boundary between two mechanically distinct zones, the dichotomous crust above and the upper mantle below, and is essential to Earth's mechanical behavior. To account for differential rotation (Section 1.5), the LVZ base must be below the depth of the continental crust, but no deeper than 350 km. Gradation in the LVZ is expected. Crucial questions center on the character of the LVZ, and how the complicated geometry of this weak zone is produced.

1.2 Interior materials and temperature constraints provided by the rock record, volcanic eruptions, and extra-terrestrial materials

Available rock samples represent only the outermost, volumetrically insignificant part of the Earth. To represent the interior, another sporadic set of samples is relied on: meteorites. Rocks, meteorites, and their constituent minerals have been studied using a wide variety of petrological and geochemical methods, so only a brief summary can be presented here. Findings from some important recent studies are emphasized. Constraints on internal temperatures from ancient molten rocks are also covered, due to their link with the chemical composition of the interior zones from whence they originated. This section attempts to distinguish what is known from what is inferred, because many of the previous deductions rest on assuming that Earth's deep interior is homogeneous and/or governed by an "adiabatic" gradient. As discussed here, the evidence requires heterogeneity and an evolving thermal gradient.

1.2.1 The well-sampled crust

1.2.1.1 Ocean sections and abyssal peridotites

Because oceanic drill holes are shallow, the vertical structure of oceanic crust is mostly based on geologic studies of ophiolites, inferred to represent sections of oceanic crust that were thrust onto continents. Ophiolites may not be typical of oceanic crust, but many characteristics of the "ophiolite sequence" summarized below agree with available information (e.g., Gregory and Taylor, 1981).

The ocean floor consists of an upper, ~ 0.4 km veneer of sediments derived from continental debris and shells of marine organisms, underlain by a ~ 0.5 km thick section of MORB basalts that have a rather uniform composition of ~ 50 wt% silica. Below these lavas is a ~ 1.5 km thick section of sheeted dikes that is slightly more mafic. Beneath this are ~ 5 km of intrusive gabbros. Gillis et al. (2014) measured chemical compositions in the upper volcanic and lower plutonic sections near the Cocos-Nazca spreading section, and ascertained that the gabbros are about ~ 48 wt% silica. The total crustal thickness of 7 km of the reconstructed ophiolite sequence agrees with the depth of the seismic MOHO discontinuity.

The underlying, compositionally distinct mantle has not been drilled, but ultramafic rocks have been dredged near ocean ridges, where magmas upwell. Abyssal peridotites found there are altered and dominated by olivine and orthopyroxene, with much less clinopyroxene and plagioclase, and have veins of gabbro and pyroxenite (Fig. 1.3 in Warren, 2016). The olivine has Fe/Mg ratios of 1/10 (Bodinier and Godard, 2014). The dominant rock types are harzburgite and lherzolite low in clinopyroxene, which is compatible with the lowermost rocks of ophiolite sections.

The mafic crust and ultramafic abyssal peridotites have very low amounts of the heat producing elements. This is consistent with the mineral structures for these rocks having few sites for large cations.

1.2.1.2 Continents

Continental rocks are highly varied but the upper continental crust is accessible and well-sampled. Available estimates of the average composition above ~ 30 km are about 67 wt% silica, and have not significantly changed in ~ 50 years, see tables in Yanagi (2011) and Henderson (1982). Several of these estimates are based on grids of rock samples from various large regions of the world.

The world's deepest drill hole, the 12 km borehole in Kola, did not reach the lower continental crust, so its composition remains poorly constrained. Andesite (57 wt% silica) is considered to represent the entirety of continents, after Taylor and McLennan (1985), based on density estimates and the compositions of abundant volcanic rocks. This average requires that the lower crust is much more mafic, ~ 54 wt% silica, than the well-sampled, overlying rocks.

The heat producing elements U and Th are abundant in the continental crust, having average concentrations of about 1 and 4 ppm, respectively. The indicated continental inventory of these elements matches that calculated for the entire Earth, assuming that the latter formed from chondritic meteorites (e.g., Henderson, 1982). The third important heat-producing radionuclide, potassium, is also concentrated upward, such that the continental crust has an average concentration of ~ 1.5 wt% K_2O. Radiogenic heat produced within the continents is lost to space, and little affects the thermal state of Earth's deep interior (Chapter 2).

1.2.1.3 Igneous rocks restricted in time

Regarding internal temperatures and their evolution over time, two special rock types, komatiites and anorthosites, warrant particular consideration. Komatiites are volcanic rocks that have very low silica ($\sim$42 wt%) but with quite high MgO ($>$18 wt%) concentrations, accompanied by low Fe (e.g., Arndt and Lesher, 2005). Consequently, their extrusion temperatures are very high, 1900–2000 K. The age of most komatiites is restricted to $\sim$3 Ga, thereby providing evidence that Earth's crust was hot and thin at that time; for comparison to today's conditions, MORB extrudes at $\sim$1400 K. Although recent arguments consider that komatiite extrusion temperature is actually much lower, due to water facilitating melting, this is questionable because the melting temperature of basalts is likewise reduced (Galenas, 2008). Komatiite lavas also lack the characteristics of explosive volcanism associated with volatiles, which are dramatically evident in eruptions of kimberlites, andesites, and rhyolites. Komatiite dikes are also very thin, and their special spinfex texture indicates rapid crystallization. These observations are consistent with very low viscosity, as is expected from their chemical composition (Arndt and Lesher, 2005).

Anorthosites (Ashwal, 1993) are volumetrically significant intrusive rocks almost entirely composed of large plagioclase crystals with subequal Na and Ca contents. Most formed between 1 and 1.8 Ga. Like komatiites, anorthosites are rich in refractory elements, compared to modern igneous rocks, but the element that has the most elevated concentration is Al, not Mg, due to plagioclase dominating the mineralogy and as reflected in their pyroxene compositions. The consequently high liquidus temperature of nearly 1700 K has fueled much debate about the origin of anorthosites. The implication is that the Archaean uppermost mantle was very hot, that significant cooling occurred between 3 and 1 Ga, and continues today.

1.2.2 Mantle samples

Informative assemblages of multiple mantle minerals are delivered in two forms. Xenoliths are rocks carried upward by volcanism, dominantly as inclusions in alkalic and potassic mafic magmas. Also available for study are small, individual minerals such as garnets and diamonds, which encase even tinier inclusions. Diamonds are carried upwards through old cratons in special, explosive eruptions of kimberlites. Some samples have been altered by metamorphism or by re-equilibration and stress during uplift. Most mantle samples represent subcontinental zones, but volcanoes in the middle of the oceanic crust also bear xenoliths. Like ophiolites, special conditions provide these samples: consistency exists, but what does this mean?

1.2.2.1 Xenoliths from the upper mantle

Xenoliths from mantle depths are known from about 3500 continental localities; see reviews by Nixon (1987) and Pearson et al. (2014). Most samples are small, typically $<$ 10 cm, but grains are coarse. The main type is peridotite, similar to samples from Alpine peridotite massifs or dredged from mid-ocean areas. Olivine with Fe/Mg $\sim$0.1 and orthopyroxene dominate. Eclogites, rocks constituted almost entirely of garnet and clinopyroxene, are rare and also occur as xenoliths. Peridotites carried by kimberlites from beneath thick, old continents are derived from below 200 km, possibly reaching 500 km (but see below), and are typically $\sim$72% olivine, $\sim$21% orthopyroxene, and $\sim$7% garnet + clinopyroxene. Isotopic ages of xenoliths range widely, but most are Precambrian, and they are generally much older than the volcanic rocks that have carried them upward. Xenoliths that origi-nated beneath thin, young continents are derived from above 150 km and are typically ~63% olivine, ~21% orthopyroxene, and ~16% garnet + clinopyroxene, and therefore are richer in Ca and Al. Many western U.S. xenoliths have been dated at 1.7 Ga.

A much smaller number of samples are provided by ocean island volcanoes. These peri-dotites tend to contain more pyroxene and iron than their continental counterparts. Hawaiian samples have been well-studied and show great variety.

Peridotites are considered to be residues after partial melting. Alternatively, the rocks may be cumulates, which are an agglomeration of minerals precipitated from a melt, as suggested by the textures. Temperatures of origin range from ~1300 to 1800 K. The *P-T* paths for xenoliths under cratons are shown in Fig. 1.3. The consistency of these *P-T* paths and xenolith origin at a depth of 190 km under Canada and 200−230 km under South Africa show that xenoliths trace the geotherms for the old, thick continental roots. Fig. 1.3 shows trends terminating at 50−60 km topside.

Many inferences of mantle composition are primarily based on xenolith data. However, the volume of the upper mantle sampled by xenoliths is miniscule, so the proportion of the residual peridotite to the parent magma in the mantle is unknown. Ringwood (1962) proposed that the upper mantle was composed of 1 part MOR basalt and 3 parts mantle olivine (pyrolite, with 43 wt% SiO_2). Other models primarily rely on meteorite data, and many assume that the compositions of the upper and lower mantles are identical. The lat-ter assumption has not been demonstrated, but rather rests on the debated hypothesis of whole mantle convection.

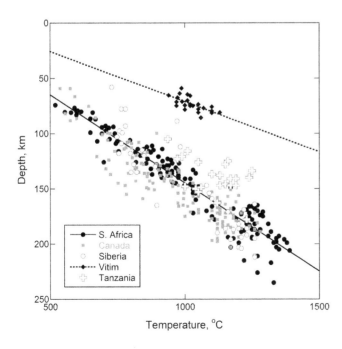

FIGURE 1.3 Depth-temperature trends from applying pyroxene Thermometry to garnet peridotite xenoliths. The least squares fit to South African material (solid line) is close to the fit for the Canadian craton. Data summarized by Pearson et al. (2014) in their 4-part figure 6 were digitized, cor-rected for a factor of 10 error in depths in their panel 6c, and for typographic errors in the y-axis for their panels 6ab, and assembled into one plot.

1.2.2.2 Inclusions in diamonds

Additional information is available from inclusions in diamonds. Large diamonds are associated with the rare eclogites, and these greatly predate their kimberlite hosts which are mostly <0.54 Ga, although the Premier pipe is ~1.5 Ga, see review by Pearson et al. (2014). Inclusions are mainly sulfides, plus the same silicate minerals observed in peridotites (see above). The underlying problem is that most, if not all inclusions, also predate their diamond host and so separate inclusions with the same age may be unrelated (Nestola et al., 2017). Thus, earlier claims that lower mantle material is preserved in microdiamonds, which were based on genetic links between the separated inclusions of (Mg,Fe)O and enstatite (Stachel et al., 2000), are unsupported. Claims that the tetragonal garnet phases are from the lower mantle have been retracted.

Regarding transition zone candidates, only one inclusion has been claimed to be a high pressure polymorph. Although Raman spectra provide a poor match, Pearson et al. (2014) argued that ringwoodite (spinel form of olivine) coexists with $CaSiO_3$ in this sample. The spectral region explored has the usual peaks associated with Si-O linkages, but does not provide direct information on cations, so the inferred chemical composition is ambiguous.

The assignments are questionable, based on existing spectra. Details are given in his paragraph for completeness. Two peaks were assigned to $CaSiO_3$. More likely, these strong peaks at 807−809 and 854−860 cm^{-1}, plus unassigned weaker peaks from 520 to 560 cm^{-1}, represent olivine stoichiometry because these correlate well with Raman spectra of Ca_2SiO_4 polymorphs by Reynard et al. (1997). The lower frequency peak is due to the olivine structure and the higher peak is present in both olivine and in high pressure larnite structures, which explains the observed intensity. A strong Raman peak at 606 cm^{-1} was not assigned. Its intensity correlates with a shoulder near 900 cm^{-1}: both are consistent with the presence of $CaSi_2O_5$ titanite (see RRUFF website). The XRD assignment, to single-phase silicate spinel, despite a missing strong diffraction peak, is compromised by the presence of multiple phases. Low transition pressure (~2 GPa), a huge volume change, back-transformation during quenching seen by Reynard et al. (1997), and the deformed nature of the diamond explains the observed large Raman peak breadths.

Multiple pure calcium silicates with the larnite and titanite structures were observed in several microdiamonds by Stachel et al. (2000), so our assignments to an upper mantle phase is compatible with previous studies. Calcium silicates are prone to hydration, which explains the OH band in IR spectra and that this signal does not match ringwoodite IR spectra.

Likewise, some individual diamonds contain separate inclusions of metal, Si-rich garnet, and $CaSiO_3$ (Smith et al., 2016). Formation depth is unconstrained.

All data on multiphase inclusions suggest that they formed at depths of ~200 to perhaps 300 km in the upper mantle, probably corresponding to the depth where the diamonds grew. The limited lower depth is connected with mechanical behavior of the productive zone, and in particular to the termination of the weak LVZ at shallow levels, suggested in Fig. 1.2 and discussed further below.

1.2.3 Information from meteorites, comets, and the sun

Earth's bulk chemical composition is ill constrained because available rock samples apparently originated in the outermost 5% of its radius, and geographic coverage is sporadic. As a consequence, estimates for the composition of Earth's interior zones below ~300 km are mostly inferred from the character of meteorites. The associated assumptions require examination.

1.2.3.1 Meteorites need not represent earth's interior

Use of meteorite chemical compositions to deduce Earth's bulk composition presumes that the early Earth formed from materials identical to extraterrestrial material delivered recently. Clearly, what is currently inside the Earth did not remain in space. In addition, meteoritic samples display a bewildering array of phases with complex chemical and isotopic relationships (e.g., Buchwald, 1975; Brearly and Jones, 1998: Mittlefehldt et al., 1998). The last authors admit that it is difficult to even organize the available descriptive information.

Meteorite samples cannot represent the proportions of material in the Earth, due to disparities in volume and mass proportions (Fig. 1.4), and the small, ~10^6 kg aggregate amount of extraterrestrial material available. The similarities of some phases to upper mantle minerals are tantalizing, but Earth is heterogeneous laterally and radially, and the common assumption that the compositions of various mantle layers are identical is unfounded. However, some information can be extracted due to common features in meteorites (Table 1.1).

1.2.3.2 Clues from isotopic data, especially oxygen

Extensive and diverse geochemical observations (e.g., Turekian and Holland, 2014) show that meteorites are unlike the rocky Earth. This section focuses on oxygen, the most abundant element in rocky material. For the planets, oxygen isotopes have well-defined trends (Fig. 1.5) connected with mass-dependent fractionation of material. Parallel trends exist for Earth plus Moon, meteorites from Mars, and the major phases in the primitive,

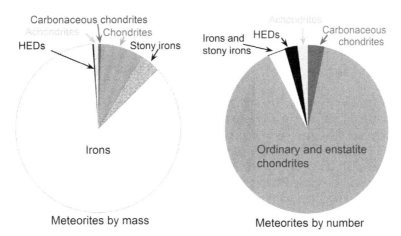

FIGURE 1.4 Large disparity between two representation of meteorite abundance. Rare carbonaceous chondrites are dominated by a single fall (Allende). *Modified after Hofmeister, A.M., Criss, R.E., 2012a. Origin of HED meteorites from the spalling of Mercury: implications for the formation and composition of the inner planets. In: Hwee-San, L. (Ed.), New Achievements in Geoscience. InTech, Croatia, pp. 153–178. http://www.intechopen.com/articles/show/title/the-case-for-hed-meteorites-originating-in-deep-spalling-of-mercury-implications-for-composition-and, which has a Creative Commons Attribution 3.0 License.*

TABLE 1.1 Gross characteristics of the three major meteorite types.

Undifferentiated chondrites	Differentiated achondrites	Irons
Old and primitive	Various ages and evolved	Old, some processed
Mostly unmelted	Previously melted	both
Chondrules[a] in a fine-grained, porous matrix of mostly silicate minerals	Large equigranular crystals are common	Coarse grained
Calcium aluminate inclusions[b] found in carbonaceous chondrites	Cumulate textures are common (gravitationally sorted from melt)	Exsolved
Presolar grains can be present[c]	Often basaltic, like surface of planets	Fe with 5–20% Ni
Some silicates are pure and iron free	Iron rich silicates and solid solutions	Sulfide rich; mostly due to FeS inclusions
Reduced	Oxidized	Reduced
Some are carbonaceous	Some are uncontestably Lunar; Others are strongly linked to Mars	Graphite common; some XN, XP, XC[d]
Aqueous alteration	Some have much metal[e]	Lack radionuclides

[a] ∼ mm balls commonly of olivine and pyroxene crystals (see Chapter 11 for discussion).
[b] Refractory oxides and silicates rich in Ca and Al, with minor Ti and Mg: hibonite, $CaTiO_3$ perovskite, anorthite, $MgAl_2O_4$ spinel, akermanite-gehlenite solid solutions, Al-Ti-rich pyroxenes (fassaite), and near end-member forsterite. Less common phases are corundum, $CaAl_4O_7$ grossite, and Al-rich diopide pyroxene. CAIs have rare inclusions of Fe-Ni alloy, sulfides, and Pt-group metal alloys, mostly encased in fassaite. These can have anomalous, mass independent oxygen isotope fractionations.
[c] Includes diamond, SiC, and graphite: it is difficult to establish other minerals as interstellar.
[d] X denotes Fe and some Ni; many other metals are present; Cu and Cr are significant.
[e] Pallasites have large olivine crystals in metal; mesosiderites are breccias of igneous rock and metal; for both the metal is like that in metallic meteorites.

chondritic meteorites, which are assigned to the asteroid belt, on the basis of reconstructed trajectories of meteorites.

The CCAM (carbonaceous chondrite anhydrous mineral) line (Clayton and Mayeda, 1975) was derived predominantly from components in the Allende meteorite, and indicates a different type of exchange than that seen in planets and most meteorite types. The trends in Fig. 1.5 show that precursor material assembled into planets was zoned, which is supported by data on refractory elements (Hofmeister and Criss, 2012a). These data motivate assignments of a large class of differentiated meteorites (the HED achondrites) to Mercury, which has an over-sized core, probably because once overlying mantle materials were largely "ball-milled" away by late-arriving impactors focused by the Sun. Hofmeister and Criss (2012a) also proposed that the asteroid belt contains both material formed in the outer Solar System, and of material ejected from large bodies of the inner Solar System during bombardment, which explains the textural and compositional dichotomy of the achondrites and chondrites (see Table 1.1).

Therefore, compositional models of Earth's mantle based on meteorite chemical compositions are only rough approximations. This finding is underscored by what reconsidered to be the most primitive meteorites showing some evidence of alteration and non-equilibrium assemblages, and by the incredible variety in mineralogy of meteorite components, down to a very fine scale (e.g., Brearly and Jones, 1998).

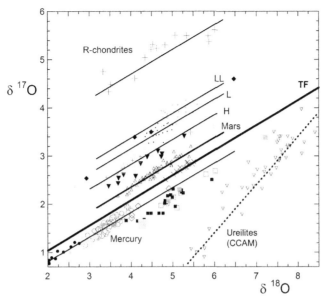

FIGURE 1.5 Oxygen isotopic data on solar system Materials. Tiny triangles = ureilites. Pen squares = winonaites. Gray square = IIICD irons. Black squares = IAB irons. Circles = HEDS. Dark gray dots = mesosiderites. Light gray dots = main group pallasites. Black dots = IIIAB irons. Open cross = angrites. Open diamond = brachinites. Light X = aubrites. Heavy X = enstatite chondrites. Open triangles = mars meteorites. Filled triangles = IIE irons. Filled diamonds = IVA irons. Small plus = H chondrites. Tiny black squares = L chondrites. Tiny gray squares = LL chondrites. Large plus = R-chondrites, altered in the desert. Light Lines = mass dependent fractionation trends for various bodies: TF (terrestrial fractionation) indicates the Earth—moon system. Dotted line = mass independent fractionation trends in certain meteorite types, as labeled (see text for details). *Modified after Hofmeister, A.M., Criss, R.E., 2012a. Origin of HED meteorites from the spalling of Mercury: implications for the formation and composition of the inner planets. In: Hwee-San, L. (Ed.), New Achievements in Geoscience. InTech, Croatia, pp. 153–178. http://www.intechopen.com/articles/show/title/the-case-for-hed-meteorites-originating-in-deep-spalling-of-mercury-implications-for-composition-and, which has a Creative Commons Attribution 3.0 License.*

1.2.3.3 Iron meteorites and the core: where is Earth's sulfur?

Oxygen isotopes of silicate inclusions in iron and stony-iron meteorites follow the same trends of non-metallic meteorites (Fig. 1.5), suggesting that inferring core compositions from irons should also be undertaken with trepidation. However, the constitution of the iron meteorites is fairly limited, in great contrast to the stony meteorites (Mittlefehldt et al., 1998; see Table 1.1), permitting some constraints to be set. A meteoritic depiction of Earth's core is particularly apt if planetary cores formed before overlying material was added (Chapter 4).

Iron meteorites suggest that Earth's core is dominated by FeNi alloy, with Ni being low. The abundance of FeS and graphite, with carbides, nitrides, and phosphides being common, indicates that the light elements in the core are likely a mixture of S, C, N, and P. However, individual grains of FeS could be density segregated into the outer core or the D'' layer. Heat producing elements are unexpected since these would provide unreasonably high temperatures (Chapter 2).

Metallic inclusions in diamonds are similar to iron meteorites, but are not identical. Smith et al. (2016) state that the assemblage in diamonds is primarily cohenite [(Fe, Ni)$_3$C], with interstitial Fe-Ni alloy, Fe$_{1-x}$S segregations, and minor Fe-phosphate, Cr-Fe-oxide, and Fe-oxide, surrounded by a thin fluid jacket of CH$_4$. Perhaps these inclusions represent raw material that was not incorporated in the core.

1.2.3.4 Comets and solar abundances: where is Earth's carbon?

Recent findings that comets are mixtures of phases respectively formed at high and low temperature (Zolensky et al., 2006) is very important. This mixture contraindicates the condensation gradient proposed by Metz (1974), and so does not fit well into currently popular models.

Based on the Solar composition (Basu and Antia, 2008), carbon should be one of the most abundant elements in the Solar System, and the pre-solar nebula probably contained copious amounts of CO. This highly stable molecule is also abundant in molecular clouds (e.g., Chiar et al., 1995) wherein stars are formed. This information, along with cometary data and the summary data of Table 1.1, suggest that Earth's carbon content has been grossly underestimated. On this basis, and assuming Earth formed as mixture of many phases, including ices, Hofmeister and Criss (2013) proposed that Earth incorporated abundant CO ice during accretion, and that subsequent heating promoted reactions between Fe metal, silicate dust, and trapped CO ice. The proposed reactions would lead to carbon being the core's major light element. Earth's core having $\sim 10\%$ C is consistent with Solar abundances and with the ubiquitous occurrence of olivine with Fe/(Mg + Fe) $\sim$ 0.1 in meteorites, mantle samples, and terrestrial basalts. These reactions also explain why the interiors of rocky planets are oxidized even though most primitive meteorites are reduced.

Alternatively, if most of the core formed first, a great amount of carbon must reside in the mantle. Diamonds are now recognized as occurring ubiquitously (Pearson et al., 2014), suggesting that the hypothesis of grossly inhomogeneous accretion has merit (Chapter 4).

1.3 Buffering at isothermal boundary layers

1.3.1 Surface temperature

The surface temperatures of bodies in our Solar System are maintained by the strong Solar flux. In addition, thermodynamic laws require that these objects shed heat by ballistic radiation to the immense and cold surroundings of space, in proportion to the surface temperature of the particular object. Therefore, boundary conditions at the surface of planetary objects are independent of their internal workings, now and in earlier times. Shapes and surface temperatures of planets are accurately measured, e.g., by images and remotely obtained spectra, allowing the rate of heat loss to space to be quantified (Chapter 2).

1.3.2 Isothermal phase boundaries, buffering, and layer thickness

The importance of phase equilibria to the physical sciences is underscored by the centigrade temperature scale being defined by the interval between the freezing and boiling points of water. Moreover, the ice bath (Fig. 1.3 in Hofmeister, 2019) illustrates the crucial effect of phase equilibria on the thermal evolution of a 2-phase system. If the bath is allowed to receive heat, ice is slowly converted to water, but the temperature remains constant as long as both ice and water are present. If the bath is placed in a freezer, more ice is made, but again, the bath temperature remains constant as long as ice and water coexist. Resistance of the bath to change is a consequence of latent heat serving as a buffer. Note that any phase transition that involves a heat of reaction will provide buffering. This discussion focusses on melting because regions of melt clearly exist in the Earth, and this provides a constraint for thermal models (Part 2).

Buffering has great consequences for a layered system with a temperature gradient that is evolving. But, a few other thermodynamic constraints need to be mentioned before buffering is discussed. Pressure increases with depth in the Earth and planets due to the force of gravity. Because phase transitions have a prescribed relationship of temperature with pressure, depths in the Earth which are melted or are undergoing some other phase transition have a specific temperature. Obviously, if the melt or transition can be identified, the temperature of the boundary can be ascertained. For this reason, much effort has been expended in exploring the behavior of mantle and core candidate phases at high temperatures and pressures (e.g., Price and Stixrude, 2015).

Temperature is expected to increase with depth since Earth is radiating to space, although the thermal gradient is affected by the position of internal heat sources (Criss and Hofmeister, 2016; Chapter 2). For T increasing with depth, as observed near the surface (Fig. 1.3), melt occupies some range of depths. If more heat flows into this area from below, this is consumed by melting more material, i.e., by expanding the region. If the supply of heat dwindles, the melt crystallizes, which releases heat. As long as a layer of melt exists, the temperature of this region is constant. Due to buffering, changes in heat flow will change the thickness and depth of the melt layer. The existence of a pressure gradient slightly modifies isothermal conditions of the melt region, depending on the thickness of the layer and whether the P gradient is rapidly changing, or not. But, this is rather inconsequential. With buffering in an evolving system, layer thickness depends on the heat flux just as much, if not more so than it does on the phase equilibria.

This situation describes the current state of Earth's core (Criss and Hofmeister, 2016). The temperature at the seismic boundary between the inner (solid) and outer (liquid) core is defined by the P-T slope of the melting curve for the metallic core material. Without knowing the exact composition, the temperature of this boundary cannot be pinpointed in the laboratory, but considerable information exists on candidate iron alloys (Li and Fei, 2014). For example, 4000 K is suggested from melting of Fe-S in the core (Chudinovskikh and Boehler, 2007). The effect of multiple alloying elements is covered in Chapter 8. The physical boundary, which is a nearly spherical shell, will move radially upon any receipt (or loss) of heat from the surroundings, or loss of heat to the surroundings. Receipt (of loss) of heat will change the relative proportions of solid and liquid phases in the whole core. The temperature where these coexist will change according to how strongly melting depends on pressure. This behavior and other observations lead Criss and Hofmeister (2016) to propose the core is progressively melting, presently at a very slow rate, because the concentrations of heat producing elements are higher, lying in the rocky, overlying mantle rather than in the metallic core.

Therefore, thickness of melt layers (along with latent heat and their temperatures) can be used as constraints on thermal evolution. Melts existing in the outer core and below the oceanic crust are utilized in Chapters 8 and 7, respectively.

Solid-solid phase transitions also have an associated heat of reaction and also provide buffering. If such regions exist in the Earth, their thickness depends on heat flux too.

1.4 Data on surface temperature gradients and heat flux

1.4.1 Methods

1.4.1.1 Surface thermal gradients in mines

Thermal gradients in mines are directly measured with thermocouples on rock faces. The world's deepest gold mine, Mponeng in S. Africa, reaches 66 °C at 4 km, averaging ~ 11 K km^{-1} (see websites). Actual temperatures and gradients are probably higher than the measurements, due to the cooling effect of mine ventilation.

1.4.1.2 Coupled subsurface measurements of heat flux and thermal gradients

Temperatures in boreholes are measured using high resolution thermistors and some non-electrical methods: see e.g., the review by Jessop (1990). The accuracy of these temperature readings is considered to be lower than those of the associated measurements of heat flux, due to the manner in which data are collected (Prensky, 1992). However, subsurface heat flux is obtained from in situ, contact measurements of both thermal conductivity (K) and temperature, using Fourier's law (Chapter 2). Because laboratory methods for K involving physical contact are uncertain by 10−20% near 298 K, due to systematic errors of losses at physical contacts and gains from ballistic radiative transfer (e.g., Hofmeister and Branlund, 2015; see also Hofmeister, 2019: chapter 4), field data have similar uncertainties. Nonetheless, these measurements should represent local conductive gradients despite possible advection (Chapter 3). Thirteen boreholes in the Venetia diamond mine in S. Africa yielded a linear gradient of 19 K km^{-1} measured at depths up to 0.6 km (Jones, 2016).

Temperatures in sediments on the ocean floor are often measured using sensors mounted on the outside of gravity-driven corers and heat flux is then determined from K measured on the recovered cores (Davis and Fisher, 2011). The reported heat flux data thus has the same 10−20% uncertainty as borehole data because conventional, contact methods are used for K. In situ measurements of K are also used, which have additional uncertainties.

1.4.1.3 Uninformative neutrino studies

Although 10^{23} U atoms decay in the Earth every second (presuming a chondritic abundance of 20 ppb), measurements of geoneutrinos only record a minute fraction of these events. Over many years, 10 ± 4 geoneutrinos were detected in Italy (Bellini et al., 2010) and 111 ± 44 in Japan (Gando et al., 2011). Not only does the colossal, $\sim 10^{30}$ fold difference between the numbers of expected and measured events require use of models to estimate flux, but these numerical values were also corrected for anthropogenic events which are far greater in number than the natural events. This correction alone is much greater than the remaining value. Geoneutrino measurements are unreliable indications of interior emissions.

In addition, to estimate the global power, Gando et al. (2011) averaged their model for the mantle contribution at the Japanese site (10 ± 10 TW) with their model for the Italian site (28 ± 19 TW), yielding 20 ± 9 TW produced by U and Th. Even accepting all of their dubious arguments and huge corrections, standard methods (Bevington, 1969) instead suggest 19 ± 12 TW. More importantly, these mantle values were obtained by subtracting model values for the continental crust, which Gando et al (2011) assert provide $2/3$ of the neutrino flux. Because crustal models range from 8 to 16 TW (e.g., Lodders and Fegley, 1998), neutrino studies do not constrain the flux from U and Th in the mantle. In addition, because potassium emissions cannot be determined by the available neutrino measurements, this approach cannot possibly constrain primordial heat.

1.4.2 Constraints on upper mantle conditions

1.4.2.1 Equivalence of continental and oceanic flux

To date, subsurface measurements have been made at over 44,000 sites on the globe (Table 1.2). Due to the sheer number of data, random errors are reduced, but systematic errors remain. The incomplete coverage (Fig. 1.6) is addressed in various approaches. Yet, the averages computed by considering the spherical geometry of the Earth have not changed in ~50 years. Lee's (1970) value is within the uncertainty of Hofmeister and Criss' (2005a,b) estimates and Vieira and Hamza's (2018) detailed computation, which involves a closely spaced grid and filling in gaps, based on age-flux correlations. The methods used in these studies differ, yet all arrive at similar global powers, despite the 12-fold increase in the number of measurements since Lee's (1970) study.

Although the estimated global power $\Pi = 31$ TW is robust, it is inexact because several regions with very high heat flow and large areas are represented by trends with age than by actual measurements: these are oceanic lithosphere west of South America and southwest of Australia and, to a lesser extent, continental lithosphere north of India (Fig. 1.6).

TABLE 1.2   Heat flux averages based on data.

Continents	Oceans	Global power	No. of data	Reference
mW m^{-2}	mW m^{-2}	TW		
62	62	31	3500	Lee (1970)[a]
61	65	31–32	22,497	Hofmeister and Criss (2005a,b)[b]
58	50–70	28–35	44,386	Vieira and Hamza (2018)[c]

[a]First quantitative estimate, where spherical harmonics were used to constrain global values.
[b]Median values presented for flux measured prior to 2002, based on three different but congruent estimates of ocean flux.
[c]The data used differ little from that of 2012 shown Fig. 1.6. Semi-empirical relations of heat flux data with age were established from binned data. These relationships were used for the bins lacking data which number 10,383 (20% of the data).

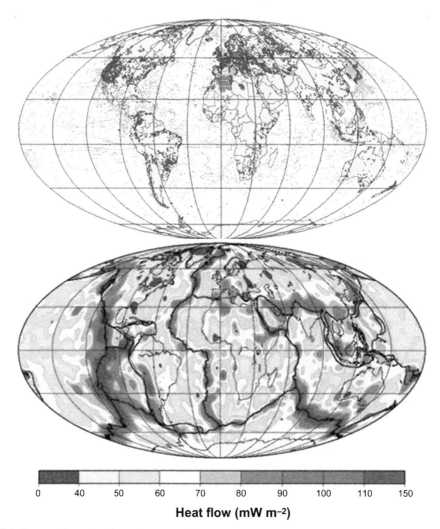

Heat flow (mW m⁻²)

FIGURE 1.6 Heat flux data. (top) Geographic locations of heat flow measurements in an updated global data set. Oceanic sites are shown as gray dots, whereas black dots are continental sites. (bottom) Global representation of heat flow based on observational data shown above, which was supplemented with estimates derived from digital maps and empirical correlation with age. Most areas of high heat flow (dark shading) are near mid-ocean ridges or convergent plate boundaries, whereas many areas of low hear flow (light gray) correspond to Precambrian cratons. Both figure parts are in color in the original article, which is open access. From Hamza, V.M., Vieira, F.P. 2012 Global distribution of the lithosphere-asthenosphere boundary: a new look. Solid Earth 3, 199−212. Available from: https://doi.org/10.5194/se-3-199-2012, which has a license level 3.0 under creative commons.

Nonetheless, these and other data-oriented studies show equivalence of oceanic and continental heat flux, within uncertainty (Table 1.2). This equivalence was noted by the earliest workers on global heat flow, and predates currently popular models (see Part 2). Fig. 1.7 shows that the position and shapes of heat flow data from continental and oceanic areas are almost identical, while illustrating the unrealism of the half-space cooling model

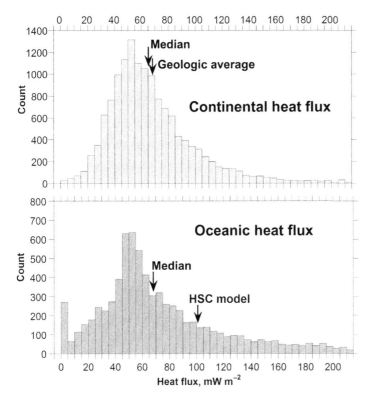

FIGURE 1.7 Histograms of continental and oceanic heat flux measurements from the interim compilation of ∼22,000 sites (Gosnold and Panda, 2002). HSC is the half-space cooling model estimate of Pollack et al. (1993) which agrees poorly with the data and is based on unsound physics, see below and Chapter 7. The continental average, also from Pollack et al. (1993), resembles the estimates in Table 1.2.

estimate (see Chapter 2). The observed median values best represent the actual averages, because heat flux cannot be negative and high flux from "hot" localities is likely overestimated due to ballistic transfer in contact methods above 298 K (Hofmeister, 2019: chapter 4).

1.4.2.2 Derivation of flux from below the Lehman discontinuity at ∼ 220 km

To address why surface heat fluxes are similar for continental and oceanic areas, temperatures at depth need to be constructed. Fig. 1.2 shows that continental crust is thicker than oceanic crust and that much anisotropy exists. However, the differences in seismic velocities of materials that underlie oceanic and continental crust are small below 200 km, which is the continental base (Fig. 1.3), and these differences become even smaller as depth increases further, such that conditions at and below ∼320 km are isotropic. Therefore, the continental geotherm can be used to estimate that the mantle temperature is 1400 K at a depth of 200 km beneath the continents. The LVZ there is a bit deeper. Extending the continental geotherm gives a temperature of 1475 K at 220 km beneath the continents, which matches the melting temperature of MORB. This is no coincidence. Mantle temperatures below the oceanic crust appear to be close to ∼1475 K over the extended depth range of 10−220 km, are thus are much higher than below the continents at an equivalent depth, as is evident in the lower sub-oceanic seismic velocities (Fig. 1.2).

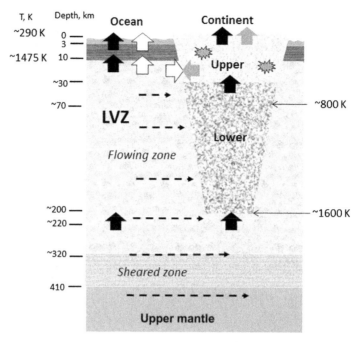

FIGURE 1.8   Data oriented model of the thermal structure beneath older and stable areas of oceanic and continental crust. Various regions are indicated by different patterns as labeled. The zone between 220 and 350 km involves the transition from seismic anisotropy near the surface to the rather uniform seismicity of the upper mantle at depth, and was designated as asthenosphere by Riguzzi et al. (2010) and others. right side, depths of layers from Seismologic data and temperatures for the surface and for the regions of melting (below basaltic crust and through the LVZ). Left side, temperatures from the geotherm defined by xenoliths, which seems high at 30 km, compared to the surface gradient due to the ages of the rocks. The continental base temperature should not depend on time, due to phase relations. Starbursts indicate confirmed location of radionuclides, which is associated with heat production (gray block arrows). Black block arrows indicate flux from the upper mantle, whereas white block arrows indicate flux arising in the LVZ, both are discussed in Section 1.5. Horizontal block arrows indicate balance of steady-state lateral flow. Dashed horizontal arrows indicate the presence of differential rotation between ~410 km and the surface, where the solid zone above the upper mantle is sheared.

If the LVZ is partially melted as assigned (Anderson, 2007), it is therefore nearly isothermal, and also flows easily under stress. A simplified version of the deduced thermal structure under the old, stable areas is shown in Fig. 1.8.

As a prelude to the thermal models in Part 2, a first-order interpretation of flux data is derived from Fourier's law:

$$\Im = -K\frac{\partial T}{\partial z} \cong -K\frac{\Delta T}{\Delta z}. \tag{1.6}$$

where $\Im$ is flux. This simple, Cartesian form is relevant due to the blocky nature of continents. The effect of pressure is neglected because P variations are small. Quasi-steady-state conditions assumed here are suitable for the older, stable crusts.

The linear thermal gradient in the lower continental crust (Fig. 1.2) shows that this region has negligible internal production (Ψ) and thus defines the flux from the underlying mantle. Thermal conductivity of rocks being $2-3\,\mathrm{W\,m^{-1}\,K^{-1}}$ at this temperature (Hofmeister, 2019: Chapter 7) gives a rough estimate of $\mathfrak{I}_{mnt} = 8-12\,\mathrm{mW\,m^{-2}}$. The surface flux is larger and the thermal gradient steepens in the upper continental crust because this region is enriched in radionuclides and thus has large $\Psi_{rad,cont}$ (Section 1.2). The inferred continental radionuclide power is consistent with geochemical estimates.

The LVZ is isothermal, and thus, the flux at the MOHO is constrained as $\mathfrak{I}_{mnt} = 8-12\,\mathrm{mW\,m^{-2}}$: these are the black block arrows in Fig. 1.8. The much larger measured surface flux requires that a substantial additional power source exist (white arrows in Fig. 1.8). This cannot be due to cooling of the solidified crust, because its average temperature is defined by the constant, isothermal boundaries. This finding does not apply to near the ridges where magma solidifies: this additional advective process is addressed in Chapter 3 and Part 2. The extra mid-oceanic flux cannot be due to radioactive heat generation because both ocean crust and abyssal peridotite are barren. Potentially large heat sources from cooling and shear are evaluated in Section 1.6. Here we note that the upper continents shed heat from radionuclides (gray block arrows in Fig. 1.8). The total flux above the continents is $58\,\mathrm{mW\,m^{-2}}$ (Table 1.2). By difference, the radiogenic flux is $46-50\,\mathrm{mW\,m^{-2}}$: Under quasi-steady state, the extra power source under the ocean will balance the radionuclide heat from the continents, as shown, so $\mathfrak{I}_{LVZ} = \sim 48\,\mathrm{mW\,m^{-2}}$.

1.4.3 No extra heat from hotspots

Much attention is given to hotspots such as Hawaii, where volcanism exists in the middle of the stable oceanic plates. Fig. 1.6 shows that the Hawaii chain is cool, like most of the middle of the Pacific. The Icelandic hotspot is hot, but this island sits on the mid-Atlantic ridge. Surveys of hot spots have shown no evidence of high conductive heat flow (Harris and McNutt, 2007) which is required for hotspots to be thermal plumes associated with mantle convection. The magma may advect heat, but this is a consequence, not a cause.

1.4.4 Problematic cooling models of oceanic lithosphere

Pollack et al. (1993) and subsequent studies report higher global power of $\Pi \sim 44\,\mathrm{TW}$ based on substituting large model values for the measured oceanic flux values, which are much smaller (Table 1.2). The HSC model is based on 1-dimensional cooling of the infinite half-space, after Stein and Stein (1992). These authors are aware of the misfit and evoke hydrothermal circulation as carrying heat away. Many problems existing with this attribution in particular and with the HSC model in general (e.g., Hofmeister and Criss, 2005a,b; Vieira and Hamza, 2018; Chapter 2), including substantial lateral heat flow, which violates model assumptions. An obvious inconsistency exists in portraying the solid oceanic lithosphere as cooling, which is not possible since isothermal upper and lower boundaries are used in model constructs. "Cooling" arises from mathematical errors, as detailed in Chapter 7. One of many conceptual errors underlying HSC and plate models is equating distance from the ridge (y) with time (t). Although the ocean floor is indeed a time sequence, the surface flux measurements represent the current state of the Earth ($t = 0$).

Quasi-steady-state conditions are indicated by the constancy of the basalts on ocean floor, which have ages up to 200 Ma over large areas and up to 340 Ma in the Mediterranean (Granot, 2016).

1.5 Earth's immense spin energy and its internal dissipation

Primary characteristics of planetary bodies are their shapes and rate of axial spin. The larger bodies are nearly round since gravity is a central force, but these are not perfect spheres, because spin supplies an additional uniaxial force, making large planets oblate. Forces and potentials associated with oblate bodies were updated and extended by Hofmeister et al. (2018). The energy in planetary spin is gargantuan, far exceeding any other energy reservoir, and furthermore is decreasing exponentially with time. The heat source derived from spin dissipation is immense but, for the most part, has been overlooked.

1.5.1 Observations connecting plate tectonics to axial spin

Ongoing tectonic activity of the Earth is popularly attributed to whole mantle convection, but problems exist with these models regarding geologic evidence (Section 1.6) and theory (Chapter 5). No compelling physical mechanism has emerged during the almost 50 years history of such models (Bercovici, 2015). The surface configuration and the motions of the plates indicate a mechanical origin that is connected with spin.

1.5.1.1 *Large scale motions indicate differential rotation*

Continental drift was proposed by meteorologist Alfred Wegener in 1912 on basis of congruent continental shapes and supporting fossil evidence, but was dismissed by geologists on the basis of rock strength. Subsequent to the discovery of the mid-oceanic rifts and parallel magnetic stripes on the ocean floor, continental drift was accepted circa 1967. Without any other viable hypothesis for the mechanism, mantle convection was also accepted. Additional studies of fossil records show that the positions of the poles are nearly time-independent, and so continental drift has been predominantly E-W over phanerozoic history (Stehli et al., 1969).

Subsequent seismological studies have focused on motions of the oceanic plates, relative to the reference frame of the globe and have shown that spin plays an important role in plate tectonics (Riguzzi et al., 2010; Doglioni and Panza, 2015; Doglioni and Anderson, 2015). Plates rotate more slowly than the interior, with an average velocity difference of ~ 6 cm year^{-1}. For a detailed analysis of diverse evidence of the importance of spin, see Doglioni and Panza (2015). The surface westward drift of the plates demonstrates that differential rotation (spin) exists between Earth's layers. Chapter 5 describes how differential rotation leads to plate tectonics, focusing on lower symmetries in the forces than uniaxial.

Strong support for differential rotation is provided by the asymmetry of the tilt of downgoing slabs, depending on whether the plate descends to the East or West (Fig. 1.9). For slabs subducting to the west, differential rotation opposes the slab motion, squeezing the bend and making the slopes steeper. For slabs subducting to the east, differential

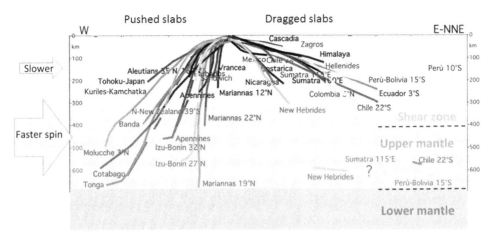

FIGURE 1.9 Tilt of Downgoing slabs. Slab dip measured along cross-sections perpendicular to the trends. Each line represents the mean trace of seismicity. Note the shorter lengths of the E- or NE-directed subduction zones, which average 27.1°. The W-directed slabs average dip is 65.6°. Mantle zones are shown as solid gray. Seismic discontinuities are shown as dashed lines. The marble pattern indicates a region which is strong, but transitions into the very weak LVZ. However, the white region rotates together, on average. Arrows on the left show how differential rotation pushes the W-dipping slabs, but drags the E-dipping slabs. *Modified after from Figure 5 of Riguzzi, F., Panza, G., Varga, P., Doglioni, C., 2010. Can Earth's rotation and tidal despinning drive plate tectonics? Tectonophysics, 484, 60—73 with permissions.*

rotation aids the slab motion, making the slopes shallow. The average values of 27° and 66° differ by a factor of ∼2 (Riguzzi et al., 2010; Doglioni and Panza, 2015), because without differential rotation, a 45° dip is expected. Because rocks are strong in compression but weak in extension, the pushed slabs remain intact, but not the dragged slabs. The breakage of the E-subducting slabs at ∼320 km correlates with the depth of seismic isotropy (Fig. 1.2): this consistent depth supports differential rotation. With the diverse properties of Earth's layers, differential rotation could be graded or abrupt or both.

1.5.1.2 Large scale fractures are from uniaxial stress of spin

The high seismic velocities in Fig. 1.1 show that the deep Earth resists compression and shear forces. The lower mantle is not only strong, but is thick, which maintains Earth's not-quite-round shape as it spins. In contrast, the lithosphere is brittle, thin, heterogeneous, and lies on the weak LVZ. With this configuration, the stresses during spin are expected to crack the lithosphere. Hofmeister and Criss (2005b) pointed out that the mid-ocean ridge system over the globe shows a fracture pattern similar to cracking of a solid cylinder subject to uniaxial stress in the laboratory. Fig. 1.10 shows the Southern Hemisphere, where the Antarctic cap is intact but fractures run out at 90°. Ridges circle Antarctica because this section of the crust possesses a disk geometry, whereas the crust at low latitudes has a geometry more similar to a hollow cylinder. The plate motions about this ring of ridges are consistent with differential rotation, such that the Antarctica cap spins congruently with the deeper regions of the Earth, as is expected.

FIGURE 1.10 Cracking of lithosphere from uniaxial stress. Heavy black lines with crossing arrows show mid-ocean ridges and heavy gray lines show subduction zones. *Orthographic projection from Weinelt, M., 2003. Create-a-map, website: http://www.aquarius.geomar.de/omc using the program of Wessel, P., Smith, W.H.F., 1995. The Generic Mapping Tools (GMT), version 3.0: Techical reference and cookbook (SOEST/NOOA), http://gmt.soest.hawaii.edu.*

The presence of thick, strong, cold continental crust in an asymmetric pattern considerably reduces the symmetry, and affects the plate motions. The origin of plate tectonics in gravitational forces, and not in convection, is discussed in Chapter 5.

1.5.1.3 Friction vs slippage and heat production

For Earth to spin differentially requires either slippage between layers or shear of layers. The crust is strong but brittle, whereas the lower mantle is strong due to high compression. The LVZ is weak, and apparently flows plastically, allowing the plates to subduct. Because this process is mechanical, the thick ~50 km elastic thickness is the appropriate depiction of the slabs, rather than the thermal thickness, obtained from models with imperfections. With slippage accompanying partial melt, the interior of the LVZ is not the zone where heat is generated. But friction would be generated at the upper and lower boundaries of the LVZ, in particular with its interface with the overlying crust as well as at its solid base. The latter seismic discontinuity is gradational rather than sharply defined.

Frictional heating would be greatest between strong layers, where resistance is provided. From the figures above, shear occurs in the vicinity of 320 km. The shear can extend upwards, but not above the continental roots at ~200 km. The shear zone can extend downwards, but probably not below the 410 discontinuity, below which isotropy exists, except near downgoing slabs. Friction produces heat and changes the large scale motions into small scale motions of vibrations, so momentum and (kinetic) energy are not conserved.

From the 1st law, friction between two sliding blocks heats both equally (Sherwood and Bernard, 1984). Therefore, heat from the shear zone diffuses around the locus of shear. The heat certainly travels to the close, cold surface, but may serve to warm the upper mantle as well, making this soft and permitting slabs to penetrate. The LVZ is certainly affected, but is currently a buffer zone. The heat leaving the LVZ can come from secular cooling of the LVZ or shearing at its boundaries.

1.5.2 Rate of spin reduction

Spin is gravitational in origin, as was known to Newton and Maclaurin. Spin exists independent of any other gravitational motions, and therefore is an independent energy reservoir. For spin energy and angular momentum to be conserved, Earth would need to rotate as a solid body, which is demonstrably untrue. This section discusses implications of the 1st law for the case of *differential* spin. We do not discuss tidal dissipation involving Earth's oceans and crust, as this is small and involves a different gravitational energy reservoir, the central force attraction between mainly the Moon and the Earth. Chapter 5 and work in preparation discuss tides, spin down, torque, and non-conservative frictional forces in detail.

1.5.2.1 Observations signifying spin loss

Deceleration of Earth's spin is evident in growth rings of corals (e.g., Scrutton, 1978). Although this has been tied to lunar recession, this attribution is in error because no torque exists in the Earth–Moon system alone. Forces from the Moon combined with Solar, can slosh the oceans, and to some extent the lithosphere, but these external forces, although central, are not co-linear, and are moreover distinct from the self-gravitational forces and centrifugal forces that are essential to spin, which balance to provide Earth's oblate shape.

The loss of spin has been projected back in time by several researchers, providing a wide range of fast initial spins (Fig. 1.11). The projections, including the exponential trend shown depends on uncertainties in the coral measurements. An alternative approach follows.

1.5.2.2 Constraints on the fast initial state

Rotational energies (R.E.) calculated from measured axial spin rates and moment of inertia are proportional to their gravitational self-potential, U_g (Fig. 1.12). For simplicity,

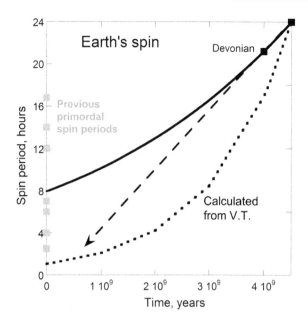

FIGURE 1.11 Earth's spin as a function of time. spin Period = solid squares and lines. Gray = post 1999 estimates from Lathe (2006). Data from Devonian corals suggests a linear extrapolation of estimates of 0.0002 s per century (see e.g., Scrutton, 1978). Exponential decay was used to provide spin down from the initial Virial state and from the Devonian coral data. Dashed line shows a linear trend. *Modified after Hofmeister, A.M., and Criss, R.E., 2012b. A thermodynamic model for formation of the solar system via 3-dimensional collapse of the Dusty Nebula. Planet. Space Sci., 62, 111–131 with permissions.*

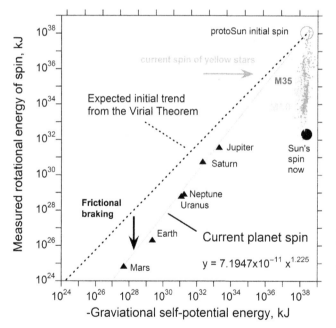

FIGURE 1.12 Dependence of measured axial spin energies of planets and stars on gravitational self-potential energy of each object. Circle = R.E. Of the newly formed Sun inferred from the Virial theorem (the dotted line). Triangles = measured spin for planets using NASA Datasheets. The fit to this data (solid line) projects to the protoSun. Large dot = Sun's present R.E. spin. Small dots = R.E. Calculated from mass, radius, and spin parameters for small stars in two clusters, using I = 0.06MR2, as modeled for the Sun. Gray dots = M50 (Irwin et al., 2009). Black dots = M35 (Meibom et al., 2009). *Reprinted with permissions from Hofmeister, A.M., Criss, R.E., 2016. Spatial and symmetry constraints as the basis of the virial theorem and astrophysical implications. Can. J. Phys. 94, 380–388.*

U_g was estimated by using the homogeneous sphere (more in Chapter 4). No other energy source of comparable magnitude exists, and thus this finding shows that spin, not primordial heat, is the consequence of accretion (Hofmeister and Criss, 2012b). Importantly, the consistency of the planetary trend points to a single cause. This cause cannot be tides, although Earth lying below the trend suggests that its oversized moon could have a slight effect. Differential rotation and energy loss has affected all bodies. Note that Mars, which was recently frozen solid, and thus now spins conservatively, lies slightly above the trend defined by the remaining planets.

The planetary trend points to the Virial value for the Sun. This being the starting point for the spin of stars is indicated by data on yellow stars like our Sun in open clusters (Fig. 1.12) and other astronomical data (Hofmeister and Criss, 2016). The Virial theorem for a bound state is not identical to energy conservation, but rather provides an additional constraint.

The Virial state should describe the planets as well. However, these are not homogeneous inside and accretion likely involved stages for the planets, as material was drawn to the Sun from unstable orbits. So our calculated initial period of 1.5 hours of Fig. 1.11 is a limiting value. To better estimate power loss in Section 1.6, we consider a range of initial spins.

1.5.3 Observational data support differential spin, not mantle convection

Researchers producing convection models acknowledge that attributing plate tectonics to whole mantle convection involves several first-order difficulties. As discussed by modeler Bercovici (2015), the most serious problem is the lack of compelling evidence for thin, thermal plumes arising from the lower mantle. Seismological evidence exists for broad features deep in the lower mantle (French and Romanowicz, 2015). However, these features are not tall and thin, as "plumes" are envisioned. From seismic studies, low-wave-speed anomalies associated with thermal upwelling are confined to the upper mantle beneath Iceland (Foulger et al., 2001; Du et al., 2006). The same is true for the Yellowstone region (Gao and Liu, 2013) and the European Cenozoic rift (Fichtner and Villasenor, 2015). Foulger (2010) discusses the plumes controversy in detail.

An uninvolved lower mantle is indicated by diverse evidence (summarized by Foulger et al., 2013), which includes flattening of many plates near 670 km (Goes et al., 2017) where earthquakes cease (Rees and Okal, 1987), and the dirth of samples that originated below 200–300 km (Section 1.3). Average plate velocities of ~4 to 6 cm year^{-1}, show that the period of whole mantle overturn must be immense (~¼ Ga) if circulation occurs. This value is a huge fraction of the ~1 Ga interval when plate tectonics clearly operated (Hamilton, 2011). Moreover, the consistent westward drift is inconsistent with mantle circulation, because the latter requires counter-circulating adjacent cells.

The inadequacy of internal heat generated by radionuclides to drive convection is another concern of Bercovici (2015). Although friction could accompany convection and could generate heat, an energy source is needed to drive the motions which produce the frictional heat. Perpetual motion machines are forbidden by thermodynamic laws. So the energy needed is the K.E. of the motions plus the heat flux. Theoretical problems with mantle convection are covered in Chapter 5.

Surface motions and surface heat flux being produced by spin degradation are consistent with the above observations. Heat is an inefficient means for producing motions; but rather is a byproduct of motions which arise from another cause. The sequence considered for differential spin:

$$potential\ energy \rightarrow kinetic\ energy \rightarrow heat \tag{1.7}$$

is theoretically and experimental verified (e.g., Halliday and Resnick, 1966).

1.6 Constrains on interior heat generation and implications

To compare possible sources of heat emissions, we compute the power emitted from a homogeneous Earth. The numerical value of flux in mW m^{-2} at the Earth's surface is conveniently close to 2× times larger than the power in TW, which is a coincidence due to Earth's particular surface area. An implicit assumption in connecting measured surface flux to the internal power is that the heat source is sufficiently close to the surface that a secular lag is not important. For conductive cooling, "close to the surface" means above 670 km (Chapter 2). Heat conduction being slow in a large body causes the internal temperature to rise. To model this, the present section provides some constraints. Also needed are the initial conditions. If accretion was cold, the surface flux should be numerically equivalent to ∼1× times the power in TW (Chapter 2).

1.6.1 Spin down

Power was calculated based on today's R.E., by assuming an exponential decay of the rotation rate from some short initial value to the current 24 hour period. Two such curves are shown in Fig. 1.11. Table 1.3 reports various powers for initial spins of 2, 4, 6.5, and 8 hours. Secular delay is not germane because differential rotation exists in the upper mantle, at least currently.

TABLE 1.3   Estimates of power (Π in TW) in the earth.

Spin down	Radionuclides	Π today	Average Π	Π at 4.55 Ga
Initial spin, hours	K (ppm); Th (ppb); U(ppb)			
2 (Virial)		7.3	211	1056
4		5.3	52	191
6.5		3.9	18.7	53
8 (Coral trend)		3.2	11.8	29
	800; 38; 11[a]	29	97	253
	200; 74; 20[b]	28	58	134
	180; 53; 15[c]	21	46	108

[a]Average of 5 chondritic meteorite types tabulated by Lodders and Fegley (1998).
[b]Used by Van Schmus (1995).
[c]Average of two bulk Earth models tabulated by Lodders and Fegley (1998).

The current power from an initial spin of 4 ± 1 hours is consistent with the estimated mantle flux delivered to the base of continents (Section 1.4.2.2; Fig. 1.3). The Devonian coral trend indicates that the current flux due to spin down can be no less than 6 mW m^{-2} whereas the geotherm suggested by xenoliths sets an upper limit of 12 mW m^{-2}. Earlier flux from shear was larger, and contributed to elevating temperatures near the surface early on (Chapter 10).

1.6.2 Radionuclides

Two well-known problems exist in meteoritic models of Earth's bulk composition. One is that the $K/U = 10^4$ ratio observed for upper mantle materials (Wasserburg et al., 1964) poorly matches that measured in chondritic meteorites (see tables in Lodders and Fegley, 1998). To address the K/U discrepancy, previous meteoritic models assume large, preferential loss of volatile elements. Table 1.3 compares an actual chondritic composition with bulk Earth models. All current powers are much larger than the power entering the base of the continents. Moreover, additional power is generated in upper continents. This brings up the well-known second problem, that given the chemical compositions of crustal rocks, all of the Th and U provided by a chondritic model now resides in continental rocks. Potassium does not contribute much current power, so if the high K values suggested by meteorites are appropriate for Earth, some of that K could reside much deeper in the Earth. Deep sequestering of K would not contribute to heat currently being conducted out of Earth's uppermost layers, but would heat the interior (Fig. 1.3; Chapter 2). Additional U in the lower mantle would have the same effect.

1.6.3 An estimate of internal power

Although meteorite models cannot accurately represent the bulk Earth, these serve as a rough but important guide. Rather than invoking unconstrained volatilization, we consider mixing of various categories of primitive meteorites. Mixing is the basis of some models (e.g., Lodders, 2000). Hofmeister and Criss (2013) showed that three reservoirs can address the large amount of metal in the core, while reconciling oxygen isotope data on meteorites with that of upper mantle materials, and providing the current surface power. Here we consider power inside the whole Earth, rather than just the surface measurements, because the latter are affected by spin dissipation and radionuclide stratification. Earth's large core requires a separate reservoir of metallic meteoritic material, because chondritic meteorites do not possess sufficiently large amounts of Fe-Ni alloy. This large metallic reservoir probably contains negligible quantities of radionuclides, given that their concentrations are tiny in iron meteorites (e.g., Mittlefehldt et al., 1998). This huge metallic reservoir must also include carbon, but the proportion of C to metal is equivocal (Section 1.2).

Oxygen isotope data suggest that Earth is a mixture of chondritic (silicate) meteorites with a component similar to their calcium aluminum inclusions (Fig. 1.5; Hofmeister and Criss, 2012a). The silicate reservoir has some average K, Th, and U composition. Since we do not know the proportions of the different primitive meteorites, the simple average of Table 1.3 is used, which is very close the common H class. The primitive C1 class lies on

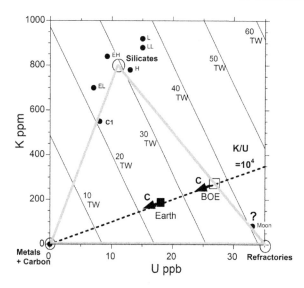

FIGURE 1.13 Computation of global radiogenic Power (diagonal contours) for a body of earth's mass having the indicated elemental compositions, presuming Th/U = 3.75. Dashed line shows the expected K/U ratio. Labeled dots show the K and U contents for various Chondritic meteorite Classes (Lodders and Fegley, 1998) and Taylor's (1982, 2014) lunar estimate. Circles are the three primordial reservoirs. The silicate and metal reservoirs are constrained by meteoritic data, whereas the refractory reservoir is estimated, based on the refractory nature of the moon. Arrows indicate the effect of substantial C on the earth's inferred composition, with permissions.

this line. Whereas this graphical representation (Fig. 1.13) directly depicts only K and U, the Th contents are effectively represented because the Th/U ratio is very uniform in diverse materials.

Calcium aluminates have substantial U, but negligible K. For simplicity, the refractory reservoir is assumed to have no K, and so must lie on the x-axis of Fig. 1.13. To place this reservoir, we utilize the fact that the Moon is more refractory than the Earth, recognizing that its composition remains equivocal since available samples are from the lunar surface only (Taylor, 2014). The Moon's position should lie on the line between the silicate and refractory reservoirs since its core is nearly negligible in radius, which places the refractory reservoir at 35 ppb U.

Earth's bulk composition lies inside the triangle defined by these three reservoirs. To place Earth, we adopt Wasserburg's ratio of K/U = 10^4. This restricts the bulk oxide Earth, but not the mantle because carbon is present (more below). The bulk Earth is placed on the diagram in accord with its huge core, but the presence of carbon dilutes the power, as indicated by the arrows in Fig. 1.13. This position is consistent with primitive carbonaceous C1 chondrites being an important component. The large amount of refractory material in the bulk Earth explains its low K/U ratio compared to chondrites, and also requires that the lower mantle is much lower in ^{18}O than +6, as pointed out by Criss (2008), and as evident in Fig. 1.5.

The resulting power for the whole Earth of ∼25 TW is now interpreted based on data for the continents, which have 16 TW of power (Section 1.4.2). Because the continents comprise only a tiny fraction of Earth's mass, their extraction from the mantle could not possibly have removed all of the U, Th, and K from great depths, even though these are preferentially partitioned in felsic rocks. The power remaining in the lower mantle is roughly estimated as 9 TW (= 25 − 16). This deep radioactive heat has not yet reached the surface due to slow cooling of large spheres but instead has warmed the interior (Criss and Hofmeister, 2016; Chapter 2).

1.6.4 A refractory, carbon-rich lower mantle

Earth's lower mantle probably has a more refractory, more carbon-rich composition than conventionally envisioned. Under ambient conditions, the density of refractory phases is intermediate between that of metal and silicate phases, so material having this composition could logically reside between the core and upper mantle. Also, refractory phases can exist at high temperatures without melting, as can diamond. Carbon is abundant in the Sun.

It follows that Earth's refractory reservoir is very substantial. On this basis, Hofmeister and Criss (2013) suggested that lower mantle is far richer in Ca, Al, and Ti oxides and poorer in silica than previous models, which assume a homogeneous mantle. This inference is supported by the presence of spinel, $CaSiO_3$, and Ti-rich inclusions in diamonds that were derived from the upper mantle (Section 1.2.2). Diamond should also be a common lower mantle phase, given the proportion of metal + carbon in the bulk Earth suggested by Solar abundances, and that the core density suggests only about 10% light elements. A refractory composition for the lower mantle suggests that it could generate significant radioactive power, but no more than 16 TW based on Fig. 1.13, core size, and the estimated position for the refractory reservoir.

References

Agee, C.B., 1998. Phase transformations and seismic structure in the Upper mantle and transition zone. Rev. Miner. 37, 165–204.

Anderson, D.L., 2007. New Theory of the Earth, second ed. Cambridge University Press, Cambridge.

Armienti, P., Gasperini, D., 2010. Isotopic evidence for chaotic imprint in upper mantle heterogeneity. Geochem. Geophys. Geosyst. 11, 0AC02. Available from: https://doi.org/10.1029/2009GC002798.

Arndt, N.T., Lesher, C.M., 2005. Komatiites. In: Selley, R.C., Cocks, L.R.M., Plimer, IR. (Eds.), Encyclopedia of Geology, 3. Elsevier, New York, pp. 260–267.

Ashwal, L.D., 1993. Anorthosites. Springer Berlin Heidelberg, Berlin, Heidelberg.

Bass, J.D., 1995. Elasticity of minerals, glasses, and melts. In: Ahrens, T.J. (Ed.), Mineral Physics and Crystallography. A Handbook of Physical Constants. American Geophysical Union, Washington, D.C, pp. 29–44.

Basu, S., Antia, H.M., 2008. Helioseismology and solar abundances. Phys. Rep. 457, 217–283.

Bellini, G., et al., 2010. Observation of geo-neutrinos. Phys. Lett. B687, 299–304.

Bercovici, D., 2015. Mantle dynamics: an introduction and overview. In: Bercovici, D. (Ed.), Treatise on Geophysics, (G. Schubert, Ed. In Chief) V. 7 Mantle Dynamics. Elsevier, The Netherlands, pp. 1–22.

Bevington, P., 1969. Data Reduction and Error Analysis for the Physical Sciences. McGraw-Hill Book Company, Saint Louis Missouri.

Bodinier, L., Godard, M., 2014. Orogenic, ophiolitic, and abyssal peridotites. In: Turekian, K., Holland, H. (Eds.), Treatise on Geochemistry, vol. 2. Elsevier, Amsterdam, pp. 103–167.

Brearly, A.J., Jones, R.H., 1998. Chondritic meteorites. Rev. Miner. 36, Ch. 3 (398pp).

Buchwald, V.F., 1975. Handbook of iron Meteorites. University of California Press, Berkeley. Available at. Available from: http://evols.library.manoa.hawaii.edu/handle/10524/33750.

Clayton, R.N., Mayeda, T.K., 1975. Genetic relations between the moon and meteorites. In: Proc. 6th Lunar Science Conf., vol. 6, pp. 1761–1769.

Chiar, J.E., Adamson, A.J., Kerr, T.H., Whittet, D.C.B., 1995. High-resolution studies of solid CO in the Taurus dark cloud: characterizing the ices in quiescent clouds. Astrophys. J. 455, 234–243.

Chudinovskikh, L., Boehler, R., 2007. Eutectic melting in the system Fe–S to 44 GPa. Earth Planet. Sci. Lett. 257, 97–103.

Criss, R.E., 2008. Terrestrial oxygen isotope variations and their implications for planetary lithospheres. Rev. Miner. Geochem. 68, 15–30.

Criss, R.E., Hofmeister, A.M., 2016. Conductive cooling of spherical bodies with emphasis on the Earth. Terra Nova 28, 101–109.

Davis, E.E., Fisher, A.T., 2011. Seafloor heat flow: methods and observations. In: Gupta, H.K. (Ed.), Encyclopedia of Solid Earth Geophysics. Springer. Available from: https://doi.org/10.1007/978-90-481-8702-7.

Doglioni, C., Anderson, D.L., 2015. Top driven asymmetric mantle convection. In: Foulger, G.R., Lustrino, M., King, S.D., (Eds.), The Interdisciplinary Earth: In Honor of Don L. Anderson. GSA Special Papers, vol. 214, pp. 51–64.

Doglioni, C., Panza, G., 2015. Polarized plate tectonics. Adv. Geophys 56, 1–167.

Du, Z., Vinnik, L.P., Foulger, G.R., 2006. Evidence from p-to-s mantle converted waves for a flat "660-km" discontinuity beneath Iceland. Earth Planet. Sci. Lett. 241, 271–280.

Duffy, T.S., Anderson, D.L.A., 1988. Seismic velocities in mantle minerals and the mineralogy of the upper mantle. J. Geophys. Res. 94, 1895–1912.

Dziewonski, A.M., Anderson, D.L., 1981. Preliminary reference earth model. Phys. Earth Planet. Inter. 25, 297–356.

Fegley Jr., B., 2015. Practical Chemical Thermodynamics for Geoscientists. Academic Press\Elsevier, Waltham Massachusetts.

Fichtner, A., Villasenor, A., 2015. Crust and Upper mantle of the western Mediterranean - constraints from full-waveform inversion. Earth Planet. Sci. Lett. 428, 52–62.

Foulger, G.R., 2010. Plates Vs Plumes: A Geological Controversy. Wiley-Blackwell, Chichester, UK.

Foulger, G.R., Pritchard, M.J., Julian, B.R., Evans, J.R., Allen, R.M., Nolet, G., et al., 2001. Seismic tomography shows that upwelling beneath Iceland is confined to the upper mantle. Geophys. J. Int. 146, 504–530.

Foulger, G.R., Panza, G.F., Artemieva, I.M., Bastow, I.D., Cammarano, F., Evans, J.R., et al., 2013. Caveats on tomographic images. Terra Nova 25, 259–281.

French, S.W., Romanowicz, B., 2015. Broad plumes rooted at the base of the earth's mantle beneath major hotspots. Nature 525, 95–99.

Galenas, M.G., 2008. Transport Properties of Mid-Ocean Ridge Basalt from the East Pacific Rise. BS Honors Thesis, University of Missouri-Columbia, 18 pp.

Gando, A., et al., 2011. Partial radiogenic heat model for earth revealed by geoneutrino measurements. Nat. Geosci. 4, 647–651.

Gando, A., Gando, Y., Ichimura, K., et al., 2011. Partial radiogenic heat model for earth revealed by geoneutrino measurements. Nat. Geosci. 4, 647–651.

Gao, S.S., Liu, K.H., 2013. Imaging mantle discontinuities using multiply-reflected p-to-s conversions. Earth Planet. Sci. Lett. 402, 99–106.

Gillis, et al., 2014. Primitive layered gabbros from fast-spreading lower oceanic crust. Nature 505, 204–208.

Goes, S., Agrusta, R., van Hunen, J., Garel, F., 2017. Subduction-transition zone interaction: a review. Geosphere 13. Available from: https://doi.org/10.1130/GES01476.1.

Gosnold, W.D., Panda, B., 2002. Interim Compilation of the International Heat Flow Commission. http//www.heatflow.org (accessed 22.09.02.).

Grand, S.P., Helmberger, D.V., 1984. Upper mantle shear structure beneath the Northwest Atlantic ocean. J. Geophys. Res. 89, 11,465–11,475.

Granot, R., 2016. Palaeozoic oceanic crust preserved beneath the eastern Mediterranean. Nat. Geosci. 9, 701–705. Available from: https://doi.org/10.1038/ngeo2784.

Gregory, R.T., Taylor Jr., H.P., 1981. An oxygen isotope profile in a section of cretaceous oceanic crust, Samail ophiolite, Oman: evidence for $\delta^{18}O$ buffering of the oceans by deep (>5 km) Seawater-hydrothermal circulation at mid-ocean ridges. J. Geophys. Res. 86, 2737–2755.

Halliday and Resnick, 1966. Physics, John Wiley & Sons Inc, New York.

Hamilton, W.B., 2011. Plate tectonics began in Neoproterozoic time, and plumes from deep mantle have never operated. Lithos 123, 1–20.

Hamza, V.M., Vieira, F.P., 2012. Global distribution of the lithosphere-asthenosphere boundary: a new look. Solid Earth 3, 199–212. Available from: https://doi.org/10.5194/se-3-199-2012.

Harris, R.N., McNutt, M.K., 2007. Heat flow on hot spot swells: evidence for fluid flow. J. Geophys. Res. 112, B03407. Available from: https://doi.org/10.1029/2006JB004299.

Henderson, G., 1982. Inorganic Geochemistry. Permagon Press, New York.

Hofmeister, A.M., 2019. Measurements, Mechanisms, and Models of Heat Transport. Elsevier, Amsterdam. 427pp. Available from: https://www.elsevier.com/books/measurements-mechanisms-and-models-of-heat-transport/hofmeister/978-0-12-809981-0.

Hofmeister, A.M., Branlund, J.M., 2015. Thermal conductivity of the Earth. Treatise in Geophysics, 2nd Edition (G. Schubert, Ed. In Chief) V. 2 Mineral Physics (G.D. Price, ed.). Elsevier, The Netherlands. pp. 584−608.

Hofmeister, A.M., Criss, R.E., 2005a. Earth's heat flux revisited and linked to chemistry. Tectonophysics 395, 159−177.

Hofmeister, A.M., Criss, R.E., 2005b. Mantle convection and heat flow in the triaxial earth. In: Foulger, G.R., Natland, J.H., Presnall, D.C., Anderson, D.L. (Eds.), Melting Anomalies: Their Nature and Origin. Geological Society of America, Boulder, CO, pp. 289−302.

Hofmeister, A.M., Criss, R.E., 2006. Comment on "Estimates of heat flow from Cenozoic seafloor using global depth and age data" by M. Wei and D. Sandwell. Tectonophysics 428, 95−100.

Hofmeister, A.M., Criss, R.E., 2012a. Origin of HED meteorites from the spalling of Mercury: implications for the formation and composition of the inner planets. In: Hwee-San, L. (Ed.), New Achievements in Geoscience. InTech, Croatia, pp. 153−178. Available from: http://www.intechopen.com/articles/show/title/the-case-for-hed-meteorites-originating-in-deep-spalling-of-mercury-implications-for-composition-and.

Hofmeister, A.M., Criss, R.E., 2012b. A thermodynamic model for formation of the solar system via 3-dimensional collapse of the dusty nebula. Planet. Space Sci. 62, 111−131.

Hofmeister, A.M., Criss, R.E., 2013. Earth's interdependent thermal, structural, and chemical evolution. Gondwana Res. 24, 490−500.

Hofmeister, A.M., Criss, R.E., 2016. Spatial and symmetry constraints as the basis of the virial theorem and astrophysical implications. Can. J. Phys. 94, 380−388.

Hofmeister, A.M., Mao, H.K., 2003. Pressure derivatives of shear and bulk moduli from the thermal Gruneisen parameter and volume-pressure data. Geochem. Cosmochem. Acta 66, 1207−1227.

Hofmeister, A.M., Criss, R.E., Criss, E.M., 2018. Verified solutions for the gravitational attraction to an oblate spheroid: implications for planet mass and satellite orbits. Planet. Space Sci. 152, 68−81. Available from: https://doi.org/10.1016/j.pss.2018.01.005.

Irwin, J., Aigrain, S., Bouvier, J., Hebb, L., Hodgkin, S., Irwin, M., et al., 2009. The monitor project: rotation periods of low-mass stars in M50. Mon. Not. R. Astron. Soc. 392, 1456−1466.

Jessop, A.M., 1990. Thermal Geophysics. Elsevier, Amsterdam.

Jones, M.Q.W., 2016. Virgin rock temperature study of Venetia diamond mine. J. South. Afr. Inst. Min. Metall. 116, 85−92. Available from: https://doi.org/10.17159/2411-9717/2016/v116n1a13.

Lambeck, K., Smither, C., Johnston, P., 1998. Sea-level change, glacial rebound and mantle viscosity for northern Europe. Geophys. J. 134, 102−144. Available from: https://doi.org/10.1046/j.1365-246x.1998.00541.x.

Lathe, R., 2006. Early tides: response to Varga et al. Icarus 180, 277−280.

Lee, W.H.K., 1970. On the global variations of terrestrial heat-flow. Physics of the Earth and Planetary Interiors 2, 332−341.

Li, J., Fei, Y., 2014. Experimental constraints on core composition. In: Turekian, K., Holland (Eds.), Treatise on Geochemistry, vol. 2. Elsevier, The Netherlands, pp. 169−251. , 527−557.

Lodders, K., 2000. An oxygen isotope mixing model for the accretion and composition of rocky planets. Space Sci. Rev. 92, 341−354.

Lodders, K., Fegley Jr., B.J., 1998. The Planetary Scientist's Companion. Oxford University Press, Oxford.

MacDonald, G.J.F., 1959. Calculations on the thermal history of the earth. J. Geophys. Res. 64, 1967−2000.

Meibom, S., Mathieu, R.D., Stassun, K.G., 2009. Stellar rotation in M35: mass−period relations, spin-down rates, and gyrochronology. Astrophys. J. 695, 679−694.

Metz, W.D., 1974. Chemical condensation sequence in the solar nebula. Science. 186, 817.

Mittlefehldt, D.W., McCoy, T.J., Goodrich, C.A., Kracher, A., 1998. Non-Chondritic meteorites from asteroidal bodies. Rev. Miner. 36, 1529−6466.

Nestola, F., Jung, H., Taylor, L.A., 2017. Mineral inclusions in diamonds may be synchronous but not Syngenetic. Nat. Commun. #14168. Available from: https://doi.org/10.1038/ncomms14168.

Nixon, P.H., 1987. Mantle Xenoliths. J. Wiley & Sons, New York.

Pearson, D.G., Canil, D., Shirey, S.B., 2014. Mantle samples included in volcanic rocks: xenoliths and diamonds. In: Turekian, K., Holland, H. (Eds.), Treatise on Geochemistry, vol. 2. Elsevier, Amsterdam, pp. 169−253.

Prensky, S., 1992. Temperature measurements in boreholes: an overview of engineering and scientific applications. Log Analyst 33, 313−333.

Prewitt, C.T., Downs, R.T., 1998. High pressure crystal chemistry. Rev. Mineral. Geochem. 37, 283−317.

Price, G.D., and Stixrude, L., 2015. Mineral physics: an introduction and overview. In: Schubert, G.editor Treatise on Geophysics, vol. 2.pp. 1-5.

Pollack, H.N., Hurter, S.J., Johnson, J.R., 1993. Heat flow from the earth's interior: analysis of the global data set. Rev. Geophys. 31, 267−280.

Rees, B.A., Okal, E.A., 1987. The depth of the deepest historical earthquakes. Pure Appl. Geophys. 125 (5). Available from: https://doi.org/10.1007/BF00878029.

Reynard, B., Remy, C., Takir, F., 1997. High-pressure Raman spectroscopic study of Mn_2GeO_4, Ca_2GeO_4, Ca_2SiO_4, and $CaMgGeO_4$ olivines. Phys. Chem. Miner. 24, 77−84.

Riguzzi, F., Panza, G., Varga, P., Doglioni, C., 2010. Can earth's rotation and tidal Despinning drive plate tectonics? Tectonophysics 484, 60−73.

Ringwood, A.E., 1962. A model for the upper mantle. J. Geophys. Res. 67, 857−867.

Scrutton, C.T., 1978. Periodic growth features in fossil organisms and the length of the day and month. In: Brosche, P., Sündermann, J. (Eds.), Tidal Friction and the Earth'S Rotation. Springer, Berlin, pp. 154−196.

Shankland, T.J., 1972. Velocity-density systematics: derivation from Debye theory and the effect of ionic size. J. Geophys. Res. 77, 3750−3758. Available from: https://doi.org/10.1029/JB077i020p03750.

Sherwood, B.A., Bernard, W.H., 1984. Work and heat transfer in the presence of sliding friction. Am. J. Phys. 52, 1001−1007.

Smith, E.M., Shirey, S.B., Nestola, F., et al., 2016. Large gem diamonds from metallic liquid in earth's deep mantle. Science 354, 1403−1405.

Stachel, T., Harris, J.W., Brey, G.P., Joswig, W., 2000. Kankan diamonds (Guinea): II. Lower mantle inclusion Parageneses. Contrib. Min. Petrol 140, 16−27.

Stehli, F.G., Douglas, R.G., Newell, N.D., 1969. Generation and maintenance of gradients in taxonomic diversity. Science 164 (3882), 947−949. Available from: https://doi.org/10.1126/science.164.3882.947.

Stein, S., Wysession, M., 2003. An Introduction to Seismology, Earthquakes, and Earth Structure. Blackwell Publishing, Oxford U.K.

Stein, C.A., Stein, S.A., 1992. A model for the global variation in oceanic depth and heat flow with lithospheric age. Nature 359, 123−128.

Taylor, S.R., 1982. Lunar and terrestrial crusts: a contrast in origin and evolution. Phys. Earth Planet. Inter. 29, 233−241.

Taylor, S.R., 2014. The moon re-examined. Geochim. Cosmochim. Acta 141, 670−676.

Taylor, S.R., McLennan, S.M., 1985. The Continental Crust: Its Composition and Evolution. Blackwell Scientific Publication, Carlton.

Tharimena, S., Rychert, C., Harmon, N., 2017. A unified continental thickness from seismology and diamonds suggests a melt-defined plate. Science 357, 580−583.

Turekian, K., Holland, 2014. Treatise on Geochemistry. Elsevier, The Netherlands.

Van Schmus, W.R., 1995. Natural radioactivity of the crust and mantle. In: Ahrens, T.J. (Ed.), Global Earth Physics. American Geophysical Union, Washington, D. C, pp. 283−291.

Vieira, F., Hamza, V., 2018. Global heat flow: new estimates using digital maps and GIS techniques. Int. J. Terres. Heat Flow Appl. Geotherm 1, 6−13. Available from: https://doi.org/10.31214/ijthfa.v1i1.6.

Walck, M.C., 1984. The P-wave upper mantle structure beneath an active spreading centre: the gulf of California. Geophys. J. R. Astron. Soc. 76, 691−723.

Warren, J.M., 2016. Global variations in abyssal peridotite compositions. Lithos. 248−251, 193−219.

Wasserburg, G.J., MacDonald, G.J.F., Hoyle, F., Fowler, W.F., 1964. Relative contributions of uranium, thorium, and potassium to heat production in the earth. Science 143, 465−467.

Weinelt, M., 2003. Create-a-map, website: http://www.aquarius.geomar.de/omc.

Wessel, P., Smith, W.H.F., 1995. The Generic Mapping Tools (GMT), version 3.0: Techical reference and cookbook (SOEST/NOOA), http://gmt.soest.hawaii.edu.

Yanagi, T., 2011. Arc Volcano of Japan: Generation of Continental Crust From the Mantle. Springer, Berlin.

Zolensky, M.E., et al., 2006. Mineralogy and petrology of comet 81P/wild 2 nucleus samples. Science 314, 1735–1739.

Websites

For data on meteorites, see the Meteoritical Bulletin database: www.lpi.usra.edu/meteor/metbull.php (accessed 16.12.18.).

Spectroscopic database: http://rruff.info/ (accessed 17.5.18.).

World's deepest mine: https://en.wikipedia.org/wiki/Mponeng_Gold_Mine (accessed 11.5.18.).

Conductive (diffusive) cooling of rocky solar system objects

Robert E. Criss

"Heat, like gravity, penetrates every substance of the universe; its rays occupy all parts of space. The object of our work is to set forth the mathematical laws which this element obeys. The theory of heat will hereafter form one of the most important branches of general physics." *Joseph Fourier* (***The Analytical Theory of Heat, 1822***).

The internal generation of heat and its loss from the surfaces of planetary objects, along with the force of gravity (Chapter 4), largely govern their magmatism, tectonism, and evolution. The importance of conductive and radiative mechanisms to heat transfer in planets was demonstrated, and quantified, using laboratory data and theory in Hofmeister (2019). However, the large length scales for planets require additional considerations, because the associated time scales are long and greatly delay the attainment of steady-state conditions. The spherical geometry of planets sets additional restrictions on the boundary conditions and on the form of differential equations governing heat flow. The present chapter quantifies ballistic radiative losses to the surroundings as well as heat conduction in planetary interiors, via macroscopic modeling.

Sections 2.1 and 2.2 review some of the most important results of previous conduction and radiation models, discuss some of their applications and limitations, and provide some new equations for radiative cooling, heat generation, and cooling scale times. Section 2.3 demonstrates the relevance of these results to rocky objects that range from individual grains to planets. Earth is discussed in detail. Importantly, shallow locations of radionuclides enhance cooling of its upper layers, while warming the interior!

2.1 Fundamental differential equations

2.1.1 Differential operators

Common problems in heat flow involve rectilinear slabs, cylindrical pipes or wires, or spherical masses, in which cases the Cartesian, cylindrical, or spherical coordinate systems are respectively convenient. The generalized equations presented below are useful for all those systems, but the various differential operators, specifically the gradient, divergence and Laplacian, have specific forms for each coordinate system. Table 2.1 illustrates the form of these operators for the cases of greatest interest to heat flow. Note that the temperature gradient is a vector, whereas the divergence and Laplacian of temperature are scalar quantities, and the Laplacian is closely related to the divergence of the flux.

2.1.2 Fourier's laws

Fourier's first law embodies the essential physics of conductive cooling:

$$\Im = -K\nabla T \tag{2.1}$$

This relationship posits a direct proportionality between the heat flux ($\Im$, J m^{-2} s^{-1}) and the temperature gradient (∇T, deg m^{-1}), via a physical property called the thermal conductivity (K, J deg^{-1}m^{-1} s^{-1}). The negative sign indicates that heat flows "down" the thermal gradient; i.e., from hot regions to cold, in accord with the 2nd law of thermodynamics. Fourier's Law is always instantaneously correct, and is very useful for problems in steady flow, where both the flux and temperature gradient remain constant.

TABLE 2.1   One-dimensional differential operators applied to temperature and flux.

		Cartesian	Cylindrical	Spherical
Gradient	∇T	$\hat{x}\dfrac{\partial T}{\partial x}$	$\hat{r}\dfrac{\partial T}{\partial r}$	$\hat{s}\dfrac{\partial T}{\partial s}$
Divergence	$\nabla \bullet \Im$	$\dfrac{\partial \Im(x)}{\partial x}$	$\dfrac{1}{r}\dfrac{\partial[r\Im(r)]}{\partial r}$	$\dfrac{1}{s^2}\dfrac{\partial[s^2\Im(s)]}{\partial s}$
Laplacian	$\nabla^2 T$	$\dfrac{\partial^2 T}{\partial x^2}$	$\dfrac{1}{r}\dfrac{\partial}{\partial r}\left[r\dfrac{\partial T}{\partial r}\right]$	$\dfrac{1}{s^2}\dfrac{\partial}{\partial s}\left[s^2\dfrac{\partial T}{\partial s}\right]$

Note: x, r and s respectively refer to position, cylindrical radius, and spherical radius, while $\hat{x}$, $\hat{r}$, and $\hat{s}$ are unit vectors in those directions. See Rojansky (1971) for multidimensional forms.

This equation is analogous to Fick's Law of diffusion, to Darcy's Law of hydrology, and to several other important physical laws.

Cases of non-steady flow are best analyzed by taking the divergence of Fourier's Law:

$$\nabla\bullet\mathfrak{I} = \nabla\bullet(K\nabla T) \tag{2.2}$$

where the divergence is another well-known differential operator. If the divergence of the flux in a spatial volume of interest is not zero, there must be a change in the heat content of that volume with time (t), because the divergence operator conveniently quantifies the net flow through the surface enveloping that volume. The loss or gain of heat equals the product of the density ρ (kg m^{-3}) and specific heat c_m (J kg^{-1} K^{-1}) of the material times the temperature change:

$$\rho c_m \frac{\partial T}{\partial t} = \nabla\bullet(K\nabla T) \tag{2.3}$$

where the subscript m for mass is used because several different heat capacities exist. If K is constant, this equation becomes what is commonly called Fourier's second Law:

$$\rho c_m \frac{\partial T}{\partial t} = K\nabla^2 T \tag{2.4}$$

where ∇^2 is the Laplacian differential operator. Note that this law was directly derived from (2.1), and that the lumped parameter $K/(\rho c_m)$ is called the thermal diffusivity, D (m^2 s^{-1}). If the physical properties are independent of T over the temperature range of interest, then D is the crucial variable. Thermal diffusivity is fundamental because it describes how fast heat moves, whereas K describes the combination of how much heat moves and how fast. Thermal conductivity is actually the more important property, but the downside is that the amount of heat and its speed cannot be separated. For further discussion, see Hofmeister (2019), Chapter 3 in particular.

If a source or sink of heat (Ψ, J m^{-3} s^{-1}) exists in the volume, it is added to the right hand side:

$$\rho c_m \frac{\partial T}{\partial t} = K\nabla^2 T + \Psi \tag{2.5}$$

In many situations, it is convenient to solve the latter equations to express temperature as a function of position and time, and then to determine the flux by differentiating those solutions, per Eq. (2.1). Also, although crafted to accommodate non-steady heat flow, the latter equations find broad application to problems in steady flow, in which case $\partial T/\partial t$ is taken as zero. For steady flow, Eqs. (2.4) and (2.5) respectively reduce to standard forms known as Laplace's and Poisson's equations.

2.2 Important solutions

A solution to any given differential equation governing a physical process must satisfy that equation, which is easily checked by taking appropriate derivatives of the proffered solution, then introducing them into the original equation and demonstrating that an identity is secured. In addition, all initial and boundary conditions set for the problem must be satisfied. If everything works out, the solution is verified and generally unique.

Modern studies commonly replace such an analytical approach with numerical methods, now that powerful computers are widely available. However, numerical results are specific to the situation examined, and in any case, the numerical algorithms can only be confirmed by comparing their results to analytical solutions. The main advantage of analytical solutions is that they reveal important physical proportionalities and relationships, thereby providing insight into the phenomenon of interest, along with generalized predictive capability. We are therefore fortunate that great historic effort has been directed toward analytically solving Fourier's equations (e.g., Carslaw and Jaeger, 1959). Results of particular importance to planetary science are provided below; in many cases these equations have been recast in ways that extends their utility. New results are also included.

Importantly, available analytical solutions are limited to K being independent of temperature and pressure. Constant K is assumed in this chapter. At high temperatures and pressures inside planets, this is fairly reasonable because their derivatives roughly offset each other; moreover, compositions of planetary interiors are too uncertain to warrant more detail. For example, samples of Earth's mantle are limited to xenoliths originating above 400 km, plus a few inclusions inferred to form near 660 km, and so the composition of the voluminous lower mantle is minimally constrained. Chapter 1 summarizes available information on Earth's interior.

2.2.1 One-dimensional flow

For one-dimensional linear flow, Cartesian coordinates are simplest, and Fourier's laws become:

$$\Im = -K\frac{\partial T}{\partial x}; \qquad \frac{\partial T}{\partial t} = D\frac{\partial^2 T}{\partial x^2} + \frac{\Psi}{\rho c_m} \tag{2.6}$$

2.2.1.1 Steady temperature

For steady temperature in a one-dimensional system, the general solution to 2.6 is:

$$T = -\Psi\frac{x^2}{2K} + b_1 X + b_2 \tag{2.7}$$

where b_1 and b_2 are constants, used to accommodate the particular boundary conditions for a problem of interest.

2.2.1.2 Fundamental solution

The "fundamental solution" to the one-dimensional heat equation describes cooling of an infinite body, initially at constant temperature T_i, after the instantaneous placement of a unit pulse of heat (delta function) into a zone of zero width. The unit pulse of heat, or Dirac delta function, embodies a finite amount of energy (one unit) represented by infinite

T in a zone of zero width, an apparent impossibility yet one whose net result is realistic and finite. The well-known solution to Eq. (2.6) for $\Psi = 0$ is:

$$T = T_i + \frac{b}{\sqrt{\pi Dt}} \exp\left(-\frac{x^2}{4Dt}\right) \tag{2.8}$$

where b is a constant. Although little used in planetary studies, an analogue of this equation known as the "thin film" solution is widely used to determine chemical diffusivities in metallurgical systems. Somewhat similarly, Criss and Winston (2003) differentiated an analogous equation to determine hydraulic flux, and used that result to craft the first theoretical hydrograph.

2.2.1.3 Half-space model

The fundamental solution can be integrated to quantify the process where an infinite half space, initially at temperature T_i, has its surface permanently maintained at temperature T_{surf}. The geometry and constraints are shown in Fig. 2.1A. The well-known solution is:

$$\frac{T - T_{surf}}{T_i - T_{surf}} = \text{erf}\left(\frac{z}{2\sqrt{Dt}}\right), \tag{2.9}$$

where erf is the error function and z represents the vertical direction. Note that cooling is a simple function of the dimensionless combination $z/(Dt)^{1/2}$, which arises in many solutions for the heat equation. For many systems, doubling the size, length scale, or position

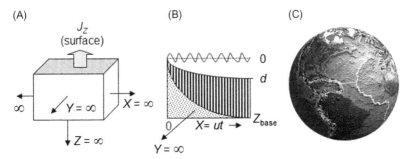

FIGURE 2.1 Geometrical constructions in various cooling problems. (A) The original 1-d half-space cooling model of Kelvin. Gray color = upper surface at $z = 0$ fixed at 0 K. The solid body extends to infinity in all directions below that surface. (B) Depiction of oceanic lithosphere in the half-space model used by Stein and Stein (1992), which is 2-d but uses the 1-d solution. The upper surface is fixed at 0 °C. At a lateral distance x from any given mid-ocean ridge, the solution is applied for time = t, using the transformation $x = ut$. Plate models are similar except that temperatures at the surface and base of the lithosphere are specified as constant. (C) The 3-d earth does not resemble a flat plate. This image shows earth's surface without oceans, with plate boundaries and volcanoes marked. *From NASA, 2018. Goddard Space Flight Center's Scientific Visualization Studio of Earth. https://svs.gsfc. nasa.gov/155 (accessed 06.02.19.). Also involved in this project are: Smithsonian Institution, Global Change Research Project, National Oceanic and Atmosphere Administration, United States Geological Survey, National Science Foundation, Defense Advanced Research Projects Agency, Dynamic Media Associates, New York Film and Animation Company, Silicon Graphics, Inc., and Hughes STX Corporation (NASA, 2018), which is open access.*

of interest yields the same temperature change when the elapsed time is quadrupled. For finite bodies, the cooling time scale is approximately: $\tau \sim z^2/D$.

Application of the half-space cooling model to determine Earth's age is historically important, as discussed in Section 2.3.1. Despite the geometry being at odds with that of the Earth (Fig. 2.1), this model is conventionally applied to tectonic plates. Alternative models for oceanic lithosphere are covered in Chapter 7.

2.2.1.4 Infinite slab

A hot infinite slab of finite width has many applications, for example it resembles an igneous dike. For a dike of initial temperature T_i and half width L, cooling into an infinite medium with initial temperature T_0, the well-known solution is:

$$\frac{T - T_0}{T_i - T_0} = \frac{1}{2}\left[\operatorname{erf}\left(\frac{L-x}{2\sqrt{Dt}}\right) + \operatorname{erf}\left(\frac{L+x}{2\sqrt{Dt}}\right)\right] \tag{2.10a}$$

For this case, the temperature of the medium near the dike increases, lengthening the cooling time. The cooling scale time can be estimated as the interval required for the object to cool by a factor of Euler's number e, i.e. for $T_{avg} - T_0 = (T_i - T_0)/e$. This condition is realized at time τ:

$$\tau = 2.011085 \frac{L^2}{D} \tag{2.10b}$$

The solution Eq. (2.10a) has the interesting property that, for short times, the surface temperature T_{surf} of the slab is maintained very near the constant value, $(T_i + T_0)/2$. This condition enables Eq. (2.10a) to be recast, for short times, to describe the cooling of a hot tabular body when the surface temperature is held constant. This solution is:

$$\frac{T - T_{surf}}{T_i - T_{surf}} = -1 + \operatorname{erf}\left(\frac{L-x}{2\sqrt{Dt}}\right) + \operatorname{erf}\left(\frac{L+x}{2\sqrt{Dt}}\right) \tag{2.11a}$$

This equation is accurate up to the cooling scale time, which for this case is:

$$\tau = 0.31688 \frac{L^2}{D} \tag{2.11b}$$

The result of Eq. (2.11b) is considerably shorter than that of Eq. (2.10b). The different time scales stem from the different boundary conditions.

2.2.2 Radial flow in spheres

The Earth and other large bodies in the Solar System are slightly flattened spheres. For radial heat flow in spherical coordinates and constant K, use of the appropriate differential operators provide the following forms for Fourier's two laws:

$$\Im = -K\frac{\partial T}{\partial s}; \qquad \frac{\partial T}{\partial t} = \frac{D}{s^2}\frac{\partial}{\partial s}\left[s^2\frac{\partial T}{\partial s}\right] \tag{2.12}$$

If there is heat generation in the body, the term $\Psi/(\rho c_m)$ is added to the RHS.

2.2.2.1 Steady radial temperature

Steady, radial flow of heat is possible only for finite bodies with internal heat generation; otherwise the body would cool. For a spherical body with outer radius s_{surf}, internal heat generation Ψ, constant surface temperature T_{surf}, and constant thermal conductivity K, the steady temperature is:

$$T = T_{surf} + \frac{\Psi}{6K}\left(s_{surf}^2 - s^2\right) \tag{2.13}$$

2.2.2.2 Steady radial temperature: two layer sphere

Steady, radial flow of heat in a two-layer sphere, for example a planet with a core and mantle, depends on the relative radii of those regions (s_{core} and s_{mtl}), on the thermal conductivities of those two regions (K_{core} and K_{mtl}) as well as on their heat productions (Ψ_{core} and Ψ_{mtl}).

For the case of heat generation in the core region alone, below a barren mantle, the solution provided by Carslaw and Jaeger (1959) is here modified in a manner that specifies external control of a constant surface temperature, T_{surf}:

$$T - T_{surf} = \frac{\Psi_{core}}{6K_{core}}\left[s_{core}^2 - s^2 + 2s_{core}^3 \frac{K_{core}}{K_{mtl}}\left(\frac{1}{s_{core}} - \frac{1}{s_{surf}}\right)\right] \quad \text{for} \quad 0 \le s \le s_{core} \tag{2.14a}$$

and

$$T - T_{surf} = \frac{\Psi_{core}s_{core}^3}{3K_{mtl}}\left(\frac{1}{s} - \frac{1}{s_{surf}}\right) \quad \text{for} \quad s \ge s_{core} \tag{2.14b}$$

Hofmeister and Criss (2013) solved the case where all the heat generation resides in the mantle, above a barren core:

$$T - T_{surf} = \frac{\Psi_{mtl}}{6K_{core}}\left[s_{surf}^2 - s^2 + 2s_{core}^3\left(\frac{1}{s_{surf}} - \frac{1}{s}\right)\right] \quad \text{for} \quad s_{surf} \ge s \ge s_{core} \tag{2.15}$$

The core temperature is everywhere constant, and takes on the value indicated by the above equation when evaluated at $s = s_{core}$. Interestingly, the core temperature depends in no way on the material characteristics of the core.

2.2.2.3 Time-dependent temperature in the sphere

Transient, radial flow of heat internal heat is described by Eq. (2.12). Even if no internal heat is generated the solution is a complicated infinite series (Carslaw and Jaeger, 1959). However, Criss and Hofmeister (2016) found a "short term" solution to this problem, namely:

$$\frac{T - T_{surf}}{T_i - T_{surf}} = 1 - \frac{s_{surf}}{s}\operatorname{erfc}\left(\frac{s_{surf} - s}{2\sqrt{Dt}}\right) \tag{2.16}$$

This result maintains the surface temperature at the value T_{surf} for all time, and is numerically identical to the series solution if the dimensionless quantity $Dt/s_{surf}^2 \le 0.028$

for $s > 0$. If the elapsed time exceeds the latter restriction, the cooling front has progressed to the center of the sphere, and the calculated results become increasingly incorrect. Even under those conditions, the solution is accurate for positions where $Dt/s^2 \leq \pi^{-1}$; i.e., for extended time in the outer part of the sphere. The average temperature of the sphere is determined by integration (Criss and Hofmeister, 2016):

$$T_{avg} = T_{surf} + (T_i - T_{surf})\left[1 - \frac{3}{\sqrt{\pi}}\frac{\sqrt{4Dt}}{s_{surf}} + \frac{3Dt}{s^2_{surf}}\right] \tag{2.17}$$

The cooling scale time can be estimated as the interval required for the object to cool by a factor of e, i.e. for $T_{avg} - T_{surf} = (T_i - T_{surf})/e$. This condition is realized at time τ:

$$\tau = \left(\frac{1}{\sqrt{\pi}} - \sqrt{\frac{1}{\pi} - \frac{1}{3} + \frac{1}{3e}}\right)^2 \frac{s^2_{surf}}{D} \sim 0.055772\,\frac{s^2_{surf}}{D} \tag{2.18}$$

Note that this scale time is much shorter than for the infinite slab. Part of this difference reflects the 3 × larger surface area to volume ratio of a spherical mass compared to an infinite slab.

Figs. 2.2 and 2.3 compare spheres to slabs, showing how their average temperatures depend on the scale time, and that their temperature profiles at unit scale time differ, when cooling under conditions of constant surface temperature.

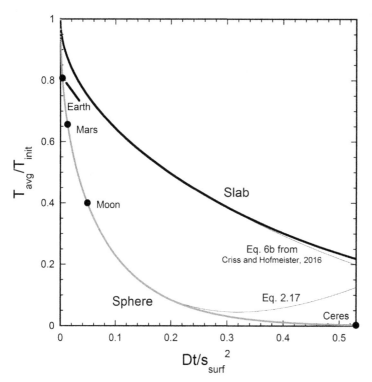

FIGURE 2.2 Plot of the average temperature of the cooling slab and sphere Vs. The dimensionless time, according to the indicated equations. Eq. (2.17) is very accurate up to a dimensionless time (Dt/s^2) of about 0.2, but becomes unrealistic afterward. *Reprinted from Fig. 2.1 in Criss, R. E., Hofmeister, A.M., 2016. Conductive cooling of spherical bodies with emphasis on the Earth. Terra Nova 28, 101–109 (Criss and Hofmeister, 2016), with permissions.*

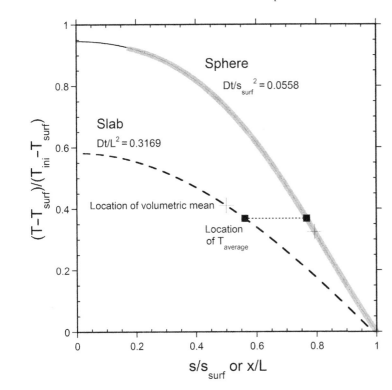

FIGURE 2.3 Temperature profiles of the sphere and infinite slab, cooling under conditions of constant surface temperature, after cooling by a factor of $1/e$, according to the indicated equations. Crosses show the location of the volumetric mean of these shapes, positions that divide the body into equal halves where temperature is hotter and colder. For the sphere, $1/2$ the total volume lies in the outermost $\sim 20\%$ of its radius. The surface area to volume ratio for a cooling sphere is $3\times$ larger than the slab, so its scale time is shorter, but cooling is concentrated in the outermost zones, preserving deep internal heat.

2.2.3 Unconventional solutions

Two modifications of Fourier's laws are useful in special circumstances. The first is venerable but the constants are commonly misconstrued; the second is largely new and has some novel applications.

2.2.3.1 Newton's law of cooling

Newton reasoned that the rate of heat loss from hot material is proportional to the temperature difference between the object and its surroundings. This proportionality requires exponential decline of temperature. This condition is accurate only if temperature gradients are unimportant, such that the object has uniform temperature as if somehow vigorously stirred or if the internal thermal conductivity is huge. Nonetheless, presuming a uniform temperature provides a crude approximation to actual cooling. In more detail, this cooling law is consistent with Fourier's law, but the divergence of the flux is considered to be equal to $\Phi\mathring{A}(T - T_{surf})$, where T_{surf} is the constant surface temperature, $\mathring{A}$ is the surface area of the object, and Φ is a descriptor of the thermal transmissivity through that surface, which has the units of J m^{-2} s^{-1} deg^{-1}. Thus, the differential equation becomes, for an object of volume V:

$$\rho c_m V \frac{\partial T}{\partial t} = -\Phi\mathring{A}\left(T - T_{surf}\right) \tag{2.19}$$

which has the solution:

$$\frac{T - T_{surf}}{T_i - T_{surf}} = \exp\left(-\frac{\Phi \mathring{A}}{\rho c_m V} t\right) \tag{2.20}$$

Note that the cooling rate depends of the surface area to volume ratio of the object.

2.2.3.2 Blackbody cooling

The Stefan-Boltzmann formula for emission of heat from a hot body into space (Hofmeister (2019), Chapter 8) provides an important way to recast Fourier's laws. For a blackbody with uniform internal temperature that is emitting heat to a much colder surroundings:

$$\rho c_m V \frac{\partial T}{\partial t} = -\mathring{A} \sigma_{SB} T^4 \tag{2.21}$$

where σ_{SB} is the Stefan-Boltzmann constant (5.670×10^{-8} W m^{-2} K^{-4}). The solution is:

$$\frac{1}{T^3} - \frac{1}{T_i^3} = 3\frac{\mathring{A}\sigma_{SB}}{\rho c_m V} t = 3\varpi t \tag{2.22}$$

where ϖ is a lumped parameter and is considered to be a constant. This equation indicates very rapid cooling, as follows. The cooling scale time can be taken as the interval required for the object to cool by a factor of e, such that $T = T_i/e$. This condition is realized when

$$\tau = \frac{e^3 - 1}{3\varpi T_i^3} \tag{2.23a}$$

For typical rocky materials, this scale time in years for s in km is about:

$$\tau = 3 \times 10^9 \frac{s}{T_i^3} \tag{2.23b}$$

Important consequences are discussed below.

2.3 Applications

2.3.1 Kelvin's half-space cooling model

Eq. (2.9) was recast by Kelvin to obtain a historically important estimate for the age of the Earth. Differentiation with respect to the spatial dimension (here z denotes the vertical) gives the thermal gradient, which near the surface is:

$$\left.\frac{\partial T}{\partial z}\right|_{surf} = \frac{T_i - T_{surf}}{\sqrt{\pi D t}} \tag{2.24}$$

As discussed by Carslaw and Jaeger (1959), Kelvin estimated that the early Earth was molten ($\sim$4100 K), that the present-day geothermal thermal gradient is 37 K km^{-1} near the surface, and that the thermal diffusivity of rocks is 1.2×10^{-6} m^2 s^{-1}. With these values, Kelvin's estimate for Earth's age was about 90 Ma, a value protested by contemporaneous geologists as being much too low. Kelvin subsequently argued for an even shorter age of 24 Ma (Carslaw and Jaeger, 1959). His pronouncement held sway for many decades.

Vigorous subsequent arguments have highlighted the dangers in oversimplifying complex problems. The subsequent discovery that radioactivity generates heat is now commonly used to explain Kelvin's low estimate. Even that argument is too simple, as the venerable question as to whether Earth's deep interior is currently cooling or warming (Takeuechi et al., 1967) remains unresolved, see Section 2.3.3.

Given how long progress in geology was impeded by Kelvin's incorrect analysis, it is surprising that the half-space cooling model and its variants (plate models) have been resurrected to explain Earth's heat flow in view of sea floor spreading. This explanation, e.g., Stein and Stein (1992), is widely accepted even though 1) the geometry is inappropriate for a spherical Earth (Fig. 2.1B and C); 2) the one-dimensional thermal conductivity needs to be corrected by a factor of $3\times$, to be volumetrically realistic; 3) huge "corrections" to most available heat flux measurements are made to make the model fit. Bathymetric data are viewed as supporting the model, but this match is disconnected from the heat flow equation because an additional free parameter is used, see Hofmeister and Criss (2005a,b) for additional discussion, and Chapter 7.

2.3.2 Cooling of rocky solar system objects

The Solar System contains rocky objects with a remarkable diversity of characteristics and geologic histories, many of which are explicable in terms of their thermal histories. While simple cooling of small objects is more rapid than for large ones, decay of radionuclides can cause temperatures to rise. The end result depends on the size of the object, the initial and surface temperatures, the concentration and interior distribution of radionuclides, and on other sources of heat. Because the complexity of the outcome increases with size, in what follows small objects are considered first. Table 2.2 provides a summary.

2.3.2.1 Meteorite chondrules

Primitive, undifferentiated meteorites are called "chondrites" because they contain abundant, small (few mm) spherical grains called chondrules. These interesting grains are often regarded as frozen drops of igneous melts. Although their origin is debated (see Chapter 11), they all formed in the earliest times of the Solar System. This section considers the hypotheses that chondrules were generated by violent impacts between the earliest accreting objects, or by localized concentrations of short-lived radioactive isotopes.

Cooling rates of these tiny objects were rapid. Estimates of the cooling scale times can be made using Eqs. (2.18) and (2.23b) for radial cooling and blackbody cooling, respectively. For plausible values of $D \sim 1 \, \text{mm}^2 \, \text{s}^{-1}$, the cooling time for an object with a 1 mm radius is only 0.06 s for the radial cooling model, and 90 s for blackbody cooling. The first estimate is far too short for the surfaces of these small grains to have attained some constant, externally-controlled temperature required by the radial cooling model, so it is unrealistic. The second estimate is superior, but nonetheless is very short.

In support of fast cooling, synthesis of ~mm balls of vitreous forsterite requires cooling rates of ~500 K s^{-1} (Tangeman et al., 2001). Thus, a ~2 s quench is needed to preserve glassy Mg_2SiO_4. Some chondrules have radiating crystals of pyroxene and olivine Mg-endmembers, which may indicate rapid cooling, but not so fast that forsterite glass formed. Our calculated time-scale of radiative cooling is consistent with this evidence. Chapter 11 suggests an entirely different origin of chondrules.

2.3.2.2 Asteroids and dwarf planets

Practically all meteorites are fragments of asteroids, and most are undifferentiated chondrites, so it follows that most asteroids are undifferentiated and have therefore never undergone melting and chemical segregation processes. Of the differentiated meteorites, many are known to have originated from much larger bodies including Mars and the Moon, that clearly have experienced magmatism and other complex planetary processes. Hofmeister and Criss (2012) argued elsewhere that any asteroids constituted of differentiated material are ejected fragments of former, much larger objects such as Mercury, where differentiation was promoted by a significant gravitational field.

Radiogenic heat production in planetary bodies is dominated by the emissions of K, U and Th. In the earliest times of the Solar System, emissions of short-lived radionuclides such as ^{26}Al may have been important, but it is likely that these isotopes had mostly decayed away before their incorporation into large objects, so they are ignored here. I used widely available information on isotopic abundances, decay energies, and half lives to develop the following equation for heat production due to K, U, and Th:

$$\Psi* = 0.00351\xi_{K,ppm}e^{0.5543t} + 0.0263\xi_{Th,ppb}e^{0.0495t} + \xi_{U,ppb}\left[0.0943e^{0.1551t} + 0.00421e^{0.9849t}\right] \quad (2.25)$$

where the heat production $\Psi*$ is reported as picowatts per kg of rock (i.e., 10^{-12} W kg^{-1}), $\xi_{K,ppm}$, $\xi_{Th,ppb}$ and $\xi_{U,ppb}$ are the present-day concentrations of the bulk elements in the rock in the indicted units, and t is the age of interest, expressed as billions of years ago. Note that $\Psi*$ must be multiplied by the rock density to obtain Ψ, defined above.

Eq. (2.25) indicates that heat generation in rocks of ordinary composition is very small. A typical chondritic meteorite would have a potassium content of about 800 ppm, thorium of 42 ppb and uranium of 14 ppb, providing a present day value for $\Psi*$ of only about 5.3 pW kg^{-1}. For an average density of 2500 kg m^{-3}, that power release would be equivalent to trying to heat a cubic kilometer of rock with a 13 Watt light bulb. The power generated would have been greater in earlier times, when radioisotope concentrations were higher, but even 4.5 billion years ago would have been only 43 pW kg^{-1}, or 8× higher.

Integrating Eq. (2.25) indicates how much energy is contributed to rocks over any time interval of interest. For example, over geologic time, 2.4 MJ kg^{-1} would be generated in chondritic material by radioactivity. Given that the heat capacity of typical rocks is about 1000 J kg^{-1}, the temperature of a typical chondrite would increase by about 2400 K, assuming no conductive loss. Temperature increases of this magnitude are possible for the deep interiors of large rocky planets (Criss and Hofmeister, 2016).

For a typical thermal diffusivity of 32 km^2 Ma^{-1}, the scale time of Eq. (2.18) for a simple spherical body to conductively cool by a factor of $1/e$ (37%) is about 0.002 s_{surf}^2 (in Ma), where surface radius is in km. Thus, a non-radioactive body slightly smaller than the Moon, which has a radius of 1737 km, would cool by a factor of $1/e$ over geologic time.

However, the largest asteroid, Ceres ($s_{surf} = 470$ km), has a scale time of only 390 Ma. Radioactive heat added over that interval could heat the body by 500 K; soon after that, conductive losses would more than offset any energy added. Nevertheless, internal heating of Ceres was probably minimally sufficient to induce partial melting, and it is the only asteroid that is approximately spherical. Fewer than 2% of known asteroids have radii greater than 50 km (Lodders and Fegley, 1998), so these would have cooling scale times of less than 5 Ma. Radiogenic heat contributed over such a short interval could have heated up these small bodies by <10 K, so conductive losses would have always outpaced any heat generated. These numerous but small objects accreted cold, and remained cold. This condition is consistent with the undifferentiated character of chondrites.

2.3.2.3 *Moon, Mars, and Mercury*

Rocky objects larger than Ceres clearly underwent internal melting, as is evident from volcanic features on their surfaces. Magmatism is consistent with the scale cooling times for the Moon ($s_{surf} = 1737$ km), Mars (3380 km) and Mercury (2438 km), which are about 5 Ga, 20 Ga, and 10 Ga, respectively. Since these times all exceed the 4.5 Ga age of the Solar System, the total heat generated would correspond to the aforementioned value of 2.4 MJ kg^{-1} for a chondritic composition, enough to heat the bodies by about 2400 K. The compositions of Mercury and Moon are more refractory than chondrites, but the bulk radionuclide compositions of these objects are not well known. Available estimates for the lunar bulk composition suggest that 1.7–3.1 MJ kg^{-1} was generated over 4.5 Ga, corresponding to temperature increases of 1700–3050 K if all this heat were retained.

Although the interiors of these objects clearly melted, practically all magmatic activity ceased about 3 billion years ago on the Moon, the age of the youngest maria. Volcanism may still continue on Mars. Heat generation is less than it was in the early days of the Solar System due to the progressive decay of radionuclides, and these objects are small enough that the current heat production rate is now more than offset by conductive losses. For the chondritic composition, heat generation has been reduced by a factor of $8\times$, or by a factor of $3.7\times$ for the suggested refractory compositions.

Interior melting promotes magmatism, which would tend to segregate the lithophile elements, including K, U and Th, upward. This process differentiated the objects into an interior "core" zone, highly depleted in radionuclides, surrounded by a more fertile mantle shell. While upward segregation of radionuclides would accelerate cooling of the entire object, the results depend on their distribution (Criss and Hofmeister, 2016). The possibilities can be considered in terms of three stages of the planetary cooling process that follows cold accretion.

During the earliest stage, interior temperatures and surface temperature gradients both increase. For small objects this stage is short, and the object then progressively cools. More interestingly, for sufficiently large objects, melting, differentiation, core formation and magmatism begin during this early stage.

During an intermediate interval that pertains only to sufficiently large bodies, planetary interiors continue to warm, and core melting and magmatism continue. However, temperatures and temperature gradients decrease in the outermost parts of the body.

In a terminal stage that applies to all objects, temperatures decrease throughout the planetary interior; for planets this rate is buffered by core freezing. Magma generation progressively decreases to nil. Mercury and smaller objects appear to be in this final stage, and Mars is likely approaching it.

2.3.2.4 Earth and Venus

Earth and Venus are the largest rocky bodies in the Solar System, and are commonly regarded as sister planets, owing to their similar size and density. However, this similarity does not extend to their surface appearance and thus not to their tectonic and magmatic histories. Earth remains very active so its surface is nearly devoid of impact craters. In contrast, Venus appears to lack large-scale tectonic movement, and its magmatism is weak or nonexistent, so its surface contains abundant craters (Hamilton, 2015). Thus, in terms of the evolutionary stages mentioned above, Earth remains well within the intermediate stage, while Venus appears to be nearing the terminal stage. An explanation for this disparity between these "sister" planets is needed.

Much more information is available for Earth than for other planetary objects, but knowledge of its internal character is routinely overstated. Venerable questions as to whether Earth is heating or cooling remain unanswered. Plausible estimates for the bulk-Earth radionuclide concentrations suggest that internal heat generation Eq. (2.25) currently approximates Earth's surface flux, but this agreement is circular because heat flow measurements are routinely used to make those compositional estimates! Even if this match were accurate, internal adjustment of internal temperatures would be ongoing, to equalize the temperatures of the buffered, two-phase core with a hotter overlying lower mantle, achieved by additional melting of the core. Such a process would continue for tens of billions of years, until the thermal gradients become outward everywhere, and the core progressively freezes.

Thus, the dramatic difference between Earth and other rocky objects in the Solar System remains an enigma. Hofmeister et al. (2019) conclude that this difference can only be explained by Earth's remnant diurnal spin and large Moon, which can respectively contribute frictional heat and tectonic force to our unique planet. See summary in Table 2.2.

2.3.3 Is the Earth heating up?

Since the discovery of heat generation by radionuclides, the possibility has been recognized that Earth may be heating up, rather than undergoing progressive cooling. This question was considered in some remarkable papers such as MacDonald (1959), with Takeuechi et al. (1967) devoting an entire chapter of their book to this subject.

Criss and Hofmeister (2016) showed that, in large bodies with a uniform chondritic composition, the interior temperatures will greatly increase over geologic time, even as the radionuclides decay away. Many geochemical processes will concentrate radionuclides upward, which will accelerate the overall cooling rate of the upper layers as the surface sheds heat to space, but still progressively increase deep interior temperatures. In the latter case, a temperature maximum is produced somewhere in the mantle (Fig. 2.4), but the particular position depends on the concentration profile, and may be near the upper

TABLE 2.2   Time scales for cooling of rocky objects in the solar system.

Object	Radius, km	Scale time[a]	Major magmatism
Chondrules	10^{-6}	90 s	N/A
Asteroids	<100	<20 Ma	?
Ceres	470	390 Ma	N/A[b]
Moon	1737	5 Ga	3.0–3.9 Ga[c]
Mercury	2438	10 Ga	~4.0–4.1 Ga[d]
Mars	3380	20 Ga	1.4 Ga[e]
Venus	6052	65 Ga	~0.3–0.6 Ga[f]
Earth	6371	70 Ga	Active

[a]For simple cooling, Eq. (2.18) pertains, except the chondrule case used Eq. (2.23b).
[b]Seasonal changes cause flow of ice (Raponi et al., 2018).
[c]Radiometric dating of lunar mare samples (Dalrymple, 1991).
[d]Estimates from crater counting (Marchi et al., 2013).
[e]Radiometric dating techniques for all Martian meteorites, except for one shergottite (Nyquist et al., 2001).
[f]Estimates from crater counting (Strom et al., 1994).

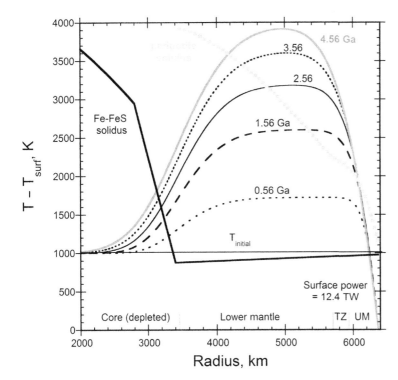

FIGURE 2.4   Evolving temperature profiles of Earth, assuming an initial temperature of 1000 K, and that almost all radionuclides are uniformly concentrated above a nearly barren core. According to this model, Earth's power and temperature gradient have changed little near the surface, while its deep interior has warmed. Note that the maximum temperature occurs in the mantle, not at the center. Stronger, upward radionuclides are more strongly concentration of radionuclides will increase the surface flux, and elevate the position of the temperature maximum. *Reprinted from Fig. 9 of Criss, R.E., Hofmeister, A.M., 2016. Conductive cooling of spherical bodies with emphasis on the Earth. Terra Nova 28, 101–109 (Criss and Hofmeister, 2016), with permissions.*

mantle-lower mantle boundary. These generalized results have many ramifications. First, Earth's core may be melting, not freezing, and most of the lower mantle may be stagnant. The hot lid will prohibit lower mantle convection, even though the outside layers are cooling. Finally, Earth is tectonically and magmatically active where internal heat is produced, which is only in its outermost zones, as summarized in Chapter 1.

Radioactivity is not the only possible source of significant internal power. Exothermic phase changes and frictional heating due to asymmetry in centrifugal and gravitation forces (Hofmeister et al., 2018, 2019) are two possibilities. Core formation was probably a passive response, so it would have contributed little or no heat (Chapter 4), contrary to current thinking. Conductive cooling models, even those that include spherical geometry and radially variable heat generation, are only a guide, albeit an important one.

2.3.4 Constraints on Earth's initial temperature

From Criss and Hofmeister (2016), cooling for the case of homogeneously distributed radionuclides is reasonably represented by adapting Eq. (2.16) to include a temperature augmentation of ΔT arising through internal heating :

$$\frac{T - T_{surf}}{T_i - T_{surf} + \Delta T} \sim 1 - \frac{s_{surf}}{s}\,\mathrm{erfc}\left(\frac{s_{surf} - s}{2\sqrt{Dt}}\right) \tag{2.26}$$

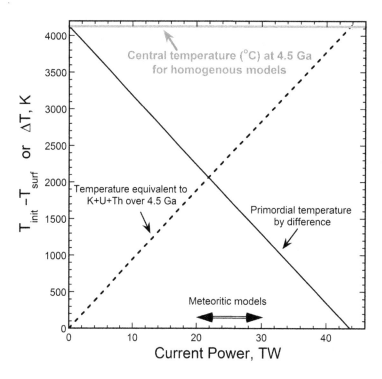

FIGURE 2.5 Trade-offs between radionuclide concentration and initial temperature allow for a given thermal Power to be radiated from earth at the present time. *Reprinted from Fig. 2.5 of Criss, R.E., Hofmeister, A.M., 2016. Conductive cooling of spherical bodies with emphasis on the Earth. Terra Nova 28, 101−109 (Criss and Hofmeister, 2016), with permissions.*

where the mean value theorem provides the effect of power Π (in J s^{-1}):

$$\Delta T = \frac{1}{\rho CV} \int_0^t \Pi(\varepsilon) d\varepsilon = \frac{\Pi_{ave} t}{\rho CV} \qquad (2.27)$$

A trade-off exists between T_{init} and ΔT. All possible conditions are represented by Fig. 2.4. Interestingly, high current power rules out high primordial heat.

The heat leaking out today is considerably lower than the internal power for homogeneous models, because most of the power goes into warming the interior, as discussed above. Neither can primordial heat "escape" because any such energy was provided during planetary assembly and would reside in the deep interior (Fig. 2.5).

References

Carslaw, H.S., Jaeger, J.C., 1959. Conduction of Heat in Solids, second ed. Oxford University Press, New York.

Criss, R.E., Hofmeister, A.M., 2016. Conductive cooling of spherical bodies with emphasis on the Earth. Terra Nova vol. 28, 101−109.

Criss, R.E., Winston, W.E., 2003. Hydrograph for small basins following intense storms. Geophys. Res. Lett. 30, 1314−1318.

Dalrymple, G.B., 1991. The Age of the Earth. Stanford University Press, Stanford, Calif.

Hamilton, W.B., 2015. Terrestrial planets fractionated synchronously with accretion, but Earth progressed through subsequent internally dynamic stages whereas Venus and Mars have been inert for more than 4 billion years. GSA Special Papers 514, pp. 123−156.

Hofmeister, A.M., 2019. Measurements, Mechanisms, and Models of Heat Transport. Amsterdam, New York, 427 pp.

Hofmeister, A.M., Criss, R.E., 2005a. Earth's heat flux revisited and linked to chemistry. Tectonophysics 395, 159−177.

Hofmeister, A.M., Criss, R.E., 2005b. Mantle convection and heat flow in the triaxial earth. In: Foulger, G.R., Natland, J.H., Presnall, D.C., Anderson, D.L. (Eds.), Melting Anomalies: Their Nature and Origin. Geological Society of America, Boulder, CO, pp. 289−302.

Hofmeister, A.M., Criss, R.E., 2012. Origin of HED meteorites from the spalling of Mercury: implications for the formation and composition of the inner planets. In: Hwee-San, L. (Ed.), New Achievements in Geoscience. InTech, pp. 153−178. Available from: http://www.intechopen.com/articles/show/title/the-case-for-hed-meteorites-originating-in-deep-spalling-of-mercury-implications-for-composition-and.

Hofmeister, A.M., Criss, R.E., 2013. Earth's interdependent thermal, structural, and chemical evolution. Gwondona Res. 24, 490−500.

Hofmeister, A.M., Criss, R.E., Criss, E.M., 2018. Verified solutions for the gravitational attraction to an oblate spheroid: implications for planet mass and satellite orbits. Planet. Space Sci. 152, 68−81. Available from: https://doi.org/10.1016/j.pss.2018.01.005.

Hofmeister, A.M., Criss, R.E., Criss, E.M., 2019. Link of planetary activity to moon size, orbit, and planet spin: a new mechanism for plate tectonics. in preparation.

Lodders, K., Fegley Jr., B.J., 1998. The Planetary Scientist's Companion. Oxford University Press, Oxford.

MacDonald, G.J.F., 1959. Calculations on the thermal history of the Earth. J. Geophys. Res. 64, 1967−2000.

Marchi, S., Chapman, C.R., Fassett, C.I., Head, J.W., Bottke, W.F., Strom, R.G., 2013. Global resurfacing of Mercury 4.0-4.1 billion years ago by heavy bombardment and volcanism. Nature 499, 59−61. Available from: https://doi.org/10.1038/nature12280.

NASA, 2018. Goddard Space Flight Center's Scientific Visualization Studio of Earth. https://svs.gsfc.nasa.gov/155 (accessed 06.02.19.). Also involved in this project are: Smithsonian Institution, Global Change Research Project, National Oceanic and Atmosphere Administration, United States Geological Survey, National Science Foundation, Defense Advanced Research Projects Agency, Dynamic Media Associates, New York Film and Animation Company, Silicon Graphics, Inc., and Hughes STX Corporation.

Nyquist, L.E., et al., 2001. Ages and geologic histories of Martian meteorites. Space Sci. Rev. 96, 105–164. Available from: https://doi.org/10.1023/A:1011993105172.

Raponi, A., et al., 2018. Variations in the amount of water ice on Ceres' surface suggest a seasonal water cycle. Sci. Adv. 4, eaao3757. Available from: https://doi.org/10.1126/sciadv.aao3757.

Rojansky, V., 1971. Electromagnetic Fields and Waves. Prentice-Hall, Englewood Cliff, New Jersey.

Stein, C.A., Stein, S.A., 1992. A model for the global variation in oceanic depth and heat flow with lithospheric age. Nature 359, 123–128.

Strom, R.G., Schaber, G.G., Dawson, D.D., 1994. The global resurfacing of venus. J. Geophys. Res. 99, 10899–10926. Available from: https://doi.org/10.1029/94JE00388.

Takeuechi, H., Uyeda, S., Kanamori, H., 1967. Debate About the Earth. Freeman, Cooper and Co, San Francisco. see Chapter 6 "Is the Earth heating or cooling?".

Tangeman, J.A., Phillips, B.L., Navrotsky, A., Weber, J.K.R., Hixson, A.D., Key, T.S., 2001. Vitreous forsterite (Mg_2SiO_4): synthesis, structure, and thermochemistry. Geophys. Res. Lett. 28, 2517–2520.

3

Heat transport processes on planetary scales

"To doubt everything or to believe everything are two equally convenient solutions; both dispense with the necessity of reflection." *Henri Poincaré (Science and Hypothesis, 1905).*

Motion of heat inside planets diverges somewhat from heat transport in the laboratory due to great differences in conditions and length and time scales, which allow processes without analogues to operate. This chapter covers the mechanisms of heat transport in planets, several of which involve simultaneous mass transport. Physical principles and theoretical analysis are the tools utilized here.

Nusselt (1915) refuted the notion that heat transport occurs by the three fundamental mechanisms denoted as conduction, radiation, and convection. Convection is not a mechanism, but rather is a combination of heat transport with the mass transport mechanism known as advection. Convection occurs when a large *fluid* system is subject to an imposed temperature gradient that is greater than that which can be sustained by conduction alone (Tritton, 1977). For *solids*, an alternative response to a substantial temperature gradient exists, namely melting, which is a focus here because its consequences have far greater importance than currently recognized in geophysics and planetary science.

The microscopic mechanism for conduction of heat at laboratory temperatures is diffusion of radiation at the low energies in the infrared (IR) region of the electromagnetic (EM) spectrum (Hofmeister, 2019a). This mechanism was revealed by accurate measurements of thermal diffusivity via laser flash analysis (LFA). The many advantages of this transient method are covered in monographs and review papers. The revised theory accounts for heat transfer resulting from *inelastic* interactions and involving the ever-present blackbody emissions: otherwise the temperature cannot evolve, and thermodynamic laws cannot be upheld, respectively.

Confusion exists because thermodynamics was developed before heat and light were recognized as being the same phenomenon and because the term radiation, as commonly used without a qualifier, generally signifies *ballistic* transfer, for which participation of the medium is insignificant. In great contrast, the medium is an essential participant in radiative *diffusion*, where its spectral characteristics exert great control. Recognition that diffusive transfer of radiation at high frequencies can be important at the high temperatures of Earth's interior is long standing (Lubimova, 1958; MacDonald, 1959), but debates still exist due to the above mentioned confusion and previous difficulties in quantifying heat transfer in weakly absorbing regions (e.g., Hofmeister, 2005).

Section 3.1 covers diffusion of radiation, extending the experimental and theoretical findings of Hofmeister (2019a,b) to planetary conditions. This section is detailed because radiative transfer is notoriously difficult to understand. Section 3.2 covers ballistic transport, which pertains to surface losses and brief heating events. Section 3.3 discusses planetary processes that transport both heat and mass, e.g., advection, magmatism, and outgassing. Open, two-phase systems are a focus. Section 3.4 summarizes theoretical problems in mantle convection models, which are based on equations for pourable fluids under constant gravity in a box being heated from below. This depiction is unlike the self-gravitating, layered, spherical Earth, with heat sources are mostly at shallow levels. Section 3.5 summarizes.

3.1 Conductive (diffusive) heat transport

All matter emits blackbody radiation at all times. This is generated internally, and becomes increasingly intense and distributed to increasingly higher frequencies (Wien's displacement law) as temperature (T) increases. One type of energy carrier acts at all temperatures as recognized by Maxwell, and this must be the electromagnetic phenomenon referred to as light. Heat is simply low frequency light.

Light is attenuated as it moves through matter, as a consequence of inelastic interactions. Inelasticity is required for temperature to evolve. The microscopic process involves stimulation of transitions from the ground state to various excited states, which is followed by the converse, with tiny losses during every atomic interaction. Inelasticity is caused by atoms deforming during collisions of any type, including with light. The outer shells of valence electrons are most affected.

Optically thick conditions describe radiative diffusion. Although metals are generally viewed as being opaque, this does not mean that radiation is prohibited from entering this material. The distinguishable colors of metals are caused by the partial penetration of light in the visible, in accord with their particular absorption characteristics. Moreover, thin films of metals transmit substantial fractions of the incident light at certain frequencies, as exemplified by 2-way mirrors. Conversely, media which appear to be fully transparent over human scales will absorb substantial amounts of light within thick masses of the same media, which describes layers in the Earth. Consequently, diffusion of radiation (conduction) in planets differs from behavior observed in the laboratory. Metals are discussed first because errors made in the 1800s still influence understanding the cores of planets. "Metals" as used here includes both elements and alloys, unless otherwise specified.

3.1.1 Electronic heat transport in metals is transient

The belief that conduction electrons carry most of the heat in metals (the Weidemann-Franz law) stems from equating electrical and thermal currents. Maxwell (1888) cautioned against taking this analogy too seriously. Actually, the analogy is invalid because charge is conserved, whereas heat is continually being lost to the surroundings, including during isothermal conditions.

The experimental and theoretical study of Criss and Hofmeister (2017) showed that metals transport heat by the same dissipative mechanism as in electrical insulators. For a succinct version, see Criss and Hofmeister (2019). Here we focus on misconceptions relevant to heat conduction in the core:

1. The tie of electrical conductivity directly to thermal conductivity at 298 K has been presented as evidence for conduction electrons carrying heat. However, the model fails at *all* other temperatures (e.g., Hust and Sparks, 1973). Moreover, a D.C. current of heat-carrying *conduction* electrons is superfluous, because the vibrating cations in a metal set up an A.C. current. Thus, the link between electrical conductivity, resistivity and

vibrational heat transfer is caused by *localized* vibrational agitations of *valence* electrons changing as heat moves down the thermal gradient (Criss and Hofmeister, 2017).

2. Misinterpretation is fueled by the ambiguities of multiplied parameters (see Transtrum et al., 2015). Thermal conductivity (K) is a convolution of specific heat (c) with average speed (u). For any given mechanism:

$$K = \rho c D \cong \rho c \langle u \rangle \Lambda \cong \rho c \langle u^2 \rangle \tau, \qquad (3.1)$$

where ρ is density, Λ is mean free path and τ is average lifetime. Thus, the fast speeds of conduction electrons (~ 100 times acoustic speeds) compensate for their low heat capacity (~ 0.01 times the measured, lattice heat capacity), making calculated values of K for electronic heat conduction fortuituously similar to measurements of K near 298 K. The original microscopic model of Drude used *lattice*, not electronic, properties (Kittel, 1971; Burns, 1990).

3. Not only do conduction electronic collisions move little heat, but this mechanism only exists for a brief time following the application of heat, due to rapid electron speeds. A transient response is evidenced by femtosecond spectroscopy (Bauer et al., 2015), and independently by monitoring the change in metal temperature as a function of time in response to a pulse of heat (Fig. 3.1A).

4. Interaction of thermally excited electrons with the lattice terminates their heat flow due to electronic velocity far outpacing lattice velocities (Criss and Hofmeister, 2017). In brief, a hot electron racing down a thermal gradient mostly interacts with other conduction electrons, due to strong Coulombic forces (Ashcroft and Mermin, 1976). These interactions change the energy levels of the conduction electron population as the front moves away from the heat source, making distant conduction electrons hot. When some traveling hot electron finally releases its extra energy to the lattice, it warms a valence electron. Other valence electrons in the distant region reached by the speeding hot conduction electrons are cold. The 2nd law prohibits transfer of heat from cold to hot entities.

Although the thermal diffusivity of the electrons is high (Fig. 3.1B and C), this mechanism operates only when conditions are far from equilibrium (Fig. 3.1A). Near-equilibrium behavior characterizes formed planets and laboratory experiments beyond several ms duration. Values for D_{lat} in Fig. 3.1B and C provide values for K (using Eq. 3.1) that are compatible with previous data on elements and alloys collected using diverse methods (e.g., Touloukian and Sarksis, 1970; Touloukian et al., 1973), all of which measure conditions close to equilibrium.

3.1.2 A model for diffusion of radiation

Radiative diffusion consists of the progressive absorption and re-emission of light down a temperature gradient inside matter. Conditions must be optically thick for diffusion, which means that light in substantial quantities does not pass through some distance where the temperature changes significantly (e.g., Siegel and Howell, 1972). This condition is obviously met in planets where gradients are circa degrees per km.

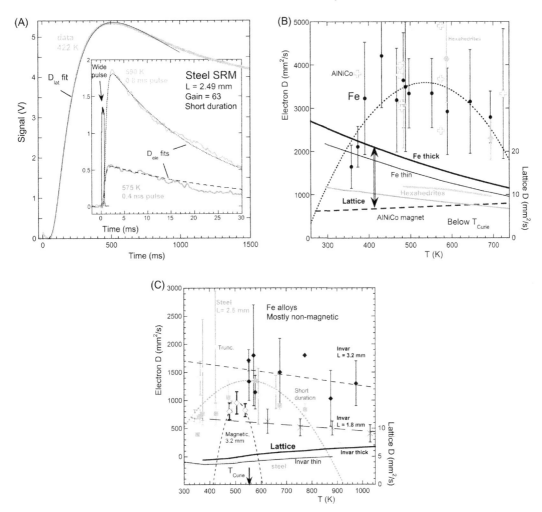

FIGURE 3.1 Thermal diffusivity of Fe and iron alloys. (A) Raw data (temperature-time curves from LFA experiments) on non-magnetic steel SRM-1461, showing collections over long time (main panel) and a short interval (inset), which describes the electronic contribution. Gray = baseline corrected data. Black = fits. The intensity (signal) depends on the software controlled gain. (B) Temperature dependence of the electronic contribution to D (symbols, dotted line, and left axis) and lattice contribution (solid and dashed lines: right axis) contributions for magnetic Fe-rich alloys (plus sign = AlNiCo8 composed of $34\%Fe + 355Co + 15\%Ni + 7\%Al$, gray diamonds = hexahedrite meteorite composed of 94% Fe + 6% Ni) compared to elemental Fe (dots), all below the Curie point. For the lattice, results from thin (<3 mm) and thick (3–7 mm) samples are shown as thin and thick lines. For D_{ele}, typical error bars are shown. (C) Mostly non-magnetic Fe-rich alloys. Solid lines are the lattice contribution. Gray = non-magnetic steel SRM-1461 (62% Fe + 20%Ni + 16%Cr). Diamonds = invar (64% Fe + 36% Ni) with $L = 3.2$ mm, where open diamonds show D_{ele} below the Curie transition. X = invar with $L = 1.9$ mm, where the linear fit well describes the data up to 1438 K. Intercepts are not constrained in fitting. *Source: Panels A, B, and C are modified after Figs. 8a, 14a, and 14b, respectively, from Criss, R.E., Hofmeister, A.M., 2017. Isolating lattice from electronic contributions in thermal transport measurements of metals and alloys and a new model. Int. J. Mod. Phys. B 31, No. 175020. Available from: https://doi:10.1142/S0217979217502058.*

To quantify the attenuation of light over a length (L), spectroscopic measurements are made. The key parameter is the absorption coefficient (A) as a function of frequency (ν):

$$A(\nu)L = -\ln\left[I_{trns}(\nu)/I_0(\nu)\right] - 2\ln[1 - R(\nu)] \tag{3.2}$$

where I is intensity of light, subscript 0 refers to incident light, subscript trns to transmitted, and R is the reflectivity $= I_{refl}/I_0$. Spectroscopy being conducted under optically thin conditions, wherein light is not necessarily thermally produced, has caused some misunderstandings in utilizing these data to ascertain heat transfer. Hofmeister (2010) discusses common mistakes (e.g., neglecting effects of refraction) leading to overestimation of A.

Thermal conductivity is obtained from an integral because the frequency of thermally produced light varies continuously, due to production by inelastic events. For parallel heat transfer across a slab, as is measured in the laboratory:

$$K_{lat,rad}(T) = \frac{V_{\text{unitcell}}\rho N_a}{MZ} \int_{\text{cuton}}^{\text{cutoff}} \frac{1}{A(\nu,T)} \frac{\nu^2}{2c^2} C_E(\nu,T) d\nu \tag{3.3}$$

where M is mass per formula unit, Z is the number of formula units in the unit cell of volume V_{unitcell}, ρ is density, N_a is Avogodro's number, and c is the speed of light (Hofmeister, 2019a). Einstein's heat capacity is:

$$C_E = k_B \left(\frac{h\nu}{k_B T}\right)^2 \exp\left(\frac{h\nu}{k_B T}\right) / \left[1 - \exp\left(\frac{h\nu}{k_B T}\right)\right]^2 \tag{3.4}$$

where ν is frequency in per seconds, h is Planck's constant, and k_B is Boltzmann's constant. Eq. (3.4) is obtained from the temperature derivative of blackbody intensity, and represents the energy carried by light at each frequency. Blackbody intensity is relevant because optically thick conditions apply. For discussion of previous formula, see Hofmeister (2019a).

To test and then to apply the above model, we need to examine the various components of Eq. (3.3). The prefactor involves well-known material properties and needs no further discussion.

3.1.2.1 Basics of light absorption

Spectra of electrical insulators (e.g., mantle minerals) contain peaks and valleys (Fig. 3.2A), such that number of peaks increases with structural complexity and the number of different atoms in the chemical compound. Metals have simple crystal structures, so peaks are few: spectra are mostly characterized by smooth changes in A with ν (Fig. 3.2B). For additional examples, Palik (1998), Hofmeister (2019c), or the websites listed at the end of this chapter.

Different mechanisms of light uptake are associated with certain spectral regions. Matter interacts strongly with infrared (IR) light, whereby optically thick conditions are met for $L > \sim 0.001$ mm in oxides (Fig. 3.2A), metals (Fig. 3.2B), and silicates (e.g., Hofmeister and Bowey, 2006). Interactions are strong because IR light stimulates interatomic motions, i.e., the fundamental vibrations of the crystal lattice. Much weaker interactions occur at slightly higher ν (the near-IR) due to interactions of light with the overtones and combination modes of these fundamental vibrations. Silicate minerals absorb more

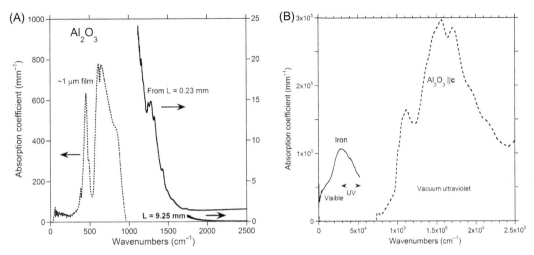

FIGURE 3.2 Comparison of spectra. (A) Absorption measurements of corundum Al_2O_3, obtained in the author's laboratory. Film thickness and derived A are approximate. Crystals are polarized along the c-axis. (B) Absorption coefficients calculated from measured reflectance spectra. Corundum absorption calculated from optical functions of from Palik (1998). Metallic iron data from our laboratory gave similar optical constants to those of Moravec et al. (1976), Ordal et al. (1983), and others (see Palik, 1998).

light in the near-infrared than do oxides (e.g., Pitman et al., 2013), because they have many more fundamentals which extend to both lower and higher ν.

Measuring A is impossible for many substances above the overtones, because lengths of meters, which exceed spectrometer bench size, are often needed: for this reason, Palik (1998) only lists index of refraction from the near-IR to visible for corundum (Fig. 3.2B).

Regarding the visible region, rock-forming minerals have Fe and other transition metal cations. The associated visible absorptions are weak for Fe/Mg ~ 0.1 that is typical of mantle samples, but increase roughly with the concentration and variety of transition metal ions. Basaltic glass, for example, is black and is optically thick in the near-IR to visible for L exceeding $\sim \frac{1}{2}$ mm (Danyushevsky et al., 1993). Metals absorb much more strongly in the visible.

Ultraviolet (UV) light stimulates transfer of charge along cation-anion bonds, so absorption strengths are very strong (Fig. 3.2B), and peaks exist in the UV for all anion-cation pairs. At even higher frequencies, interactions become weaker, until the energy of light is sufficient to stimulate the nucleus. Nuclear transitions, however, are not part of the thermal energy of planets, although these are relevant at the extraordinary temperatures deep in stars.

3.1.2.2 Behavior of heat capacity

Heat capacity affects how much heat can be moved. Both ν and T strongly affect C_E (Fig. 3.3A). As temperature increases, the blackbody curves move to higher frequency and thus stimulate higher energy transitions: Fig. 3.3 lists peaks in electrical insulators (e.g., mantle minerals). A competition exists: as temperature increases, higher energy processes increasingly participate in diffusion of heat, but these mechanisms generally carry little heat at planetary temperatures.

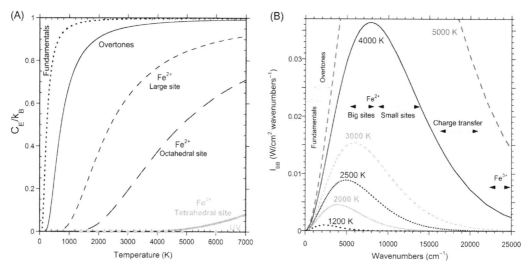

FIGURE 3.3 Energies and heat capacities associated with vibrations. (A) Temperature dependence of the heat capacity associated with light of different frequencies. The curve marked fundamentals was calculated for a frequency of 500 cm^{-1} and represents light created during inelastic interactions of fundamental vibrational modes, including acoustic. For the overtones, a frequency of 1500 cm^{-1} was used. The electronic transitions considered assume that the material has the same Fe content for each species, where frequencies considered are 5000, 10,000, and 30,000 cm^{-1} from left to right. (B) Planck curves, with temperatures labeled. Horizontal double arrows indicated ranges of absorption for common minerals.

3.1.2.3 Why frequency limits are essential

Previously, integration included the transparent, near IR region, where $A = 0$ within uncertainty, providing large and uncertain values for K. Averaging or incorporating a scattering parameter (Lee and Kingery, 1960; Shankland et al., 1979) was used to circumvent infinities. These approaches are somewhat arbitrary. Problems also arise from the unattainable limits of 0 and ∞ in the integrals used. These limits are part of the mathematical idealization known as the blackbody spectrum. However, perfect absorbers, emitters, reflectors, or transmitters do not exist.

Inelasticity is the source of an upper cutoff (Hofmeister, 2019a). The frequency of the emissions must be less than the energy of the most energetic of the possible transitions or collisional exchanges. Small losses prevent attainment of very high velocities or highly excited states. Importantly, the concept of a cut-off frequency underlies Debye's model for heat capacity, which reasonably describes many solids (e.g., Burns, 1990).

Regarding the lower cutoff, we note that the integral of Eq. (3.4) is a simple summation. However, when multiple carriers are involved (the frequencies), a sum rule is germane. The component K's must be weighted in direct proportion to the volumetric heat capacity C of each component (Criss and Hofmeister, 2017):

$$K_{single} = \frac{\sum C_i K_i}{\sum C_i} \tag{3.5}$$

An equivalent summation is required to calculate the thermodynamic Grüneisen parameter from its microscopic vibrational components (e.g., Krishnan et al., 1979). Thus, the radiative diffusion model must account for differences for heat capacity among the different spectral regions (Fig. 3.3A).

The $C_i K_i$ terms in the numerator are tiny for frequencies that are not excited, whereas the sums in the denominator are dominated by the transitions with large C_i at the temperature of interest. Thus, thermal conduction is dominated by a particular type of transition for a certain frequency range, and temperature. We can describe K for the various transitions by evaluating the integral over the associated frequency range. Although one mechanism exists (absorption/reemission), not all carriers are equally effective at any given temperature or frequency range.

For example, the heat capacity of a d-d transition in the visible is negligibly small at laboratory and lithospheric conditions (Fig. 3.3A), and thus the calculations of thermal conductivity at low temperatures need only involve low frequencies. This finding is tested below.

3.1.3 Laboratory evidence for conduction by radiative diffusion

3.1.3.1 Implications of thermal diffusivity depending on sample size in laboratory measurements

Laser flash experiments on Al_2O_3 (Fig. 3.4), silica glass, and many other electrically insulating crystalline solids (Hofmeister, 2019d) show that $D \to 0$ as $L \to 0$. This limit is mandated in a radiative diffusion model, because a medium is required for that mechanism to operate. Experiments on metals show that their lattice K and D also increase with

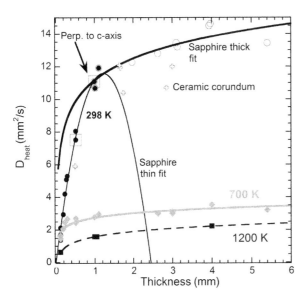

FIGURE 3.4 Thermal diffusivity of synthetic, pure Al_2O_3 at various temperatures for different length samples. Heat flow was parallel to the c-axis, except as indicated. Ceramics have isotropic heat flow, but these samples have varying amounts of porosity.

sample thickness (Fig. 3.2B and C), but the dependence of D on L has not been quantified. Therefore, we focus on insulators.

Twelve different crystal structures and glass are reasonably represented by (Hofmeister, 2019d):

$$D_{heat}(L) = D_\infty \left[1 - \exp(-bL)\right] \tag{3.6}$$

Eq. (3.6) closely describes isotropic material. Parameter D_∞ represents thermal diffusivity for a very long sample whereas the inverse of parameter b represents attenuation, a hallmark of diffusion.

The steep rise in $D(L)$ from the origin and the nearly constant D at large L (Fig. 3.4) is consistent with absorption of overtones being much weaker than A of the fundamental IR modes. Specifically, thick samples are optically thick in the IR and near-IR, whereas thin samples are only optically thick in the IR. For small L, only the fundamentals participate in diffusion while the overtones transmit ballistically. The value of D is low when only the fundamentals participate because their A is large. For large L, both overtones and fundamentals participate. Thermal diffusivity is larger mostly because A is smaller. The nearly flat region persists to large L because the samples studied absorb negligibly in the visible and the temperatures were insufficient to stimulate UV transitions.

This length scale dependence is a likely source of variation between laboratories. Contact losses and ballistic gains also existing makes systematic differences appear as random errors. These experimental difficulties coupled with beliefs that heat is transported elastically (Clausius' and Debye's kinetic theories) has prevented recognition that all heat transport is radiative.

3.1.3.2 Evaluation of the model by comparison with cryogenic thermal conductivity data

Spectra on planetary materials collected over wide ranges of both frequency and temperature are needed inputs for the radiative diffusion integral, but such data are in short supply (see the websites). Consequently, comparing Eq. (3.3) to measurements requires approximating spectra. Because integration is a form of averaging, simple functions such as boxcars and ramps were used for A. This approach permitted analytical evaluation of the integral (Hofmeister, 2019a), as follows:

The series of peaks in the IR spectrum of an electrical insulator (Fig. 3.2A) can be represented on average by constant A, so data on corundum were fit to:

$$K_{IR}(T) \propto \left[\frac{24}{b^3} - \frac{e^{-b\nu}}{b^3}\left(24 + 24b\nu + 12b^2\nu^2 + 4b^3\nu^3 + b^4\nu^4\right)\right]; \quad b = \frac{1.44}{T} \tag{3.7}$$

The combination of constant A with a cutoff frequency is identical to a boxcar function. Such a distribution was used by Kieffer (1979) to model heat capacity of insulators.

For metallic elements, A increasing proportional to ν^2 was assumed, based on Fig. 3.2B and reflectivity data. The fitting equation for metals is:

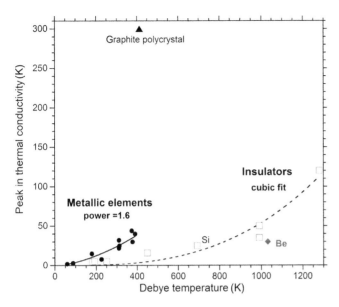

FIGURE 3.5 Link of the peak in thermal conductivity with the Debye temperature. Dots = metals. Plus = electrical insulators, including Si, which is a single-crystal. Triangle = semi-conductors. Diamond = Be metal, which has Raman activity (Feldman et al., 1968) and a Debye temperature more than double that of any other metallic element. Fits are power laws, as indicated. Debye temperatures mostly from Burns (1990). Peaks in K mostly from Ventura and Perfetti (2014). *For details see the source: Figure 11.6 and Table 11.1 in Hofmeister, 2019a. Modelling diffusion of heat in solids. In: Hofmeister, A. M. (Ed.), Measurements, Mechanisms, and Models of Heat Transport, vol. 427. Elsevier, New York, pp. 359–398, with permissions.*

$$K_{IR}(T) \propto \left[\frac{2}{b} - \frac{e^{-b\nu}}{b} \left(2 + 2b\nu + b^2\nu^2 \right) \right]; \quad b = \frac{1.44}{T} \quad (3.8)$$

The combination of A increasing with ν and a cutoff frequency is related to Debye's model for heat capacity, which fits metals well.

Eqs. (3.7) and (3.8) provide non-dimensionallized thermal conductivity. The form chosen for $A(\nu)$ along with the cutoff frequency determines the shape for $K(T)/K_{max}$: this choice constitutes one fitting parameter. Both formulae provide a peak where its position in T depends largely on the cut-off frequency.

A strong peak in $K(T)$ is the hallmark of cryogenic experiments. Its position in T depends directly on the Debye temperature for both metals and insulators, but in two different trends (Fig. 3.5). Different trends in Fig. 3.5 for metals and insulators are connected with the need for different forms for $A(\nu)$ to reproduce both their C_P and K.

Data on some metals required a second integral to represent the overtones. A more gradual increase in A with ν exists at high ν, so we used $A \sim \nu$ for the additional integral, yielding:

$$K_{near-IR}(T) = const. \left[\frac{6}{b^2} - \frac{e^{-b\nu}}{b^2} \left(6 + 6b\nu + 3b^2\nu^2 + b^3\nu^3 \right) \right]; \quad b = \frac{1.44}{T} \quad (3.9)$$

The function for A of the overtones is a simple triangle, constituting one parameter. But because Eq. (3.9) for the overtones is summed with Eq. (3.8) which represents the fundamentals, a constant is needed to represent their relative contributions. These are three parameter fits.

Cryogenic data for K of corundum could be fit reasonably well with Eq. (3.7) for a boxcar spectrum, using a cut-off frequency of 108.6 cm^{-1} (Fig. 3.6A), which is low compared

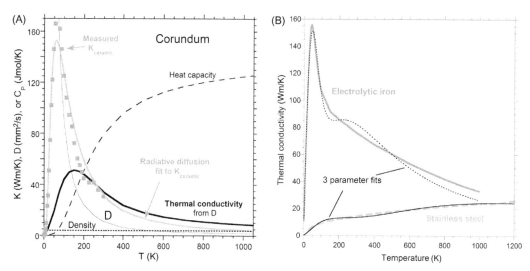

FIGURE 3.6 Transport and associated thermodynamic properties of a partially transparent insulator and a metal. The x- and y-axes in both panels are similar. The radiative diffusion model used 1 free parameter to fit the insulator and 3 to fit the metals. (A) Insulating sapphire, pure Al_2O_3. Dotted line = density (Fiquet et al., 1999). Dashed line = heat capacity (Ditmars et al., 1982). Gray quares = measured K on ceramic from Berman et al. (1960), which had large amounts of ballistic transfer. Thermal diffusivity from Hofmeister (2014), extrapolated below 298 K. (B) Polycrystalline electrolytic iron, Fe, which is NIST standard SRM-8421. Dark gray curve = measured K from Hust and Lankford (1984), which agrees with LFA measurements above 298 K. Light gray curve is K for stainless steel, SRM-1641 from Hust and Lankford (1984). *Source: Modified after Figures 11.5a and 11.10 in Hofmeister, 2019a. Modelling diffusion of heat in solids. In: Hofmeister, A.M. (Ed.), Measurements, Mechanisms, and Models of Heat Transport, vol. 427. Elsevier, New York, pp. 359–398, with permissions.*

to the spectra (Fig. 3.2A). However, Berman et al.'s (1960) data on K contain systematic errors with opposing signs (contact losses and ballistic gains) which affect $\partial K / \partial T$ and therefore the shape and placement of the peak. Limitations of periodic and steady-state methods when applied to electrical insulators are covered in Hofmeister et al. (2007) and Hofmeister (2019e). Despite substantial experimental uncertainties, the placement of the peak is roughly correct. The low cutoff frequency indicates that the weak absorption in the far-IR controls cryogenic thermal transport, which is in accord with the blackbody intensity peaking in the far-IR at cryogenic temperatures, and weak A enhancing radiative diffusion (Eq. 3.3).

Data on $K(T)$ of metals are more accurate than for insulators, but minor problems arise in merging datasets and how the physical contacts are made. Nonetheless, K on pure Fe metal and a non-magnetic steel alloy were fit reasonably well (Fig. 3.6B) by summing Eqs. (3.8) and (3.9).

Iron is used as an example due to its planetary relevance. However, this element is strongly magnetic with a Curie point at 1043 K. Transport properties are affected over a several hundred K by this transition in weakly magnetic insulators (yttrium iron garnet and magnetite: see Hofmeister, 2006, 2007), so data on Fe^0 could be impacted over a wider range. Note that non-magnetic steel was accurately modeled over the entire range of measurements (Fig. 3.6B), whereas other metallic elements and graphite were reasonably fit with our model (Hofmeister, 2019a).

Summing the two integrals provides two peaks in K data. The breadth of each peak in calculated K is related to the form for $A(\nu)$, whereas the strength is related to C_E/A of the overtones, relative to C_E/A of the fundamental modes. The lowest T peak in K is the largest, because overtones have much lower C_E than fundamentals (Fig. 3.2A), which compensates for differences in A.

3.1.3.3 Evaluation of the model by comparison with accurate high temperature D-data

Thermal diffusivity on ~300 electrically insulating solids from 298 K upwards, sometimes to above melting, obtained using accurate laser-flash analysis, and provided in 33 journal publications, are accurately described by:

$$D = FT^{-G} + HT \tag{3.10}$$

where G and H are positive (Hofmeister, 2019d). Coefficient G is on the order of unity whereas H is small: resolving its value requires T exceeding ~800 K. This simple form for D also applies to high temperature K, which are related through Eq. (3.1), because heat capacity and density slowly vary above ~800 K.

Hofmeister et al. (2014) devised Eq. (3.10) based on experimental data, and attributed the power law term to phonon scattering, which had long been considered the mechanism for heat transport. The high temperature term, HT, was proposed to arise in diffusive radiative transfer in the infrared.

Actually, *both* terms in Eq. (3.10) are due to radiative diffusion. The first term is caused by interaction of light with the fundamental modes, which provide very high K (and thus high D) at cryogenic temperatures, and then declines rapidly as T increases (Fig. 3.6A). The second term is assigned to the overtones.

For metals and alloys, D depends linearly on T. At high temperatures, the slope is frequently positive. Negative slopes exist for some of the elements, mostly at low temperature (see Figure 9.1 in Criss and Hofmeister, 2019). This behavior is consistent with radiative diffusion of fundamentals at low T but of overtones at high T, recognizing that the spectra of metals change gradually with frequency well into the UV, and that individual vibrational modes are rarely resolved (Fig. 3.2B).

3.1.4 Effect of pressure

The radiative diffusion model provides:

$$\frac{\partial \ln(K_{lat,rad})}{\partial P}\bigg|_T \cong \text{const.} \frac{\partial \ln(\nu_{cutoff})}{\partial P} \cong \text{const.} \frac{\gamma_{th}}{B_T} \tag{3.11}$$

Due to taking a logarithmic derivative, the proportionality constant is just a numerical factor, equaling 4.6 for insulators, but 2.65 for metals, due to spectral differences. The cutoff frequency is related to the upper frequency of the fundamental vibrations. The thermodynamic relationship on the RHS is based on the thermal Grüneisen parameter (γ_{th}) being

3. Heat transport processes on planetary scales

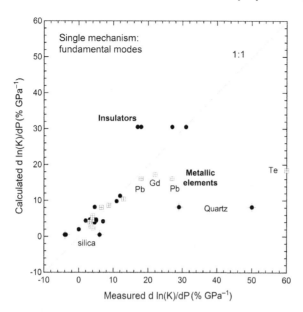

FIGURE 3.7 Comparison of pressure derivatives calculated for insulators (dots) and metals (squares) from Eq. (3.13) to measurements. Data on SiO_2 involve cracking, so are suspect. Data on Te are approximate. *Source: Figure 11.12 in Hofmeister, 2019a. Modelling diffusion of heat in solids. In: Hofmeister, A.M. (Ed.), Measurements, Mechanisms, and Models of Heat Transport, vol. 427. Elsevier, New York, pp. 359–398, with permissions.*

proportional to the average of the mode Grüneisen parameters, which are defined as $\gamma_i = B_T^{-1}[\partial \ln(\nu_i)/\partial P]$, where B_T is the isothermal bulk modulus. Because the thermal Grüneisen parameter weakly depends on T or P (e.g., Anderson and Isaak, 1995), the bulk modulus dictates how thermal conductivity depends on pressure.

Pressure derivatives near ambient conditions are compared here, due to problems in the high P experiments (Hofmeister, 2010). In particular, the dependence of D on thickness means that referencing the diamond anvil cell experiments on ~ 0.1 mm samples to ambient pressure measurements on samples of ~ 3 mm, is inappropriate.

Calculated derivatives from Eq. (3.11) agree with multiply confirmed, low pressure experimental determinations (Fig. 3.7). Both insulators and metals are accurately modeled. Although previous equations are similar to Eq. (3.11), no other model predicts derivatives for both insulators and metals with this level of agreement (see the review of Hofmeister et al., 2007).

3.1.5 Conduction/diffusion involving layers and mixtures

3.1.5.1 *Heat flow across layers (in series)*

The time-evolution of heat flow for two and three layers was modeled by Lee et al. (1978) and is utilized in laser flash measurements. Here we are interested in some equivalent thermal conductivity which represents a multi-layer sample under steady-state conditions. For simplicity, we assume the properties K_i and C_i are independent of temperature over any given layer i. Sum rules lead to:

$$\frac{L}{K} = \frac{L_j^2 C_j}{K_j} \sum \frac{1}{L_i C_i} = \sum \frac{L_i}{K_i} \qquad (3.12)$$

For a series configuration with temperature dependent properties, the layer width can be made small to account for this behavior. In a layered geometry, the temperature field is constrained and independent of time for each layer. Hence, poor thermal conductors greatly impede cooling in a layered system.

3.1.5.2 Heat flow along layers (in parallel)

This case is identical to the sum rule for multiple mechanisms, where heat capacity is a weighting factor (Eq. 3.5). At the high temperatures inside Earth, heat capacity varies little, i.e., the Dulong-Petit limit is germane. Consequently, heat conduction in parallel at high T is an average:

$$K \cong \frac{\sum K_i}{n} \qquad (3.13)$$

where n is the number of layers. Thermal transport properties of rocks vary over only a limited range: mafic rocks have diffusivities of $0.8-2$ mm^2/s near 298 K, whereas felsic rocks vary from 1.5 to 3.6 mm^2/s (Whittington et al., 2009; Merriman et al., 2013). This difference is connected with the amount of quartz, which has high D, and decreases as T increases. Hence, parallel heat flow is not so important on large planetary scales. Parallel flow is germane in the heterogeneous crust, particularly the continents. For cold temperatures in the outer Solar System, weighting is needed.

3.1.5.3 Multicomponent mixtures (rocks) at high temperature

For rocks, the series formula (Eq. 3.12) is relevant, but in this case, the sum is over the number of different phases, K_i is the thermal conductivity of each phase and L_i is the proportion of any given phase, rather than the grain size. The sum of L_i in this situation is unity. The value of K_i should represent the grain size of the ith phase in the rock. Our approach is based on Eq. (3.12) representing rocks, which is demonstrated by our LFA database (Hofmeister, 2019d). Porosity can be represented by this approach. Testing this formulation against rock data is underway (Merriman et al., in prep.)

3.1.6 Radiative diffusion inside planets at high temperatures

Laboratory experiments cover the temperature range of the lithosphere and into the upper mantle. The data and model described above and detailed in Hofmeister (2019a,d) and our website can be extrapolated to transition zone conditions, for any given mineralogy. However, temperatures above ~ 2500 K in the lower mantle can bring about diffusion of light in the visible, on the basis of Fig. 3.3. Radiative diffusion at these conditions is not directly measured in the laboratory, and must be calculated from Eq. (3.4).

Regarding pressure, its effect is quasi-harmonic and independent of the temperature response. Since Eq. (3.11) pertains to $K_{rad,vis}$, the pressure derivative should be taken at a relevant high frequency, e.g., of the peak maximum in the visible. If the boxcar approximation is used to describe the average visible absorption, then the numerical constant in

Eq. (3.11) is ~ 4.6. If $A \propto \nu$ is assumed, the constant is ~ 3.6, and if A is proportional to ν^2, then the constant should be 2.65. Compression can be quantified by spectroscopy since peak positions are the most readily measured spectral parameter and are not terribly affected by baseline corrections.

3.1.6.1 *Electronic transitions of insulators in the visible*

Eq. (3.4) gives heat capacity per atom. Thus, the C_E curves in Fig. 3.3a need to be divided by ~ 10 to be appropriate for the ratio of 9 Mg atoms per one Fe^{2+} observed in upper mantle minerals. Whether the lower mantle has the same composition of the upper is unknown, since diamond inclusions are limited to ~ 300 km depths of the upper mantle, and these differ from lithospheric minerals (Chapter 1). Even so, higher concentrations of chromophores increase the absorption coefficient which balances the increase in heat capacity, as in Eq. (3.4), and thus the division by ~ 10 holds for a wide range of iron contents.

To gauge $K_{rad,vis}$, we consider ratios of C_E/A, the span of the integral (Eq. 3.3) and its internal factor of ν^2. Note that Eq. (3.7) for a boxcar (average absorption) centers on a ν^2 term, so this factor and average values of A's should suffice. For large samples above 298 K, the overtone region ($\nu \sim 2000$ cm^{-1}) controls $K_{rad,IR}$ that is measured in the laboratory, which has $A \sim 10$ mm^{-1}, and $C_E \sim 0.7$ k$_B$ (Fig. 3.3A). Olivine absorbance in the visible is ~ 1 mm^{-1} near 10,000 cm^{-1}, and at 3000 K, $C_E \sim 0.02$ k$_B$ (Fig. 3.3a, considering 1 Fe per 10 atoms Mg in the formula unit). Roughly speaking, $K_{rad,vis}$ at 3000 K should be $30 \times K_{rad,IR}$ measured in the laboratory at ~ 1000 K. An important factor is ν^2 in the integral. As temperature climbs, heat transfer climbs, as is consistent with the HT term for thermal diffusivity measured above about 800 K. By about 5000 K, $K_{rad,vis}$ should be $\sim 100 \times K_{rad,IR}$.

A more definitive estimate for conduction in the lower mantle requires modeling its mineralogy (Chapter 8). In any case, the strong increase in K with T and P means that the mantle becomes increasing efficient at diffusing heat with depth. A diamond rich lower mantle would have very high K.

3.1.6.2 *Continuum behavior of metals and alloys in the visible*

The linear increase in $K(T)$ for steel alloy with T above ~ 200 K (Fig. 3.6B) is associated with absorption coefficients of metals being high from ~ 200 cm^{-1} upwards and increasing with ν. Thus, the linear increase in $K(T)$ measured in the laboratory for Fe-Ni alloys and steel can be extrapolated to the high temperatures of Earth's core. The compressed core is very efficient at conducting heat.

3.1.6.3 *The UV region and beyond*

For metals, absorption climbs from the IR to a broad peak in the UV whereas for insulators a steep increase in absorption occurs deeper in the UV (Fig. 3.2). For planetary problems, the UV region is unlikely to participate due to the combination of high absorption and low heat capacity.

Beyond the UV, light does not interact as in optical measurements. Nuclear processes are not germane to heat transfer in planets, which underlies use of a cut-off frequency in radiative transfer. However, nuclear processes are important in stars, as are extreme temperatures and the UV region.

3.2 Ballistic radiative transfer

Due to the presence of iron and other transition metal cations and slight changes in temperature over the large scales of planets, ballistic transfer of radiation in their interiors is unexpected. In contrast, release of heat from a planetary surface to the surroundings is ballistic.

The Stefan-Boltzmann formula for emission of heat from a hot body into space is well-known. This law presumes emissivity of unity (i.e., a blackbody). Formulae for cooling of a blackbody given in Section 2.2.3.2 can be adjusted for lower emissivity, but the findings are little affected: cooling by radiation is very fast. For this reason, surface temperatures are maintained by a balance with the Solar flux, which appears to be rather constant from the historical record (Coddington et al., 2016). This boundary condition clearly represents the inner planets, given that the flux of heat from Earth's interior is $\sim 10^{-3}$ of the Solar influx, and Earth is by far the most active planet. Emission of heat in the outer Solar System can potentially overtake the Solar influx. As regards the rocky bodies, only Io may be affected.

3.3 Advection

Mass elements in a flowing medium have finite temperature and carry heat in accord with their heat capacity. This process, called advection, can occur on any scale from the microscopic, as in elastic or plastic deformation, up to the size of the system. Movement of suspended sediments in a river, or movement of rocks on a tectonic plate, are examples of advective transport (Fig. 3.8).

Advection differs fundamentally from mass diffusion because the associated velocity (u) results from an *applied* external force. Gravitational buoyancy is the key source in planets. Consequently, advection is not associated with a characteristic, intrinsic speed. Furthermore, the process of advection depends on whether the system is closed or open, and whether the system contains phases with disparate properties. The open, two-phase system of fluid imbedded in a solid is of great importance, and the fluid can be molten rock, percolating water, or a gas such as CO_2. Moreover, fluids themselves can be two phase systems (Section 3.3.4).

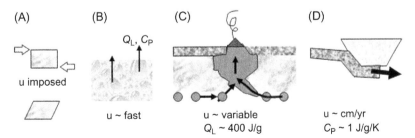

FIGURE 3.8 Schematic of various advection processes transferring heat and mass inside the Earth on various scales. (A) Deformation of small grains, which occurs at a predetermined rate in the laboratory. (B) Outgassing, which depends on bouyancy. (C) Magma generation and motion. (D) Advection of the lithosphere.

3.3.1 Plate motions

The vertical flux across the surface of oceanic plates is a consequence of the isothermal boundaries, not cooling of these solids (Chapters 1 and 7). Maintenance of these isothermal boundaries is therefore of interest, and is pursued in thermal models in Part 2.

Also of interest are the lateral motions of the lithospheric plates which are slow, averaging ∼6 cm/yr (Doglioni and Panza, 2015). Regions of subduction decrease radial emissions (relative to the middle of the plate) because the cold, descending plates displace warmer material. Behavior near the ridges, where hot magma ascends, has the opposite effect. Conservation of mass requires that the amount of heat associated with the heat capacity of the moving material is approximately conserved in a large, slowly evolving system. This approximation becomes exact at high temperatures where heat capacity approaches the constant Dulong-Petit limit. However, latent heat release is not balanced during the circuit, see Section 3.3.3.

3.3.2 Laboratory experiments differ from planetary interiors

To mimic conditions inside the Earth and planets, experimenters seal various materials in a capsule, which is then compressed and heated, and sometimes subjected to differential stress. Probing the sample during such testing is difficult, but rewarding. For example, diamond anvil cells are transparent to many types of electromagnetic radiation, permitting in-situ observations that facilitate diverse studies. A summary of techniques used in mineral physics is given by Price and Stixrude (2015).

Recovered samples can be examined to obtain information on small scales along with details of chemical reactions. Experimental containers are made of various materials, generally chosen to prevent unwanted reactions. Although concerns were once expressed as to whether platinum capsules are permeable to hydrogen, the current belief is that mass diffusion is unimportant over the short time-scales of experiments.

If sample chambers are truly impermeable, no material can escape, including volatiles. Such systems are closed to the exchange of matter but not of energy with the surroundings. The resulting assemblages thus entirely depend on the experimental material and the run conditions. Although such control is desired, the closed systems in the laboratory differ from the Earth and planets, which are open systems with a gravitational gradient. Thermal and chemical gradients also exist, and the time-scales are extremely long. Of concern is behavior of low density, low viscosity phases mixed with denser solids. Water is a particular concern, due to its unusual physical properties such as its thermal expansivity increasing with pressure (Chaplin, 2018): see Section 3.3.3.

To estimate the time required for the low-density phase to cross the sample chamber, the terminal velocity of Stokes is considered:

$$u_{Stokes} = \frac{g \Delta \rho d^2}{18\eta} = \frac{g d^2}{18 v}\left(1 - \frac{\rho_{particle}}{\rho_{medium}}\right) \tag{3.14}$$

where η is dynamic viscosity and v is kinematic viscosity. The density factor in the brackets is on the order of unity and can be neglected. Chamber sizes are below 2 mm, so belbs of fluid or melt must be small, generally <0.2 mm, which sets an upper limit on the speed. Using the kinematic viscosity of silica glass, 10^{12} mm/s (Doremus, 2002) also overestimates speed, giving 10^{-10} mm/s. A lower limit for the time for a low-density phase to cross the sample is ~ 30 yr, a time interval not used in experiments. Moreover, minerals and rocks are difficult to deform, so their effective viscosities are enormous: estimates for the mantle are $\sim 10^{21}$ mm/s (e.g., Schubert et al., 2001). An upper limit for time of settling in the chamber for this case would be enormous, ~ 30 Ga. Both time-scales are sufficiently long that mass diffusion or chemical reactions, rather than advection, are the likely transport process in laboratory situations where fluid is interspersed with compressed solids. Thermal gradients that might exist in the chamber are also relevant.

The above estimates of Stoke's velocity and settling time reveal that laboratory experiments do not depict gravitationally driven motions of matter inside planets. Volatile behavior in particular is not correctly portrayed, because volatiles can move at much higher speeds due to the great density contrast and the small molecule size, both of which promote percolation and mass diffusion. Water converting to very low density steam at relatively low temperatures (400 K) greatly enhances mobility of H_2O. An important consequence is sealed capsule experiments that feature high concentrations of H_2O and OH (e.g., Murakami et al., 2002) need not describe Earth's interior.

Low permeability and low porosity of certain regions may impede advection of volatiles, but over time, slow percolation and even slower mass diffusion promote stable density stratification of multiple phases in a gravity gradient. Ascent of melts is slower than that of volatiles due to their lower density contrast, much larger "molecule" size, and higher viscosity, but Eq. (3.14) shows that these effects are far less important than the resistance applied by the medium.

3.3.3 Importance of magma and volatile ascent

Magmatism and outgassing are especially effective means of cooling planets for several reasons: these processes are fast, carry latent heat upwards, and alter the compositions of planetary interiors.

First, melt and gas are buoyant and flow under any stress, which results in their rapid vertical speeds near the surface, since gravitational acceleration is large ($g \sim 9$ m/s throughout the mantle) and density contrasts are substantial ($\sim 10\%$ difference for silicates). Emplacement times of only ~ 8 hr for kimberlites that originate at ~ 200 km depths (McGetchin et al., 1973) suggest ascent speeds of $\sim 10^{10}$ cm/yr. Dimensional analysis supports the reasonableness of this estimate, which dwarfs the speed of plates. Such explosive magmatism involving volatiles is the fastest process. The next fastest would be percolation of gas and fluids such as water, which can ascend molecule by molecule through a permeable rock matrix. Stiffer silicate magmas ascend slowly by processes such as tectonic dilation at the mid-ocean ridges, which is described by velocities similar to u of plates, or by destabilizing overlying rocks which fall into the magma, or by chemical reactions as the hot melts eat their way to the surface. The latter two modes can be rapid or slow but in any case are independent of plate motions.

Second, moving melts not only carry heat described by their heat capacity, but also transport latent heat (Q_{melt}) up to the time of crystallization. When and where Q_{melt} of silicate magmas is released is important because it is large ($\sim 400\,J/g$). Furthermore, Q_{melt} is taken up at the source region, which need not be identical to the region of solidification. If buoyant magmas reach the surface, Q_{melt} is radiated to space along with advected heat, so both are permanently removed from the global heat budget.

Eruptions at the mid-ocean ridges provide $15-25\,km^3$ of new basalt per year (Elderfield and Schultz, 1996). This estimate does not include continental, island and arc eruptions. Flood basalts are the next largest class of volcanics, being produced at $\sim 1\,km^3$ per yr (White et al., 2006), a rate that is two orders of magnitude larger than any other class, including the Hawaiian volcanoes. Therefore, mid-ocean ridge volcanism well describes Earth's heat losses due to volcanism. From this, and the discussion above and in Chapter 1, the magmatic heat currently being lost by the Earth is only 0.8 ± 0.2 TW.

Third, preferential partitioning of incompatible elements and large ion lithophiles into silica-rich melts (e.g., Henderson, 1982) means that these magmas carry the long-lived radioactive species (U, Th, K) upward. Buoyant motion of heat-producing elements towards the surface hastens Earth's cooling while creating chemical layering, specifically the special, continental crust (Chapter 6).

3.3.4 Two-phase systems

3.3.4.1 Circulation of water in rock

Water has unique properties. The heat capacity of water is uniquely high, suggesting that hydrothermal circulation of water and steam through rocks in active magmatic areas could move substantial amounts of heat, thereby enhancing the heat flux over conductive rates. However, this is not the case for large and small hydrothermal systems, as documented by oxygen isotopic measurements of continental rocks (e.g., Singleton and Criss, 2004), or for ophiolites (Gregory and Taylor, 1981) which were once part of the ocean floor. Advection plays a limited role in heat transfer near the surface because the temperature gradient is controlled by other processes, namely those processes which create the magmatic intrusion. A mathematical analysis was present by Hofmeister and Criss (2005), who considered 1-dimensional, steady-state flow.

The total heat flux is the sum of the conductive flux and the advective flux:

$$\Im_{heat,total} = -K_{rock}\frac{\partial T}{\partial z} + \Im_{mass,fluid}C_{fluid}T \qquad (3.15)$$

where the fluid flux has units of kg/m^2-s. The Nusselt number (Nu) is obtained by dividing Eq. (3.15) by the conductive flux, which reveals that the important quantity is the lumped parameter:

$$b = \frac{\Im_{mass,fluid}C_{fluid}}{K_{rock}}. \qquad (3.16)$$

The solution for a fixed upper temperature (T_0) and a fixed basal temperature (T_L) where $z = L$ is:

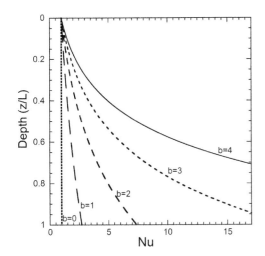

FIGURE 3.9 Relationship of normalized depth to the Nusselt number for vertical transport of heat by conduction and advection. The line $b = 0$ represents conduction only. *After Figure 5 in Hofmeister, A.M, Criss, R.E., 2005. Earth's heat flux revisited and linked to chemistry. Tectonophysics 395, 159−177, with permissions.*

$$\frac{T - T_0}{T_L - T_0} = \frac{1 - e^{-bz}}{1 - e^{-bL}}. \tag{3.17}$$

But the more revealing quantity is Nu (Fig. 3.9). High rates of fluid flow (large b) affect the temperature profile and enhance the flux at depth, but this augmentation is not evident at the surface, where heat flux is measured.

The above analysis shows that hydrothermal circulation cannot provide the additional heat flux needed for 1-dimensional cooling models for the plates to explain available measurements of the oceanic lithosphere. See Hofmeister and Criss (2005) and Chapter 7 for additional discussion.

3.3.4.2 Behavior of water in silicate liquids

The current view in petrology is that water and OH^- are dissolved in the silicate liquid, and that the physical properties measured in the laboratory describe a single phase. This view is exemplified by oxide summation models for density and α of melts (e.g., Lange, 1997), and stems from reliance in geoscience on equilibrium thermodynamics, which really should be termed thermostatics (Nordstrom and Munoz, 1986). The Earth is *not* a static environment, as it has both thermal and gravitational gradients, and is cooling and evolving.

Many examples of fluid mixtures exist where the system does not behave as a single phase (e.g., Li et al., 2010). Oil and vinegar mixtures provide a familiar example of *2-phase* (biphasic) behavior, where these liquid components separate in the gravity field on Earth's surface, unless forcibly agitated. Aqueous biphasic systems are common and have been investigated since the 1950s (Wang et al., 2016). Notably, static compression is insufficient to homogenize certain mixtures, such as ionic liquid $+ CO_2$ (Blanchard et al., 2001). The latter system is similar to magma, where the solute is a small gas molecule incorporated in a liquid solvent with strong internal bonding.

Biphasic behavior is a consequence of the much different physical properties of the solvent and solute. Water (steam at high T) is a component of many biphasic systems due to its bizarre physical properties compared to virtually all other substances. For example, thermal expansivity of water or steam increase as pressure is increased (e.g., Chaplin, 2018). Water being polar and small is a contributing factor. Lack of bonding between components, which leads to immiscibility, is also important.

Regarding magmas, the density of water is much less than that of a silicate melt. Importantly, the expansivity of water/steam is about $1000\times$ that of silicates at ~ 400 K, and strongly increases as T climbs further (Kell, 1975). Near 945 K, the silicate glass component has a density near ~ 3000 kg/m^3, whereas the steam component will have a density of ~ 0.225 kg/m^3. Per the polyhedral bulk moduli model, this difference exists on both microscopic and macroscopic scales. Earth's gravitational field requires that water rise through silicate melt, whether as individual molecules or droplets (fluid inclusions). The only impediment is when the melt has extremely high viscosity as in rhyolites, due to the rigidity of the 3-dimensional SiO$_2$ framework (e.g., Whittington, 2019).

Several observations support that 2-phase behavior describes H$_2$O molecules in many different magma compositions: (1) Magmas have fluid inclusions and bubbles, showing that phases have separated. Pumice is an extreme case. (2) The Pala pegmatites of California testify to the separation and buoyancy of water-rich melt on a large scale from granitic bodies. Earthquakes permitted rapid escape of the watery fluid, so large gemstones were preserved. Note that gem pockets over the world occur in tectonically active areas (Simmons et al., 2013). (3) Explosive volcanism is associated with water retention, e.g., the high-silica dacite dome of Mount St. Helens; see also Fowler and Spera, 2008. (4) Explosive volcanism is uncharacteristic of basalts, which corroborates relative ease of water migration in this chemical composition.

3.3.4.3 Irreversiblity is important to advection of heat

Two-phase behavior in the Earth is complex and depends on time. Regarding flow of heat, 2-phase behavior in aqueous systems may affect heat transfer because the heat capacity of water is rather large. Hydrothermal systems differ somewhat, as discussed above, a consequence of slow *circulation*, which approximates reversible behavior. The ocean's temperature is regulated by the Solar flux. Thus, the ocean not only provides a reservoir of water, but also serves as a regulatory temperature bath for the oceanic lithosphere.

One-way motion of water from the deep interior to the surface is an entirely different story. Such transport is irreversible, where water that is in contact with hot rocks far beneath the surface is released at the surface, but little is carried back down to deep source zones, although subduction of altered material provides a reverse path for part of the material. The presence of primordial, deeply sequestered H$_2$O, and CO$_2$ also needs to be addressed in time-dependent thermal models of the Earth and other planets.

3.4 The physics of convection: how planets differ from the laboratory

Whole mantle convection is not supported by most observations, nor does mantle tomography provide compelling support (see Chapter 1 for discussion and references). Even proponents of whole mantle convection acknowledge that first order difficulties remain with their models, despite ~ 50 years of effort. For example, Bercovici (2015) noted that:

- the mechanism moving plates has not been identified;
- known energy sources are too weak to drive convection;
- evidence for thin, thermal plumes is lacking; and
- persistence of the known chemical heterogeneities in the upper mantle is not explained.

Evidence for whole mantle convection rests on the large magnitude of the dimensionless Rayleigh number (Ra $\sim 10^8$) for Earth's ~ 3000 km thick mantle. However, the historic derivation of the Rayleigh number assumes conditions quite different from those inside the rocky Earth, which motivated Hofmeister and Criss (2018) to probe the underlying physics. This section summarizes their theoretical assessment of stability criteria and arguments for stability of the lower mantle.

3.4.1 The Rayleigh number

The dimensionless number Ra derived by Rayleigh (1916) describes fluids in a box in a uniform gravity field, where:

$$Ra = \frac{\alpha_{vol}\Delta T g h^3}{Dv}, \tag{3.18}$$

α_{vol} is volumetric thermal expansivity, ΔT is the temperature difference across the system, g is the constant gravitational acceleration, D is thermal diffusivity, and v is kinematic viscosity. Minor modifications to Ra made in geodynamic studies are described below. Non-dimensional numerical simulations of mantle convection use Ra as a variable (e.g., Zhong et al., 2015).

Estimates of supercritical Ra are insufficient evidence for whole mantle convection, because applying Eq. (3.18) to the continental lithosphere suggests that this region should convect (Hofmeister and Criss, 2018), contrary to geologic evidence. Regarding the mantle, the very slow motions of the plates are unlike the strong, time-dependent, turbulent conditions suggested by its high (10^8) values for Ra. Experimentally, convective flow has only been observed in rather inviscid materials (v up to 22000 mm^2/s as in syrup: White, 1988). That the experimental constraints on stability involve v that is 17 orders-of-magnitude lower than estimates of mantle viscosity has not been questioned when using Ra to describe Earth's mantle. In contrast, a much smaller discrepancy between strain rates of 10^{-3} to 10^{-6} s^{-1} in laboratory experiments with those of 10^{-12} to 10^{-16} s^{-1} estimated from perceived mantle circulation has been a long-standing concern in deformation studies (e.g., Cordier et al., 2012).

3.4.2 Existing adaptations of the Ra to planets and their pitfalls

3.4.2.1 Accounting for the adiabat?

In atmospheric studies, the adiabatic gradient is substracted from ΔT (e.g., Tritton, 1977). The equation used [$\Delta T^* = \Delta T(1 - \gamma_{th} \, B_T^{-1}\Delta P)$] actually depicts an isentrope.

Adiabats and isentropes are equivalent for ideal gas and reasonable for real gas (Hofmeister and Criss, 2019a,b), but not for solid planets. This correction amounts to about 30% of Ra and is thus unimportant to stability arguments.

Instead, thermal stability hinges on whether the material remains solid under the applied temperature gradient. Solids conduct heat up to their melting temperature. This response is irrespective of the thermal gradient, but rather depends on the thermodynamic characteristics of the material. The classical Ra number presumes constant physical properties and thus cannot describe a two-phase system, e.g., a system with co-existing solids and fluids.

3.4.2.2 Incorrect assessment of internal heating

The current adaptation of Ra in geodynamics for internal heating depends more strongly on h than on its third power as in Eq. (3.18) (e.g., Schubert et al., 2001). That the dependence is stronger signals an obvious problem, because internal heating supplies heat towards the top of a box, rather than just at its bottom, and this distribution is stabilizing against convection, not invigorating.

In detail, over a long time, the middle of an internally heated slab becomes increasingly hotter than its bottom, which makes it difficult for the lower half of the slab to convect. Short times are more relevant due to the immensity of Earth (Chapter 2). For this case, internal heating would destabilize the upper half of the slab, at the expense of stabilizing the lower half. Because the internal heating of the Earth is weak, this destablizing effect on the upper half should be neglected, especially insofar as ΔT is an estimate.

3.4.3 Mass conservation violation due to spherical geometry

The classical derivation of the Rayleigh number considers vertical instability of a tabular system, which greatly differs from radial instability in a planet. The different shapes of the volume elements in Cartesian, cylindrical, and spherical-polar coordinate systems lead to different rules for mass conservation during motions of these elements.

Equal thickness (δh) layers in a box have the same mass, obviously. However, concentric shells of equal thickness (δs) in a sphere have different mass because shell area varies with radius (Fig. 3.10). For other geometries, length scales require amending to conserve mass during its transfer between layers. For the sphere, h in Eq. (3.18) must be replaced with L:

$$L = h \frac{R_{in}^2}{R_{out}^2} = h \frac{(R_{out}-h)^2}{R_{out}^2} = h \left(1 - \frac{h}{R_{out}}\right)^2 \tag{3.19}$$

To conserve momentum and heat during flow (the continuity equations) requires the same scaling (Hofmeister and Criss, 2018). Thus, for planets, which are nearly spherical, Ra must be multiplied by $(1 - h/R_{out})^6$. This correction is ~ 0.01 which changes the value of Ra by only two orders of magnitude. More importantly, the correction of Eq. (3.19) changes the character of non-dimensionalized mantle convection equations, which do not currently incorporate the size of the planet.

Size greatly affects heat conduction in spheres (Criss and Hofmeister, 2016; Chapter 2) and therefore must greatly affect convection. Not considering the shapes of the elements has led to excluding a key descriptor of a celestial body, its size, from convection models.

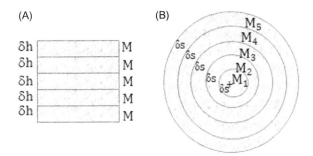

FIGURE 3.10 Schematics in 2-dimensions illustrating how geometry affects scale length in dimensional analysis. (a) Mass balance of parallel layers with equal thickness and density in Cartesian geometry. (b) Imbalanced mass in concentric layers with equal thickness and density. M_1 is the smallest mass, progressing to M_5, the largest mass. If heat energy is considered in terms of heat density (energy per volume), then for equal thickness layers, the same imbalance occurs as for mass. For cylindrical symmetry, the missing factor is $(1 - h/R_{out})$, rather than that given in Eq. (3.19).

3.4.4 Stability criteria for flow of plastic solids

Infinitesimal stress causes liquids to flow, but the response of solids to stress is much more complicated and has been studied in great detail in engineering and other fields. The Ra number assumes that flow is viscous, but for solids, viscous behavior is limited to certain materials above the glass transition. The much different process of creep can occurs in solids that are hot, but below melting (Meyers and Chawla, 2009). Certain mathematical descriptions of creep (e.g., Nabarro-Herring or Coble) can be recast to mimic equations that depict Newtonian viscous behavior (e.g., Kohlstedt and Hansen, 2015), lending support to models that treat the solid mantle as if it were fluid. However, more realistic equations for creep cannot be made equivalent to viscous behavior. On this basis, Hofmeister and Criss (2018) considered the idealized behavior of a Bingham plastic to develop criteria for flow in the mantle.

Bingham's model requires a minimum shear stress (σ_{yield}) for plasticity, in order to describe materials with both liquid- and solid-like behavior (see Davis and Selvadurai, 2005; Selhke et al., 2014 for applications to mud and lavas). Regarding mantle materials, multiple types of plasticity are observed, including motion of dislocations, atomic diffusion, and dynamic recrystallization (Kohlstedt and Hansen, 2015). That substantial energy is needed to break and reform atomic bonds in these mechanisms, explains why significant differential stress ($\delta\sigma$) is needed to deform oxides and silicates in the laboratory. For example, Jin et al. (2001) applied $\sim 90-200$ MPa at hydroscatic pressures of $\sim 30-50,000$ MPa to deform synthetic ecolgites and gabbros.

Thus, for solids to exhibit the liquid-like behavior that underlies the Rayleigh number requires:

$$\frac{\delta\sigma}{\sigma_{yield}} > 1. \tag{3.20}$$

Except for locations close to the moving slabs, conditions in the mantle are hydrostatic ($\delta\sigma \sim 0$), so creep is unexpected.

3.4.5 Derivation of Gr and Ra from force balance

Assuming Newtonian flow allowed Hofmeister and Criss (2018) to evaluate the ratio of the buoyancy to drag forces under constant gravity (g) and physical properties using dimensional analysis as:

$$\frac{F_{\text{buoyancy}}}{F_{\text{drag}}} \cong \frac{\rho_0 \alpha_V \Delta T g h \mathring{A}}{\mathring{A} \rho_0 \upsilon (\partial u_{\text{flow}}/\partial y)} = \frac{\alpha_V \Delta T g h}{\upsilon(u_{\text{flow}}/h)} = \frac{\alpha_V \Delta T g h^3}{\upsilon(u_{\text{flow}} h)} = \frac{\alpha_V \Delta T g h^3}{\upsilon^2}. \tag{3.21}$$

where $\mathring{A}$ is area and the adiabatic gradient is substracted from ΔT. The RHS is identical to Grashof's dimensionless number, Gr, which equals $\upsilon Ra/D$.

Criteria for instability were developed by recognizing that heat and mass flow are decoupled in highly viscous liquids (Hofmeister and Criss, 2018). Hence, each force in Eq. (3.21) is mitigated by the relevant diffusion coefficient. Instability exists when:

$$\frac{F_{\text{bouyancy}}}{F_{\text{drag}}} \frac{D_{\text{mass}}}{D_{heat}} > 1; \quad \text{i.e.,} \quad \frac{\text{Gr}}{\text{Le}} > 1; \quad \text{or} \quad \frac{\text{Ra}}{\text{Sc}} > 1 \tag{3.22}$$

where $Sc = \upsilon/D_{\text{mass}}$ is the Schmidt number and $Le = D_{\text{heat}}/D_{\text{mass}}$ is the Lewis number. Presumably, the minimum stress of Eq. (3.20) is met. Eq. (3.22) provides $\text{Ra}_{\text{critical}} \sim 2000$ from Sc for water. For gas, $\text{Le} \sim \text{Sc} \sim 1$ provides critical numbers of unity for both Gr and Ra. Both deductions are supported by experiments: gas convects when the adiabatic gradient is exceeded (i.e., near Ra of unity) whereas water has a critical number of ~ 1760.

For a solid close to the yield point, the ratio of buoyant to resistive forces is near unity. Thus, the flow criterion simplifies to $Le < 1$, which is not observed in solids, since mass diffusion is miniscule. The much faster process of thermal diffusion overwhelms creep in solids.

Although mantle convection is unexpected, an alternative exists to instability in a high thermal gradient, namely, when a solid becomes too hot, it melts. In planets, melting produces a phase that is highly buoyant and far less viscous than the solid. Thus, the key heat-transfer processes expected in planets are conduction (radiative diffusion: Section 3.1) and the advective, one-way ascent of magma and volatiles (Section 3.3).

3.5 Summary

Solids differ from fluids in two important ways that govern how they transport heat. The first key property is rigidity, which allows solids to resist stresses up to a certain point. This includes thermal stresses, such as an imposed temperature gradient: as long as a body remains solid for the range of temperatures experienced, it will conduct heat. Second, the conductive response can be explained in terms of the diffusivities as well: mass diffusivity is negligible compared to thermal diffusivity because displacing atoms requires that their bonds are broken and reformed. The third transport property, viscosity, represents resistance to flow, which is immense in solids. Hofmeister and Criss (2018) cast this resistance in terms of differential stress, and argued that the mantle being essentially hydrostatic cannot convect due to rigidity of solids. Several errors in the formulation and implementation of the Rayleigh criterion that were elucidated in this report are summarized in this chapter, including system wide violations of mass and energy conservation. Due to the geometry of the sphere and the fact that solids melt, both the size of a planet and its maximum temperature are essential to realistic models of convection models, but these factors have not to date been included. To address first-order observations (Chapter 1), such as plate tectonics, a mechanism involving gravitational forces in the Earth-Moon-Sun system is discussed in Chapter 5.

That melting, and not convection, is the response of the Earth to high temperatures is evident in the oceanic lithosphere, which is melted at its base, and in the core, which is substantially melted. Isothermal boundaries constrain not only temperature, but control flux (Chapter 1). To utilize these findings in ascertaining the thermal states and evolution of planets in Part 2, conduction in large spheres must be understood (Chapter 2) and initial conditions also need to be specified, which necessarily involve some assumptions (Chapter 4).

References

Anderson, O.L., Isaak, D.G., 1995. Elastic constants of mantle minerals at high temperature. In: Ahrens, T.J. (Ed.), A Handbook of Physical Constants. American Geophysical Union, Washington D.C., pp. 64−96.

Ashcroft, N.W., Mermin, N.D., 1976. Solid State Physics. Holt, Rinehart and Winston, New York.

Bauer, M., Marienfeld, A., Aeschlimann, M., 2015. Hot electron lifetimes in metals probed by time-resolved two-photon photoemission. Prog. Surf. Sci. 90, 319−376.

Bercovici, D., 2015. Mantle dynamics: an introduction and overview. In: second ed. Schubert, G. (Ed.), Treatise on Geophysics, vol. 1. Elsevier, Amsterdam, The Netherlands, pp. 1−22.

Berman, R., Foster, E.L., Schneidmesser, B., Tirmizi, S.M.A., 1960. Effects of irradiation on the thermal conductivity of synthetic sapphire. J. Appl. Phys. 31, 2156−2159.

Blanchard, L., Gu, Z., Brennecke, J., 2001. High-pressure phase behavior of Ionic liquid/CO_2 systems. J. Phys. Chem. B 105, 2437−2444.

Burns, G., 1990. Solid State Physics. Academic Press, San Diego.

Chaplin, M., 2018. Water structure and science. < http://www1.lsbu.ac.uk/water/water_structure_science.html > (accessed February 10, 2019).

Coddington, O., Lean, J.L., Pilewskie, P., Snow, M., Lindholm, D., 2016. A solar irradiance climate data record. Bull. Am. Meteorol. Soc. 96, 1265−1282. Available from: https://doi.org/10.1175/BAMS-D-14-00265.12.

Cordier, P., Amodeo, J., Carrez, P., 2012. Modelling the rheology of MgO under Earth's mantle pressure, temperature and strain rates. Nature 481, 177−180.

Criss, R.E., Hofmeister, A.M., 2016. Conductive cooling of spherical bodies with emphasis on the Earth. Terra Nova 28, 101−109.

Criss, R.E., Hofmeister, A.M., 2017. Isolating lattice from electronic contributions in thermal transport measurements of metals and alloys and a new model. Int. J. Mod. Phys. B 31, No. 175020. Available from: https://doi.org/10.1142/S0217979217502058.

Criss, E.M., Hofmeister, A.M., 2019. Transport properties of metals, alloys, and their melts from LFA measurements. In: Hofmeister, A.M. (Ed.), Measurements, Mechanisms, and Models of Heat Transport. Elsevier, Amsterdam, pp. 295−325.

Danyushevsky, L., Falloon, T., Sobolev, A., Crawford, M., Price, R., 1993. The H_2O content of basalt glasses from Southwest Pacific back-arc basins. Earth Planet. Sci. Lett. 117, 347−362.

Davis, R.O., Selvadurai, A.P.S., 2005. Plasticity and Geomechanics. Cambridge University Press, Cambridge.

Ditmars, D.A., Ishihara, S., Chang, S.S., Bernstein, G., West, E.D., 1982. Enthalpy and heat-capacity standard reference material: synthetic sapphire (α-A1203) from 10 to 2250 K. J. Res. Natl. Bur. Stand. 87, 159−163.

Doglioni, C., Panza, G., 2015. Polarized plate tectonics. Adv. Geophys. 56, 1−167.

Doremus, R.H., 2002. Viscosity of silica. J. Appl. Phys. 92, 7619−7629.

Elderfield, H., Schultz, A., 1996. Mid-Ocean ridge hydrothermal fluxes and the chemical composition of the ocean. Ann. Rev. Earth Planet. Sci. 24, 191–224.

Feldman, D.W., Parker Jr., J.H., Ashkin, M., 1968. Raman scattering by optical modes of metals. Phys. Rev. Lett. 21, 607–609.

Fiquet, G., Richet, P., Montagnac, G., 1999. High-temperature thermal expansion of lime, periclase, corundum and spinel. Phys. Chem. Miner. 27, 103–111.

Fowler, S.J., Spera, F.J., 2008. Phase equilibria trigger for explosive volcanic eruptions. Geophys. Res. Lett. 35, Paper L08309. Available from: https://doi.org/10.1029/2008GL033665.

Gregory, R.T., Taylor, H.P., 1981. An oxygen isotope profile in a section of cretaceous oceanic crust, Samail ophiolite, Oman: evidence for $\partial^{18}O$-buffering of the oceans by deep (>5 km) seawater-hydrothermal circulation at mid-ocean ridges. J. Geophys. Res. 86, 2737–2755.

Henderson, G., 1982. Inorganic Geochemistry. Permagon Press, New York.

Hofmeister, A.M., 2005. The dependence of radiative transfer on grain-size, temperature, and pressure: implications for mantle processes. J. Geodyn. 40, 51–72.

Hofmeister, A.M., 2006. Thermal diffusivity of garnets at high temperature. Phys. Chem. Miner. 33, 45–62.

Hofmeister, A.M., 2007. Thermal diffusivity of aluminous spinels and magnetite at elevated temperature with implications for heat transport in Earth's transition zone. Am. Miner. 92, 1899–1911.

Hofmeister, A.M., 2010. Scale aspects of heat transport in the diamond anvil cell, in spectroscopic modeling, and in Earth's mantle: implications for secular cooling. Phys. Earth Planet. Inter. 180, 138–147.

Hofmeister, A.M., 2014. Thermal diffusivity and thermal conductivity of single-crystal MgO and Al_2O_3 as a function of temperature. Phys. Chem. Miner. 41, 361–371.

Hofmeister, 2019a. Modelling diffusion of heat in solids. In: Hofmeister, A.M. (Ed.), Measurements, Mechanisms, and Models of Heat Transport, 427. Elsevier, New York, pp. 359–398.

Hofmeister, A.M., 2019b. Reconciling the kinetic theory of gas with gas transport data. In: Hofmeister, A.M. (Ed.), Measurements, Mechanisms, and Models of Heat Transport. Elsevier, Amsterdam, pp. 143–179.

Hofmeister, A.M., 2019c. Macroscopic analysis of the flow of energy into and through matter from spectroscopic measurements and electromagnetic theory. In: Hofmeister, A.M. (Ed.), Measurements, Mechanisms, and Models of Heat Transport. Elsevier, Amsterdam, pp. 35–73.

Hofmeister, A.M., 2019d. Thermal diffusivity data on non-metallic crystalline solids from laser flash analysis. In: Hofmeister, A.M. (Ed.), Measurements, Mechanisms, and Models of Heat Transport. Elsevier, Amsterdam, pp. 201–250.

Hofmeister, A.M., 2019e. Methods used to determine heat transport and related properties, with comparisons. In: Hofmeister, A.M. (Ed.), Measurements, Mechanisms, and Models of Heat Transport. Elsevier, Amsterdam, pp. 99–142.

Hofmeister, A.M., Bowey, J.E., 2006. Quantitative IR spectra of hydrosilicates and related minerals. Mon. Not. R. Astron. Soc. 367, 577–591.

Hofmeister, A.M., Criss, E.M., 2018. How properties that distinguish solids from fluids and constraints of spherical geometry suppress lower mantle convection. J. Earth Sci. 29, 1–20. Available from: https://doi.org/10.1007/s12583-017-0819-4.

Hofmeister, A.M., Criss, R.E., 2005. Earth's heat flux revisited and linked to chemistry. Tectonophysics 395, 159–177.

Hofmeister, A.M., Criss, R.E., 2019a. Reconciling classical thermodynamics with heat and mass transfer: implications for the kinetic theory of gas. Can. J. Phys.

Hofmeister, A.M., Criss, R.E., 2019b. The macroscopic picture of heat retained and heat emitted: thermodynamics and its historical development. In: Hofmeister, A.M. (Ed.), Measurements, Mechanisms, and Models of Heat Transport. Elsevier, Amsterdam, pp. 1–34.

Hofmeister, A.M., Dong, J.J., Branlund, J.M., 2014. Thermal diffusivity of electrical insulators at high temperatures: evidence for diffusion of phonon-polaritons at infrared frequencies augmenting phonon heat conduction. J. Appl. Phys. 115, 163517. Available from: https://doi.org/10.1063/1.4873295.

Hofmeister, A.M., Pertermann, M., Branlund, J.M., 2007. Thermal conductivity of the Earth. In: Schubert, G., Price, G.D. (Eds.), Treatise in Geophysics, V. 2 Mineral Physics. Elsevier, The Netherlands, pp. 543–578.

Hust, J.G., Lankford, A.B., 1984. Update of thermal conductivity and electrical resistivity of electrolytic iron, tungsten, and stainless steel. National Bureau of Standards Spec. Pub. 260-290, 1–71.

Hust, J.G., Sparks, L.L., 1973. Lorenz ratios of technically important metals and alloys. NBS Technical Note 634 (U.S. Government Printing Office, Washington, DC).

Jin, Z.-M., Zhang, J., Green II, H.W., Jin, S., 2001. Eclogite rheology: implications for subducted lithosphere. Geology 29, 669–670.

Kell, G.S., 1975. Density, thermal expansivity, and compressibility of liquid water from 0″ to 150°C: correlations and tables for atmospheric pressure and saturation reviewed and expressed on 1968 temperature scale. J. Chem. Eng. Data 20, 97–105.

Kieffer, S.W., 1979. Thermodynamics and lattice vibrations of minerals: 3. Lattice dynamics and an approximation for minerals with application to simple substances and framework silicates. Rev. Geophys. Space Phys. 17, 20–34.

Kittel, C., 1971. Introduction to Solid State Physics, fourth ed. John Wiley and Sons, New York, pp. 224-226, 239–265.

Kohlstedt, D.L., Hansen, L.N., 2015. Constitutive behavior, rheological behavior, and viscosity of rocks. In: second ed. Schubert, G. (Ed.), Treatise in Geophysics, vol. 2. Elsevier, The Netherlands, pp. 389–427.

Krishnan, R.S., Srinivasan, R., Devanarayanan, S., 1979. Thermal Expansion of Crystals. Pergamon Press, New York.

Lange, R.A., 1997. A revised model for the density and thermal expansivity of $K_2O-Na_2O-CaO-MgO-Al_2O_3-SiO_2$ liquids from 700 to 1900 K: extension to crustal magmatic temperatures. Contrib. Miner. Petrol 130, 1–11.

Lee, D.W., Kingery, W.D., 1960. Radiation energy transfer and thermal conductivity of ceramic oxides. J. Am. Ceram. Soc. 43, 594–607.

Lee, T.Y.R., Donaldson, A.B., Taylor, R.E., 1978. Thermal diffusivity of layered composites. Therm. Conduct. 15, 135–148.

Li, C., Han, J., Wang, Y., Yan, Y., Pan, J., Xu, X., et al., 2010. Phase behavior for the aqueous two-phase systems containing the ionic liquid 1-butyl-3-methylimidazolium tetrafluoroborate and Kosmotropic salts. J. Chem. Eng. Data 55, 1087–1092.

Lubimova, H., 1958. Thermal history of the earth with consideration of the variable thermal conductivity of the mantle. Geophys. J. Roy. Astron. Soc. 1, 115–134.

Maxwell, J.C., 1888. An Elementary Treatise on Electricity. Henry Frowde, London.

MacDonald, G.J.F., 1959. Calculations on the thermal history of the earth. J. Geophys. Res. 64, 1967–2000.

McGetchin, T.R., Nikhanj, Y.S., Chodos, A.A., 1973. Carbonatite-kimberlite relations in the Cane Valley diatreme, San Juan County, Utah. J. Geophys. Res. 78, 1854–1869.

Merriman, J.D., Alan, G.W., Hofmeister, A.M., Nabelek, P.I., Benn, K., 2013. Thermal transport properties of major Archean rock types to high temperature and implications for cratonic geotherms. Precambrian Res. 233, 358–372.

Meyers, M.A., Chawla, K.K., 2009. Mechanical Behavior of Materials. Cambridge University Press, Cambridge.

Moravec, T.J., Rife, J.C., Dexter, R.N., 1976. Optical constants of nickel, iron, and nickel-iron alloys in the vacuum ultraviolet. Phys. Rev. B13, 3297–3306.

Murakami, M., Hirose, K., Yurimoto, H., Nakashima, S., Takafuji, N., 2002. Water in Earth's lower mantle. Science 295, 1885–1887.

Nordstrom, D.K., Munoz, J.L., 1986. Geochemical Thermodynamics. Blackwell Scientific, Palo Alto, CA.

Nusselt, W., 1915. Das Grundgesetz des Wärmeüberganges. Susndh. Ing. Bd. 38, 477-482, 490-496.

Ordal, M.A., Bell, R.J., Alexander Jr., R.W., Long, L.L., Querry, M.R., 1983. Optical properties of fourteen metals in the infrared and far infrared; Al, Co, Cu, Au, Fe, Pb, Mo, Ni, Pd, Pt, Ag, Ti, V, and W. Appl. Opt. 22, 1099–1119.

Palik, E.D., 1998. Handbook of Optical Constants of Solids. Academic Press, San Diego.

Pitman, K.M., Hofmeister, A.M., Speck, A.K., 2013. Revisiting astronomical crystalline forsterite in the UV to Near-IR. Earth Planets Space. 65, 129–138.

Poincaré, H., 1905. Science and Hypothesis, English Translation. The Walter Scott Publishing Co., London.

Price, G.D., Stixrude, L., 2015. Mineral physics: an introduction and overview. In: 2nd edition Schubert, G. (Ed.), Treatise in Geophysics, vol. 2. Elsevier, The Netherlands, pp. 1–5. Available from: https://doi.org/10.1016/B978-0-444-53802-4.00029-4.

Rayleigh, L., 1916. On convection currents in a horizontal layer of fluid when the higher temperature is on the underside. Phil. Mag. 32, 529–546.

Schubert, G., Turcotte, D.L., Olson, P., 2001. Mantle Convection in the Earth and Planets. Cambridge University Press, Cambridge.

Selhke, A., Whittington, A., Robert, B., Harris, A., Gurioli, L., Médard, E., 2014. Pahoehoe to 'a'a transition of Hawaiian lavas: an experimental study. Bull. Volcanol. 76, 876–896.

Shankland, T.J., Nitsan, U., Duba, A.G., 1979. Optical absorption and radiative heat transport in olivine at high temperature. J. Geophys. Res. 84, 1603–1610.

Siegel, R., Howell, J.R., 1972. Thermal Radiation Heat Transfer. McGraw-Hill, New York.

Simmons, W.B., Pezzotta, F., Shigley, J.E., Beurlen, H., 2012. Granitic pegmatites as sources of colored gemstones. Elements 8, 281–287. Available from: https://doi.org/10.2113/gselements.8.4.281.

Singleton, M.J., Criss, R.E., 2004. Symmetry of flow in the Comstock Lode hydrothermal system: evidence for longitudinal convective rolls in geologic systems. J. Geophys. Res. 109, B03205.

Touloukian, Y.S., Powell, R.W., Ho, C.Y., Nicolaou, M.S., 1973. Thermal Diffusivity. IFI/Plenum, New York.

Touloukian, Y.S., Sarksis, Y., 1970. Thermal Conductivity: Metallic Elements and Alloys. IFI/Plenum, New York.

Transtrum, M.K., Machta, B.B., Brown, K.S., Daniels, B.C., Myers, C.R., Sethna, J.P., 2015. Perspective: sloppiness and emergent theories in physics, biology, and beyond. J. Chem. Phys. 143, 010901.

Tritton, D.J., 1977. Physical Fluid Dynamics. Van Nostrand Reinhold, New York.

Ventura, G., Perfetti, M., 2014. Thermal Properties of Solids at Room and Cryogenic Temperatures. Springer, Heidelburg.

Wang, Y., Li, Y., Han, J., Xia, J., Tang, X., Chen, T., et al., 2016. Cloudy behavior and equilibrium phase behavior of triblock copolymer L64 + salt + water two-phase systems. Fluid Phase Equilibria 409, 439–446. Available from: https://doi.org/10.1016/j.fluid.2015.10.046.

White, D.B., 1988. The planforms and onset of convection with a temperature-dependent viscosity. J. Fluid Mech. 191, 247–286.

White, S.M., Crisp, J.A., Spera, F.J., 2006. Long-term volumetric eruption rates and magma budgets. Geochem. Geophys. Geosyst. 7, Q03010. Available from: https://doi.org/10.1029/2005GC001002.

Whittington, A.G., 2019. Heat and mass transfer in glassy and molten silicates. In: Hofmeister, A.M. (Ed.), Measurements, Mechanisms, and Models of Heat Transport. Elsevier, New York, pp. 327–357.

Whittington, A.G., Hofmeister, A.M., Nabelek, P.I., 2009. Temperature-dependent thermal diffusivity of Earth's crust: implications for crustal Anatexis. Nature 458, 319–321.

Zhong, S.J., Yuen, D.A., Moresi, L.M., Knepley, M.G., 2015. Numerical method for mantle convection. In: second ed. Schubert, G. (Ed.), Treatise in Geophysics, vol. 7. Elsevier, The Netherlands, pp. 197–222.

Websites

Spectra, infrared and visible (accessed 22.0219.).
http://www.astro.uni-jena.de/Laboratory/Database/jpdoc/f-dbase.html
http://minerals.gps.caltech.edu/
Thermal diffusivity (accessed 22.02.19.).
http://epsc.wustl.edu/~hofmeist/thermal_data/
Cryogenic data on thermal conductivity (accessed 14.04.18.).
https://ws680.nist.gov/publication/get_pdf.cfm?pub_id = 913059
https://www.nist.gov/mml/acmd/nist-cryogenic-materials-property-database-index
https://www.lakeshore.com/Documents/LSTC_appendixI_l.pdf
Heat capacity (accessed 22.02.19.).
https://pubs.er.usgs.gov/publication/b2131.
Debye temperatures (accessed 22.02.19.).
http://www.knowledgedoor.com/2/elements_handbook/debye_temperature.html
Properties of elements (accessed 20.04.18.). This site includes citations.
http://www.knowledgedoor.com/
Steam properties (accessed 22.02.19.).
http://www.thermopedia.com/content/1150/

Physical constraints on the initial conditions and early evolution of the solar system

Anne M. Hofmeister and Robert E. Criss

"Into the eternal darkness, into fire and into ice." Dante Alighieri (The Divine Comedy, 1320).

Currently popular models of accretion of bodies in our Solar System stem from Safronov's (1969) description of the process as consisting of random and fractal collisions, where particles clumped into progressively larger aggregates until rocky bodies and ultimately planets were formed. One premise is the historical notion of Laplace that the Solar System formed from material ejected from the equator of the spinning Sun, while another premise is based on the erroneous notion of Kelvin that gravitational contraction produces heat, including the radiant power of stars. These hot fractal models fail to explain most first-order features of the Solar System, which are:

- All planets spin upright and similarly
- Orbits are concentric and nearly circular
- $> 99\%$ of the angular momentum of the Solar System resides in the planets
- The axial orientation and handedness of spin, orbits, and satellite orbits are nearly the same
- Regular trends exist in rotational and spin energies
- Regular trends exist in planetary mass for each of gassy and rocky planets
- Planets of the inner and outer sub-systems differ greatly in size, chemical composition, and their number of satellites

Several serious flaws with the fractal models are acknowledged by the researchers involved. (1) Angular momentum is not conserved (Armitage, 2011), a major problem for Laplace's solar ejection hypothesis. (2) The first step of grain formation is a mystery (Wetherill, 1988). No subsequent fractal model has addressed this problem. (3) The chaos envisioned in the final, giant impact stage conflicts with the observed regularities listed above. (4) Evidence is lacking for high-temperature nebular condensates being located near the Sun (Lodders, 2003) because the innermost planet Mercury, like the chondrites attributed to the distant asteroid belt, is composed of silicates and iron which condense at similar temperatures. (5) The discovery that comets are a mix of ice and refractory minerals (Zolensky et al., 2006) contradicts the hot accretion idea, and shows that the condensation sequence is not an important factor. In short, ejection of material from the Sun to form the planets lacks supporting evidence. Likewise, the main component of chondrites, the roundish chondules, are popularly depicted as melt droplets, although a viable heat sources is absent. Chapter 11 provides further discussion and an alternative origin.

As an alternative to assuming hot initial conditions and fractal accretion, Hofmeister and Criss (2012a) proposed that the nebula collapsed and contracted in two steps, forming the planets and Sun in accord with thermodynamics and conservation laws. Since then new data have appeared that pertain to formation of stars and planets, and our own ideas have been developed further.

However, this book is not about accretion: rather, the purpose is to better understand the thermal state and evolution in the Earth and other large rocky bodies. The importance of the initial boundary conditions to solving time-dependent problems in heat transfer is the motivating reason. Thus, the present chapter focuses on constraining temperatures upon formation and shortly thereafter. However, what should be an important tool (classical thermodynamics) has two serious flaws. First, the effect of gravity is ignored in the classical equations presented in modern thermodynamics books. Second, the dynamics of heat and mass transfer are not a part of classical thermodynamics, although they should have been (Truesdell, 1980). To address heat transfer and heat reservoirs in planets requires considering basic principles of Newtonian gravitation, since these bodies are compressed under their own weight.

This chapter is organized as follows. Section 4.1 provides a theoretical discussion of basic principles. Since gravity is a conservative, mechanical force, we begin with the Virial Theorem which conveniently incorporates the additional special requirements of a bound state. Heat and mass transfer during contraction of a cloud to create a spinning body are discussed. Section 4.2 summarizes observations that point to a cold start, and observations that point to an orderly, not chaotic, accretionary process. The next sections discuss steps of planetary formation and differentiation, while distinguishing between external and internal events. Section 4.3 covers how our Solar System initially formed from a nebula, addressing the known shortcomings in previous models mentioned above. We present arguments for iron cores accreting first. Section 4.4 discusses the much different heating and cooling processes that operate after protoplanetary bodies are formed, but have not yet reached full size. Giant impacts and large planetary drift are dismissed on the basis of elementary physics. Section 4.5 discusses gravitational segregation in planetary interiors, and why a minimum size for a formed body is needed to induce this process and add to the protocore. We explain how incorporation of ice nucleated on dust grains into accreting planetary bodies can affect their initial temperatures and early evolution. Section 4.6 summarizes.

4.1 Basic Principles

4.1.1 Virial Theorem - the key to uniting mechanics and thermostatics

Orbits in our Solar System were described long ago by Kepler. These bound states adhere to the Virial Theorem (VT) of Clausius, which for gravitational forces stipulates:

$$RE = -\frac{U_g}{2} \tag{4.1}$$

where U_g is the gravitational potential energy (PE), and RE is rotational energy. Eq. (4.1) and the definition of total energy (TE = RE + PE) yield:

$$TE = -RE = \frac{U_g}{2} \tag{4.2}$$

Clausius derived the VT from a mathematical identity, without invoking any conservation law (Hofmeister and Criss, 2016). Bound states are obviously conservative, so the laws of energy and momentum conservation also governed the formation of the bound state. The presolar nebula would have been vast, so from Eqs. (4.1) and (4.2), its total energy would have initially been nearly zero. Thus, energy was lost while forming the bound state of the Sun and its planets. From the VT, $\frac{1}{2}U_g$ was clearly lost, which meets second law requirements since, during any real process, the energy of the system must

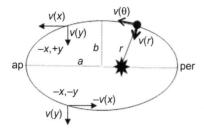

FIGURE 4.1 Schematics of the bound state of an elliptical orbit. A small mass (dot) orbits about a large mass (Sun) which sits at a focal point. At perihelion, aphelion, and the crossing of the minor axis, the radial velocity = 0, but the tangential velocity is finite. Spatially symmetrical pairs of xv_x and yv_y each balance: one example is shown. Radial and tangential decompositions also balance over the cycle. *After Figure 1b in Hofmeister, A.M., Criss, R.E., 2016. Spatial and symmetry constraints as the basis of the Virial Theorem and astrophysical implications. Can. J. Phys. 94, 380–388 with permissions.*

decrease while the energy and entropy of the surroundings increase. The energy loss from the system is achieved by radiative transfer because light is generated in collisions of atoms and molecules, due to inelasticity (Hofmeister, 2019a).

Angular momentum conservation is evident in planetary orbits. Momentum conservation was not invoked by Clausius, so it provides a constraint in addition to that provided by the VT. However, momentum can be conserved in states that are not bound, such as hyperbolic orbits. The special condition for the VT is that linear momentum cancels in all directions over a cycle in the restricted space that defines any particular bound state (Hofmeister and Criss, 2016). This balance involves the canonical conjugate pairs of position and linear momentum (Fig. 4.1). Symmetry is the source of this balance, whereby the return path mirrors the forward motion. It is this balance which allows the dynamic problems of orbit and spin to be conveniently evaluated as static averages.

Existence of bound states in the Solar System (e.g., Keplerian orbits) requires that angular momentum was conserved during its formation. Hofmeister and Criss (2012a) incorporated this principle in their conservative model, which utilized the first and second laws, but not the VT. Additional restrictions exist (Criss and Hofmeister, 2018): see below. The many restrictions underscore the inappropriateness of retaining heat during formation, when energy must instead be lost to the surroundings.

Lastly, the Solar System clearly formed as a single entity, as evidenced in Fig. 1.12. Additional evidence is provided in Section 4.2. Much of the original character is retained today, including the congruent rotational sense of both spin and orbits, and the approximately parallel spin axes of the planets and the Sun.

4.1.2 Heat and mass transfer principles

Inside planets, mass and heat can move independently or together. Diffusion of mass in response to a chemical gradient is appropriately modeled as independent from heat transfer (Fick's laws). In condensed matter, this process is very short range, and is overshadowed by mechanical motions over large scales driven by gravity. In the latter case, when matter moves in response to a gravitational field, heat is carried with it. This process is called advection and is important in planets because gravitational sorting can produce internal layering. Advective transfer of heat is proportional to the material properties of heat capacity and latent heat, in view of the mechanics. It matters how much matter is being moved, how fast it moves, and what are its contents of sensible and latent heat (Section 3.3).

Heat also moves through matter as a response to thermal gradients, a process called conduction, in a manner that is independent of the physical motions of matter. The mathematical physics is covered in Chapter 2. This process is always radiative, i.e., light is involved (Section 3.1; Hofmeister, 2019b). Conduction is diffusion of low frequency (infrared) light. Heat transfer being radiative and ever present originates in collisions of atoms being inelastic, regardless of whether the matter is a gas or a condensed phase. Inelastic interactions provide a gas of light that is intermingled with atoms, which cannot be contained as no perfect reflector exists. However, how fast the photon gas can move inside the medium (i.e., how rapidly temperature evolves) depends greatly on the density of the medium. In detail, the interaction is measured in spectroscopic experiments and is quantified by the medium's absorption coefficient. Density, chemical composition, and thus the absorption coefficients vary greatly between that of the gas of the Solar nebulae and condensed matter in the formed planets, and so the discussion of radiative diffusion inside planets (Section 3.1) needs extending to include gas.

4.1.2.1 Radiative transfer in very rarified media

If interactions of light with matter are strong, conditions are *optically thick*, where the medium participates and the process is diffusive, but if interactions are weak, conditions are *optically thin*, where involvement of the medium is minimal to negligible, and the process is ballistic (e.g., energy is transferred at great speed from boundary to boundary). Which of these conditions applies depends on the absorption characteristics of the particular medium. (Section 3.1; Hofmeister, 2019c).

Of great interest is the change in absorption of the Solar nebula as it contracts, which can be gauged from changes in its density. Star-forming environments (molecular clouds) are mostly H and H_2 gas with some other molecules, ices, and dust. Earth's N_2 atmosphere is a reasonable analogue material, since monatomic and diatomic molecules lack infrared fundamental vibrations. It is mostly optically thin, so ballistic transfer describes the Solar nebula up until its density is similar to Earth's atmosphere. Based on this criterion, Hofmeister and Criss (2012a) showed that optically thin conditions persisted up until the Solar nebula contracted to inside the orbit of Mercury, i.e. when the protosun formed.

The hallmark of ballistic processes is rapidity. The characteristic speed is c/n, where c is lightspeed and n is the index of refraction, and so ballistic heat transfer is far faster than particle velocities in the Solar System during accretion. The presence of the medium slows the travel by only a small amount, since n is typically ~ 1 for gas but can reach ~ 4 into the visible (e.g., elemental Ge).

Consequently, the effect of ballistic radiation in problems of planetary interest are readily accounted for. A quantitative description of cooling of a blackbody via ballistic radiation is provided by the Stephan-Boltzmann law. As shown in Sections 2.2 and 2.3, such cooling proceeds as T^3 and is very fast. Changes in temperature for particulates of matter, large or small, therefore depend on other processes, such as internally generated heat. Extremely rapid cooling in response to applied heat thus describes the earliest stages of planetary accretion.

Due to the nearly instantaneous nature of ballistic energy transfer, the energetics of the initial and final states can be approximated via the well-worn approaches of thermostatics

(Section 4.1.3). Furthermore, this approach can be applied to understanding heat evolution during atomic collisions and surficial impacts on planets.

4.1.2.2 Radiative transfer in dense gas and condensed matter

Gases significantly denser than Earth's atmosphere, as would have occurred during the later stages of accretion or occur now in the atmospheres of the giant planets, are optically thick and are governed by radiative diffusion. Laboratory measurements at high temperature and pressure apply.

Ample data exist on dense monatomic and diatomic gas in the laboratory, where the power laws for measured K cover a narrow range of $T^{1.69}$ to $T^{1.84}$ over all temperatures accessed, sometimes from cryogenic temperatures to ~ 2000 K (Hofmeister, 2019a). For some gases, vibrational overtones exist into the ultraviolet region, such that the absorption coefficient decreases with frequency while the bands broaden with temperature (Hofmeister, 2019c). Hence, on average, gas absorption increases with T. The combination provides trends in K for gas that are weaker than the well-known T^3 law (e.g., Kellett, 1952; Clark, 1957) derived for radiative diffusion (Eq. 3.8), by assuming A is neither a function of ν nor of T.

At low to moderate temperatures, the most accurate description of heat diffusion is thermal diffusivity (D) as best determined from accurate laser flash measurements. Multiplication of D by the density (ρ) and heat capacity (C_P), which are generally well known, provides thermal conductivity:

$$K = \rho C_P D. \tag{4.3}$$

But by ~ 2500 K, laboratory data are no longer reliable. Much of the problem is excessive ballistic transport across the small samples, which does not occur inside the Earth due to the much lower thermal gradient. For these reasons, the spectroscopic integral of Eq. (3.4) must be used to portray the efficient heat conduction of mantle insulators by transitions near visible frequencies, as well as the interesting mechanism in core metals which involves valence (not conduction) electrons interacting with light (Section 3.1).

In short, thermal evolution of dense media due to heat diffusion (conduction) is described by Fourier's equations (Chapter 2), using the material properties of heat capacity, density, and thermal diffusivity and/or thermal conductivity, as described in this section and Chapter 3.

4.1.3 Thermostatics and the thermodynamics of self-gravitating systems

Each thermodynamic state function explicitly depends on two of the four fundamental state variables, whose full set embodies entropy (S), pressure (P), volume (V), and temperature. For example, the Helmholtz free energy is the characteristic function of T and V, whereas the Gibbs free energy is the characteristic function of P and T; similarly, internal energy depends on S and V while enthalpy depends on S and P. For each choice of state function, the remaining two state variables depend on the two independent variables. Gravitation being important to planets and accretion has ramifications.

4.1.3.1 *Why volume and temperature are the key thermodynamic variables*

For large-scale astronomical problems, which are fundamentally governed by gravitation and heat transfer, the Helmholtz free energy is the function of choice (e.g., Müller, 2003). Volume and temperature are the independent variables for several reasons. First, Newton's law of gravitation, and gravitational potential energy, depend on radius, because radius is clearly related to the distance of particle separation, and to the size and potential energy of an accreted object. The relationship between radius and volume is direct and obvious. Second, temperature and distance are the essential variables of heat transfer, as embodied in Fourier's laws (Chapter 2).

The underlying, fundamental reason that temperature is singularly important to thermodynamics and evolutionary behavior is that this one variable describes blackbody emissions. All objects emit blackbody radiation, albeit modified by spectral characteristics of real material (Bates, 1978). As a consequence, the laws of thermodynamics can be recast in terms of the behavior of light (Hofmeister, 2019a), which explains why reversibility is impossible and why absolute zero is unachievable.

The use of any other state functions than Helmholtz free energy in a thermostatic evaluation of problems involving gravitation can only lead to unnecessary trouble and ambiguity. Hofmeister and Criss (2012a, 2019) provide further discussion.

4.1.3.2 *Evolutionary equations*

As discussed above, thermostatics can be used when heat transport is ballistic transport. Otherwise, the role of time needs to be considered in the evolution of the system. To address this problem, we presume that Newton's classical laws are valid and that the reservoirs of mass and energy are separately conserved. Interiors of stars, relativity, and cosmology are not covered in this book. Our Solar System is a late comer, having been produced from recycled material in a rather isolated pocket in the galaxy, as indicated by the large distance to the nearest star (300,000 a.u.).

Because the mass (m) and heat content (q_V) are separately conserved, these quantities describe the system through:

$$\frac{\text{System inventory}}{\text{Independent variable}} = \text{Key material property.} \tag{4.4}$$

The key parameter of a large body, i.e. its mass, is determined at constant temperature implicitly (Table 4.1). One explicit reason is that conversion of condensed matter to gas, when no container walls are present, permits mass loss. If diffusional processes are

TABLE 4.1   Governance of planetary and astronomical scale thermodynamic systems.

Description of the system	Dependent variables	Evolution	Transport properties
$\frac{dm}{dV} = \rho$	$P = f(V,T,m)$	Newton's laws	G, u
$\frac{dq_V}{dT} = C_V$	$S = F(V,T,m)$	Fourier's laws	K, D

Notes: In classical thermodynamics, mass and energy are separately conserved. Historically, these two independent inventories were used to ascertain which variables were extensive and which were intensive, because it was not recognized that four state variables existed. F and f indicate functions specific to the problem of interest.

important, then Fick's laws apply, as is the case for laboratory studies. However, for the cases of interest in this book, gravitation governs the retention or loss of mass as gauged by the escape velocity. In either depiction, the important material property is density.

Density is often viewed as being a function of pressure, which results from familiarity with laboratory experiments, where pressure is controlled mechanically, as in a piston-cylinder apparatus. For a self-gravitating body, pressure is instead dictated by the distribution of mass and temperature in a volume of interest. Pressure is therefore a *dependent* variable in such situations, rather than a controlled, *independent* variable. Here the bulk modulus or compressibility of the material is relevant: these are not explicitly included in Table 4.1, because these properties are secondary, being derivatives of V or ρ. The same holds for thermal expansivity.

Table 4.1 explicitly states that heat is determined at constant volume because much confusion has been caused by describing heat transfer in terms of reversibility, which is impossible. As pointed out by Hofmeister and Criss (2019), reversible heat transfer is the fundamental oxymoron of thermostatics! Consequently, C_V is of utmost importance in gravitational systems (Table 4.1). This finding should not be surprising, as heat capacity is intimately tied to heat transport through sum rules (Criss and Hofmeister, 2017). It links the rate of heat transport (D) with the amount of heat being transported (K) via Eq. (4.3), and so is central to radiative diffusion (Eq. 3.4).

Separate conservation of mass and energy permits the evolution of a system to be evaluated using Newton's and Fourier's laws. Some additional properties and variables are relevant, such as the speed of light (c) and the speed of moving matter (u), where the latter depends on solving Newton's equations. Although speed is a derived parameter, it is important when advection (Section 3.3) supplements diffusion of heat.

4.1.4 Gravitation provides motions, which may then create heat

Every conservative force is the gradient of a specific potential function. In conservative systems, potential energy and kinetic energy can be interchanged, but the positions and motions allowed in such exchanges are restricted if the system remains conservative. As an example, the orbital velocity of a planet or particle clearly depends on its position, but is not a function of its temperature, electric charge, magnetization, etc. Hofmeister and Criss (2016) showed that every type of conservative motion has its own Virial Theorem. This restriction stems from the stringent requirements for transformation of energy between reservoirs. The need for a distinct Virial Theorem for each separate energy inventory is fundamental, but is greatly misunderstood: its neglect has created great scientific problems, many of which are not even unrecognized.

When heating is possible, constraints from thermodynamics must be satisfied in addition to conservation of energy and momentum (e.g., Sherwood and Bernard, 1984). Heat is produced when external work is performed on a system by the surroundings: for example, when gas in a piston-cylinder device is compressed. No external work was applied during the gravitational contraction that forms stars and planets. Beliefs that such self-gravitating systems produce heat, which is furthermore internally retained, stem from Kelvin's disproven hypothesis for the origin of starlight. In Kelvin's historical analysis, stars radiate heat to space as they contract. Nuclear processes are now known to produce starlight.

Contraction cannot directly provide internal heating, but rather creates a state of high kinetic energy (KE) on the macroscopic scale of the system. Graviational PE must initially be converted into the KE associated with axial spin, but that KE can subsequently dissipate, generating heat via internal friction stemming from differential rotation of the constituent layers. So in the end, most planetary heat has a gravitational origin, but it is released over a long time interval via an additional, non-conservative force. Spin-down of planets over geologic time was covered in Sections 1.5 and 1.6, which showed that the current surface emissions arise from near surface, ongoing processes. Exponential loss of spin was suggested, and will be used in Chapter 10. An obvious corollary is that very slow accretion could involve heating of the planets, and so this chapter discusses ideal and more realistic cases, with reference to Figs. 1.11 and 1.12 and those presented below.

4.1.5 Accretion as a phase transition

An appropriate thermodynamic analogy to formation of the Solar System is the phase transition of condensation. Clearly, the nebula is a gas whereas the planets and Sun are solids with some layers that are liquid (e.g., oceans) and/or gas (atmospheres). So a phase transition indeed occurred during formation of these large bodies.

For more ice to form in an ice bath, the water must eject all latent heat released to the surroundings. The temperature of the bath thus remains constant during progressive freezing, but the proportions of the phases change, as is observed. Moreover, because ice is structurally ordered but water is disordered, formation of more ice decreases the entropy of the bath. The entropy of the surroundings (outside of the bath) is increased by casting off the evolved latent heat, so the overall entropy of the universe increases. Note that *none* of the released energy is internally retained in the progressively freezing, ice-water system.

The Solar System is obviously more ordered than a nebula (Fig. 4.2). It follows that the entropy of the Solar System decreased during its formation, while S of the Universe increased by a greater amount! Entropy of real nebular gas likewise decreases during contraction, because contraction reduces entropy of an ideal gas according to:

$$\Delta S = N R_{gc} \ln\left(\frac{V_f}{V_i}\right) \tag{4.5}$$

where N is the number of moles and R_{gc} is the gas constant.

If real nebular gas were to heat up during self-contraction, then this gas would expand, opposing the contraction. Thus, the gas must shed heat for contraction to proceed. To model these incremental changes in view of Eq. (4.5), Hofmeister and Criss (2012a) invoked conservation of energy. Because the governing thermodynamic variables in gravitational problems are T and V, Helmholtz' free energy (F) is relevant:

$$F = E - TS = U_g + RE - TS \tag{4.6}$$

where E is internal energy (e.g., Müller, 2003). The resulting changes and the second law require that the surroundings of the nebula become slightly warmer.

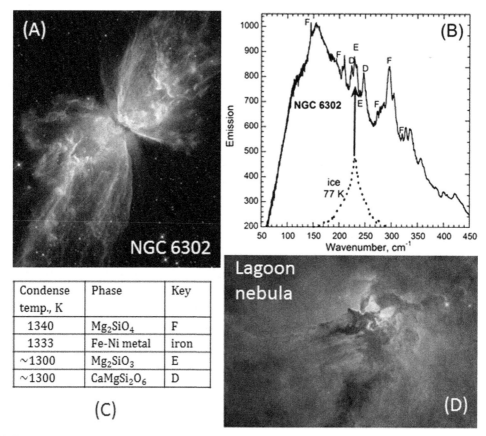

FIGURE 4.2 Astronomical environments. (A) Old NCG 6302, about 1 light year across. At the center of the nebula is a dying but hot star that was once about five times more massive than the Sun. The envelope of ejected gas and dust is moving at $\sim 10^6$ km/h, and is bathed in ultraviolet radiation, warming the cast-off material and making it glow. Temperature of the dust is about 50 K. (B) Emission spectrum of part (A), originally obtained by Molster et al. (2001), with definitive assignments compared to ice spectrum. (C) Key to spectral assignments of dust, with condensation temperatures from Lodders and Fegley (1998). (D) The young Lagoon nebula, which is a star forming region about 4 light years across. *(A) and (D) Publically available composite image from NASA, using the Hubble telescope. (B) Modified after Figure 7 in Hofmeister, A.M., Wopenka, B., Locock, A., 2004. Spectroscopy and structure of hibonite, grossite, and CaAl$_2$O$_4$: implications for astronomical environments. Geochim. Cosmochim. Acta 68, 4485–4503 with permissions.*

Warming in the gas stage can only proceed via radiative transfer. Specifically, ideal gas behavior exists because well-separated molecules interact negligibly, and so frictional heating is not possible until densities have increased far above the very low densities of molecular clouds ($\sim 10^{-18}$ g/cm^3; Bergin and Tafalla, 2007). Before the required high density ($\rho \gg 10^{-3}$ g/cm^3, which describes Earth's atmosphere) is attained, heat produced during inelastic collisions is shed in accord with blackbody cooling (Section 4.1.3).

Blackbody radiation is always present. Hence an isolated nebula without an internal heat source will slowly contract. The thermal energy balances the gravitational. Equations based on V and T being the independent variables are provided by Hofmeister and Criss (2012a). However, this slow collapse differs from the catastrophic collapse discussed below.

Thus, like the ice bath, thermostatic formation of condensed bodies from a nebula cannot involve production of heat that is retained. However, frictional heat can be produced once the forming bodies became sufficiently dense, and so some limited amount of "primordial" heat could have been generated via collisions and compaction of particulates during the latest stages of accretion (see Chapter 11).

4.2 Observations supporting cold accretion

4.2.1 Raw materials and stellar recycling

Meteorites possess rare, small, but telling grains of isotopically distinctive, pre-solar material (e.g., Bernatowicz and Zinner, 1997). The presence of these grains proves that the nebula which formed the Solar System included at least some dust predating our Sun. It is currently popular to assume that most material in the Earth and other planets was ejected from the already formed Sun, as suggested by Laplace. However, radial ejection of such proto-planetary matter cannot produce our Solar System, in which the planets have more angular momentum than the Sun. A more reasonable hypothesis is that the planets and Sun formed simultaneously as a spinning nebula contracted (Hofmeister and Criss, 2012a, b). Unlike prevailing models, this latter process can preserve angular momentum.

It is well known from astronomy and astrophysics that stars have generations, and that the younger generations condense from nebulae that contain high atomic number elements formed by nucleosynthesis in precursor stars and novae. That the Sun and planets formed from a common nebula is supported by similar elemental abundances in the Sun, terrestrial rocks, and chondritic meteorites, and by the location of the Solar System in a gas-poor region of the Milky Way. Accretion sucked in the gas and debris. Stars and their planetary systems can form individually from small clouds known as Bok globules, but more commonly form within gigantic molecular clouds that are the locus of a succession of stars that continuously form and evolve on a grand scale (Yun and Clemens, 1990).

Phases and the chemical composition of the Solar nebula should closely resemble those observed in star forming regions. Silicate dust has been detected (van Breemen et al., 2011). Simple ices (CO, CO_2, H_2O) and hydrocarbons have also been spectroscopically identified in the environments of young stellar objects: see Bowey and Hofmeister (2005) for a list of studies. These ices are considered to nucleate on mineral grains. Iron is also expected to exist, but its bland spectrum (Hofmeister, 2019b) does not permit remote identification. Assignments of spectral peaks to polyaromatic hydrocarbons (PAHs) are ubiquitous in astronomical literature, but this assignment is probably in error as the intensity vs frequency pattern matches spectra of solid hydrogen (Schaefer, 2007). Hydrogen is the main component of all astronomical environments, and will freeze at the very cold temperatures of the interstellar medium (~ 4 K: Bergin and Tafalla, 2007).

Aging stars are categorized either as carbon stars or oxygen stars on the basis of solid phases condensed during mass loss. This division exists due to the stability of the CO molecule. If any given star has more O than C, then oxide and silicate minerals are formed. Conversely, SiC forms around carbon stars, as shown from spectroscopic identification (e.g., Thompson et al., 2006). Thus, identification of SiC along with oxides in presolar grains points to the Bok globule preceding our Solar System having multiple sources of material.

Multiple sources of dust should come as no surprise. Some stars terminate in explosions (novae). The ejected cloud NGC 6302 (Fig. 4.2) has spectral patterns clearly indicative of endmember forsterite and enstatite, i.e., of pure Mg_2SiO_4 and $MgSiO_3$ (Molster et al., 2001). Assignments of other peaks are less certain. Possibilities are diopside (Kemper et al., 2002) and high temperature refractories similar to phases in meteorites (Hofmeister et al., 2004).

4.2.2 Cold temperatures in nebulae

Ballistic radiative transfer is an effective means of dissipating heat because it occurs at light speed. As shown in Chapter 2, $\sim$mm sized balls of molten forsterite formed at ~ 1800 K cool to room temperature in only ~ 90 s. Smaller grains of condensed dust suspended in a cloud would cool even faster (see Fig. 4.2 for the main phases and their cooler temperatures of formation).

Cold temperatures of nebulae are confirmed by emission spectra. The broad spectral peak near 160 cm^{-1} in Fig. 4.2 indicates a temperature of ~ 47 K for much of this dust. Regarding the nebula as a whole, radiative transfer would actually be diffusive, from grain to grain. Between the grains, the speed is near that of light, and so steady-state is rapidly achieved. Hence the grains in NGC-6302 are in radiative equilibrium with each other. The dust is warmed by UV light from the bright, hot central star but each grain radiates this heat to its neighbors. Such a cold temperature for dust around a hot star shows that radiative transfer in a dust cloud is quite efficient, as calculated in Section 2.2.

Even colder temperatures (8–20 K) are typical of molecular clouds (Bergin and Tafalla, 2007). Yet, these star-forming regions contain young stars and planets, whose emissions should heat their host nebulae. The low temperatures associated with extended regions of star formation further support that diffusive radiative transfer is quite efficient in nebulae. Icy moon properties imply cold accretion (Chapter 11).

From the above subsections, the following sequence during formation is inferred:

1. Stars synthesize higher atomic number elements in their deep interiors.
2. Aging stars eject ions (e.g., H^+, O^{2-}) and atoms (e.g., Fe^0), sometimes during novae.
3. The hot ejecta cool and condense into molecules (e.g., H_2, CO) and mineral dust, which further cools and drifts elsewhere.
4. Eventually sufficient material (gas, ice, and dust) gathers that new star(s) form.

Lastly, recent work associates exoplanets with dusty regions, which is consistent with the above.

4.2.3 Evidence for orderly accretion in Solar System trends

Many commonalities exist in the properties of the planets (see the bullet list in the introduction). Several trends in energy and mass across the Solar System follow simple power laws, as demonstrated by Hofmeister and Criss (2012a). The most important of these trends are comparisons of current axial spin (Fig. 1.12) or of orbital rotational energy (Fig. 4.3) to the gravitational self-potential energy. Both types of RE depend nearly linearly on PE, which was computed for a sphere of constant density with mass (M) and body radius (R_b):

$$U_{g,self} = -\frac{3}{5}\frac{GM^2}{R_b},\tag{4.7}$$

where $G = 6.67 \times 10^{-11}$ m^3/(kg s^2) is the gravitational constant. Eq. (4.7) is known as the self-gravitational potential (e.g., Eddington, 1926). The specific volume contracting to form any object is unimportant due to the inverse dependence of $U_{g,self}$ on radius. Rotational energy of the orbits is given by:

$$RE_{orbit} = \frac{1}{2}\frac{GMM_{Sun}}{r_{orbit}}\tag{4.8}$$

Spin energy is provided by $\frac{1}{2}I\omega^2$ where I is the moment of inertia and ω is the spin rate. Fig. 4.3 used measured parameters tabulated by NASA: see Hofmeister and Criss (2012a) for details.

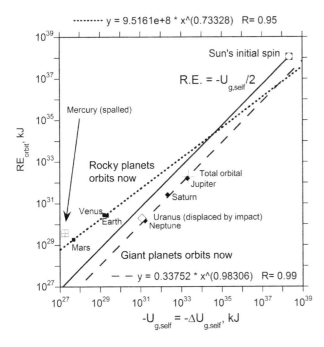

FIGURE 4.3 Dependence of orbital rotational energies on gravitational self-potential energy in the solar system. Gravitational potential estimated for homogenous density. Squares and dotted line = current RE$_{orbit}$ of inner planets. Mercury (open square) was not included because it was spalled to the extent its mantle was lost, plus it probably moved inwards during multiple impacts, as evidenced by its tilted and eccentric orbit. Diamonds and dashed line = gas giants. Uranus (open diamond) was not used in the fit because its tilted spin indicates a rather strong impact. + = total RE$_{orbit}$ of all planets, which is dominated by Jupiter. Circle in square and solid line = Virial Theorem for spin. Least squares fits are shown. Modified after Figure 2a in Hofmeister and Criss (2012a) with permissions.

Current spin and orbital energies of the planets are immense! In fact, RE_{orbit} of the inner planets exceeds their gravitational self-potential energy, which can be explained by contraction of the Solar nebula during its vertical collapse, which conserves angular momentum (Section 4.3). Regardless of model details, the data show that copious amounts of heat did not accompany formation of the planets and Sun. Rather, gravitational PE became two types of rotational kinetic energy, and afterwards heat was produced by friction during the vigorous spin (Sections 1.5 and 4.1.4). The orbital energy is directly tied to formation conditions because no dissipative mechanism exists, except impacts on formed planets (see below).

Production of both spin and orbital RE complicates models of accretion. Two stages of accretion are suggested by the presence of extensive satellite systems around the gas giants. This deduction is supported by the masses of rocky and gassy bodies each defining a well-defined trend (Fig. 4.4). These two trends suggest an initial, vertical collapse of the dusty components of the nebula, which are less stable than gas due to larger particle mass and slower velocities, followed by the contraction of gas around local nuclei, but predominantly around the center, forming the Sun.

Local contraction provided the outer planets with huge atmospheres and large satellite systems, as explained by Hofmeister and Criss (2012a) in terms of gravitational

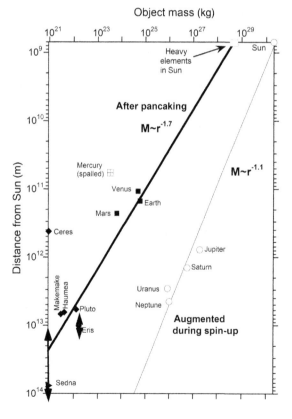

FIGURE 4.4 Dependence of object mass and compositional diffusion on distance from the center. Circles and light diagonal line = Sun and gas giants, with least squares fit. Squares and heavy line = least squares fits to rocky planets. The fit projects to the outer dwarf planets (diamonds). Mercury (cross in square) was not fit because its mass was greatly reduced by spalling. Sedna (large triangle) is likely a dwarf planet, but Ceres (also a diamond) is too small to be primary in this region. Double arrow = the range of orbital radii for Eris and Sedna. *Modified after Figure 11 in Hofmeister, A.M., Criss, R.E., 2012a. A thermodynamic model for formation of the solar system via 3-dimesional collapse of the Dusty Nebula. Planet. Space Sci. 62, 111–131 with permissions.*

competition with the simultaneously growing Sun. For a particle within a dust cloud to be incorporated in a growing body orbiting the growing Sun:

$$\left(\frac{r_{cloud}}{r_{orbit}}\right)^2 < \frac{M_{cloud}}{M_{Sun}} \tag{4.9}$$

The cloud mass is that of the fully formed planet, since moon mass is comparatively small. Fig. 4.5 shows that Eq. (4.9) describes the gas giant systems. Mars apparently captured its tiny moons. That the Earth-Moon system agrees with Eq. (4.9) suggests this region of the Solar system had dust only, whereas gas occurred in the more distant surrounding shell that now includes the gas giants. Likewise Pluto and the other dwarf planets seem to have formed in a region where much less gas was available. Hofmeister and Criss (2012a) discuss zonation in terms of kinetic energy of the material in the nebula, gravitational stability, and timing.

From Eq. (4.9), the growing Sun limited Mercury's draw to within 5×10^7 m. This small distance supports that the mass of Mercury was limited by gravitational competition

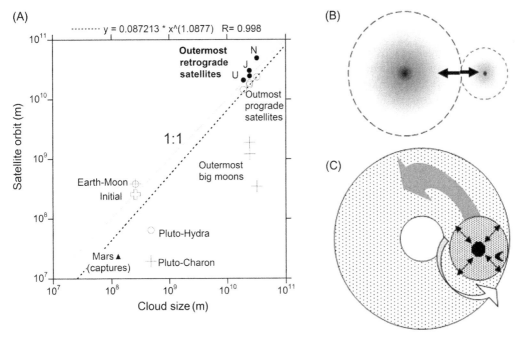

FIGURE 4.5 Data and models on satellite systems. (A) Comparison of satellite orbits to the cloud size set by balancing gravitational force from the Sun against that of each planet (Eq. 4.9). Circles and dotted line with least squares fit = outermost prograde satellite for the planets as labeled: note the dot representing Saturn is beneath Jupiter. Dots = retrograde satellites. + = large moons. Triangle = Mars, which is capturing its tiny moons. Open plus = initial position of the moon from back projecting lunar drift. (B) Schematic of gravitational competition. (C) Schematic of the second stage of collapse and contraction, which requires expansion of the tiny clouds as the planet spins up, and results in formation of the satellite systems. The width of the planet forming ring is exaggerated. *Parts (A), (B), and (C) are modified after Figures 10, 6 g and 6 h, respectively, in Hofmeister, A.M., Criss, R.E., 2012a. A thermodynamic model for formation of the solar system via 3-dimesional collapse of the Dusty Nebula. Planet. Space Sci. 62, 111–131 with permissions.*

(Fig. 4.5B). Subsequent loss of its mantle is indicated by spalling, as suggested by Mercury's overly large core. For both reasons, Mercury departs from the trends of the other rocky bodies in Figs. 4.3 and 4.4.

Gravitational competition describing the satellite systems provides strong evidence that the Sun, its planets, and their moons essentially co-accreted. The vertical collapse that provides dusty clumps and precedes their growth and spin-up into planets has recently been quantified in a conservative model discussing the formation of spiral galaxies (Criss and Hofmeister, 2018; Sect. 4.3).

4.3 Beginnings: accretion of the Sun and planets from a nebula

The presence of peculiar stable isotope anomalies, and the former presence of short-lived isotopes such as ^{26}Al, has been conclusively proven in several primitive meteorites, signifying that a dramatic nuclear event attended the earliest stages of the Solar System. Several possibilities come to mind. Many stars in the solar vicinity have variable proper motions, so close passes between two or more stars are possible over vast intervals of time. Mass transfer between proximal stars is known, including jets streaming from the spin axis (see website list). Explosions (novae) are also well-documented. Addition of mass destabilizes a cloud in which the gravitational pull to the center is balanced by centrifugal (inertial) forces.

Although the process that triggered formation of the Solar System may never be known, a destabilized large cloud of gas will contract and/or collapse because the attractive force of gravity draws mass towards the center, until the resistance inherent to condensed matter manifests itself. The dynamics governing the processes after triggering can be inferred from physical principles, as follows:

4.3.1 Contraction and collapse of a nebula to form a central star and planets

4.3.1.1 Energy minimization and angular momentum conservation

The physical conditions during the birth of the Solar System will always be uncertain, but a fundamental thermodynamic precept for any real, formative process must be that each finite step resulted in a decrease in the inventory of energy. An equally important concept is that angular momentum should be conserved, since both the precursor nebula and the Solar System are bound states. Criss and Hofmeister (2018) used these concepts to suggest a sequence of evolutionary stages in the formation of spiral galaxies. Although these galaxies are many billions of times more immense and massive than the Solar System, and have some notable dynamic differences, their highly flattened character and congruent sense of rotation have obvious affinities to the Solar System. They also both appear to have formed in a process that involved condensation and dynamic evolution of an enormous, roughly equant, slowly spinning "molecular cloud, or in the case of the Solar System, a precursor "Solar Nebula." The first order question is, why have these dispersed, roughly equant precursor "clouds" become condensed and highly flattened?

4.3.1.2 Why is the Solar System "flat"? - the pancaking process

Hofmeister and Criss (2012a) demonstrated that the rotational energy associated with the spin and orbits of the planets define regular trends that closely match the gravitational self-potential of the various bodies. Importantly, these coherent trend lines project to a plausible, yet very fast and energetic, initial spin rate of the protoSun (Figs. 1.12 and 4.3). This paper proposed an intuitive process, called "pancaking," to describe an early stage where the Solar Nebula rapidly collapsed into a highly flattened shape, allowing it to preserve much of its angular momentum.

Subsequently, Criss and Hofmeister (2018) quantified the "pancaking" process using energy minimization. A roughly spherical "cloud" of low density can reduce its gravitational self-potential, while preserving all of its angular momentum, by progressively collapsing into a denser, highly flattened spheroid cloud, if it simultaneously undergoes a modest contraction, down to ~²/₃ of the original equatorial radius (Fig. 4.6). For example, an equant Solar Nebula with a radius of ~ 100 AU (astronomical units) could collapse into a highly flattened spheroid, whose shape resembles spiral galaxies, with a radius of ~66 AU.

4.3.1.3 Inward densification, round-up and planet formation

Following the production of flattened oblate cloud by the "pancaking" process, Criss and Hofmeister (2018) quantitatively demonstrated that the homogeneous nebula can continue to reduce its inventory of total energy by densifying its interior zones, while maintaining its shape, volume and mass. However, some fraction of the angular momentum of the contracting zone must be lost, which depends on the form of the densification. If we assume that the original nebula has a homogenous density ρ_0 and an initial angular momentum of L_0, and describe the densification during contraction as a power law, then:

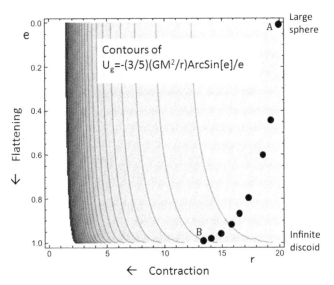

FIGURE 4.6 Contours of self-potential $U_{g,self}$ on a graph of ellipticity vs radius (non-dimensionalized). A large, equant, low density nebula (point A) can conserve its mass and angular momentum if it flattens while contracting its equatorial radius by a factor of ~2/3, which leads to a ~20× increase in density (point B). Dots show the path taken during flattening, which minimizes the energy.

$$\frac{\rho}{\rho_0} = \frac{3-n}{3}\left(\frac{A}{r}\right)^n \tag{4.10}$$

To probe densification independent of round-up, we presume the shape is unchanged, while conserving volume, and mass. For a differentially rotating body, Criss and Hofmeister (2018) derived:

$$\frac{L}{L_0} = \frac{10}{3}\frac{3-n}{(10-3n)} \tag{4.11}$$

For example, for the $n = 2$ power law, which seems to describe spiral galaxies and thus their precursor gas cloud, $1/6^{th}$ of the original inventory of angular momentum must be lost during the densification of the object's central zones. Yet, as is well known, angular momentum must somehow be conserved! Resolution is as follows:

In spiral galaxies, angular momentum balance is realized by outward movement of the spiral arms, which gain angular momentum, while the central portions become denser and lose angular momentum. In the Solar System, the analogous process is formation of the planets, where these bodies move away from the densifying Sun, thereby "carrying off" this momentum. At this time the Sun is also rounding up (see Criss and Hofmeister, 2018 for the similar process of forming a bulge in galaxies).

This hypothesis is supported by Solar system data. The orbital angular momentum of the planets is close to $\sim 3 \times 10^{43}$ kg-m^2/s, of which 60% represents Jupiter and almost 40% is provided by the other giants (Fig. 4.3). Six times this amount is expected for $n = 2$ in Eq. (4.11) which gives 2×10^{44} kg-m^2/s. A protoSun with the latter angular momentum, an inverse-square density distribution, and the size of the modern Sun would be rotating every few hours. This rapid rotation rate corresponds very closely to that predicted from the projection of the planetary trend lines on Fig. 1.12, and data on young yellow stars, as discussed earlier. Importantly, these two methods of estimating the Sun's initial rotation rate are independent.

Fig. 4.3 provides another important clue: the inner planets have more orbital RE than RE $= -U_{g,self}/2$ (Eq. 4.1) which describes contraction of a cloud from infinity to form a central object, whereas the outer planets have less orbital RE (Note that RE increases when L decreases, and vice versa, due to the RE going as ω^2 and L as ω.) Although Eq. (4.1) was derived considering spin, the same should hold for orbits, due to the reference state of $U_{g,self} = 0$ being at infinity, so off-center positions for the planets in the Solar system makes little difference, and because the relationship embodies conservation of energy and angular momentum, but does not specify the type of angular momentum. The inner planets being drawn towards the center with the Sun negligibly changes L, due to the large mass of the giants (see above). During this stage, the planets need not be fully formed: motion of shells or rings of gas have the same effect. Consolidation of material in a ring conserves both orbital angular momentum and energy. Spin of planets is another issue, which is covered in Section 4.4.

Notably, this division in radial motions apparently created the region now known as the asteroid belt, as indicated by the following observations. First, insufficient matter existed near 2–4 AU for a planet to form. Second, distinct trends exist for the masses of the inner and of the outer planets on orbital radius (Fig. 4.4). Two trends show that matter was redistributed after pancaking. Third, the gas giant trend points to the origin, the Sun, whereas the rocky planet trend points to the "dust" in the Sun, represented by its inventory of elements heavier than H and He. This behavior is consistent with the outer Solar system forming directly from the nebula which moved away from the origin, whereas additional processing occurred in the inner Solar system in order to form the Sun. Specifically, the Sun out-competed the inner planets for gas (Eq. 4.9) as it rounded up and densified. Although, it is not possible that the inner planets formed without H_2 and He, based on abundances, clearly the amounts were not large.

A crucial question remains: *Why do the rocky planets exist?* For the inner planets to not have been swallowed up by the Sun during its final round up, these must have been in stable orbits. Consequently, formation of planets predates or is simultaneous with the contraction which yielded the Sun. The above momentum balance actually applies to rings or shells of nebulae, and does not address why the rocky planets formed.

4.3.2 Planets began with formation of proto-cores via several processes

Planets and stars form because (point) masses attract each other in accord with Newton's law:

$$F_g = -G\frac{Mm}{s^2}$$ (4.12)

Other attractive forces exist: Coulomb's law describing charged particles has an analogous mathematical form. Over short distances Coulombic attraction is powerful; this force must have contributed to the production of solids from ions in material ejected from aging stars.

Another strong, attractive, short-range force is magnetism. Although iron has not been detected in molecular clouds due to its lack of a distinctive IR spectrum, metallic Fe particulates are an important product in models of nucleosynthesis. Regarding our Solar System, meteorite constitution and the large size of planetary cores demonstrate that iron was abundant in the precursor nebula.

The behavior of Fe in the collapsing and contracting nebula depends on its physical properties, which are:

$$\text{high density} + \text{high atomic mass} + \text{ferromagnetic} + \text{ductile} + \text{durable}$$ (4.13)

The element Ni has similar properties and readily alloys with Fe. For simplicity, the term "iron" is used to represent both metals and their blends. Based on these properties, we deduce that iron cores form first, as follows:

Ferromagnetism is the strongest type of magnetism and results from exchange coupling between iron atoms. Iron particles are magnetic dipoles, even if very small. Opposite poles attract each other. The Curie temperature is 1043 K at ambient pressure, so even a warm cloud and energetic collisions would have been affected by the above properties of iron alloy. Finally, magnetic forces are extremely strong at short range.

We postulate that iron dust in a nebula will accrete first due to this extra attractive force between even the most minute particles of metal. This extra short-range force promotes adhering during collisions (Fig. 4.7a), which is the crucial first step. The tiny dipoles will align during the interaction, forming the domains that are typically of macroscopic iron,

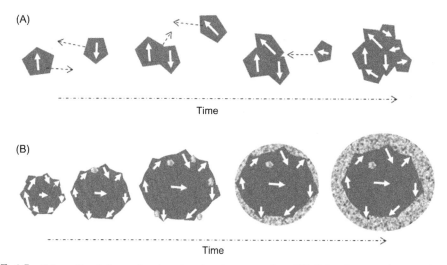

FIGURE 4.7   Schematic of the early steps in planetary accretion. (A) Role of magnetic attraction in grain agglomeration and production of planetary nuclei. Collisions of small iron particulates (gray hexagons with the dipoles as white arrows) result in agglomeration, due to magnetic attraction coupled with gravitational attraction and slow but random velocities (dashed arrows). As the grains grow, left to right, their larger mass attracts more of the tiny grains, which stick to surface grains of compatible orientations, due to welding, magnetism, and ductility. Mineral grains break up rather than sticking like iron grains, so planetary nuclei are mostly iron. (B) Gravitational growth of a protocore into a planet. Time is from left to right. The conglomerates once formed provide a central attracting mass, which further grows, depleting the local region of iron. For this reason, and because of their large and growing size, the protocores also attract mineral dust, and hold on to it. As the iron content of the nebula is progressively depleted, accreting matter becomes increasingly dominated by silicates and other non-metallic solids.

which is polycrystalline. The growing protoplanet core has no net magnetism, but its surface has domains which would attract opposite dipoles.

Collisions between iron particles are less destructive than collisions between oxide and/or silicate particles, because iron is ductile and durable. These properties, the force of magnetism, and the ferromagnetic properties of iron thus promote core growth, which in turn enhances its gravitational attraction. Moreover, once an iron particle impacts the growing core, it is stuck for three reasons. (1) Considerable force is needed to pry apart magnets, even the weak ones used for convenience on refrigerators. (2) Friction welding (like the hammering of a blacksmith) will accompany the proposed gravitational collapse which is augmented by magnetic attraction, forming strong material with substantial mass. High temperature is not necessary for friction stir welding, which process is most like the gravitational indentation described here (Reinhardt Criss, personal communication, 2018; see website list). (3) Cold welding of metal is expected (Everett Criss, personal communication, 2019). This process has caused mechanical problems with satellites in space (Merstallinger et al., 2009). Magnetism and welding are crucial initially, but gravity is the governing force overall.

It is also important that single atoms of iron are heavy compared to other abundant atoms and therefore have lower thermal velocities. Hofmeister and Criss (2012a) describe how the thermal velocities of gas oppose gravitational collapse. Molecules of minerals are more massive than atoms of iron and could accrete. However, collisions of silicate dust particles are damaging and fracturing, so that some average grain size is maintained via their mutual collisions (~1 mm, see Chapter 11). In contrast, collisions of Fe particles, no matter how tiny, will result in grain growth (Fig. 4.7A).

Once a good-sized protocore is formed, mineral dust can then accrete due to the growing force of gravity. Fracturing of the grains does not hamper growth during this stage (Fig. 4.7B). Hence, planets start with a small iron core, which is then overlain by a mixture that is more representative of the nebula composition. Essentially no heat is evolved in this process of small impacts on the surface, because of ballistic emissions. Hence, ices are also incorporated, but magnetism and welding may limit ice proportions in the core.

A minimum body size of ~1200 km (~6×10^{22} kg) for the rocky planets is required to augment and purify the protocore, as is suggested by the trends of current core size with planetary radius and surface gravity (Fig. 4.8). The moons have different compositions than planets (Section 4.4) and need not have *large* iron cores. It is expected that the moons have tiny protocores, which is consistent with the small core deduced for our Moon and other bodies (see e.g., Lodders and Fegley, 1998). Growth of protocores subsequent to accretion of the essentially full-sized body is covered in Section 4.5.

The above proposal addresses how grains form, the initial steps of accretion, and the approximately equal sizes of the cores of three planets closest to the Sun, which are not explained by any previous model. Mars has both a smaller mantle and core, which is

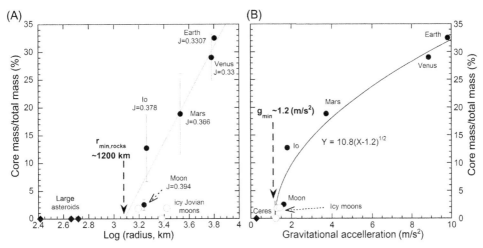

FIGURE 4.8 Dependence of core size, expressed as the ratio of core mass to total planetary mass, on (A) the logarithm of the surface radius and (B) the surface gravitational acceleration. Diamonds = asteroids. Dots (with error bars) = rocky moons. Open circles = icy moons with highly uncertain core sizes. Mercury was not included in the linear regression, although the x-intercept of the fits would change little if it were. Measured axial moments of inertia ($J = I/MR^2$ where R is equatorial radius and M is total mass) are indicated. Bodies with cores have $J < 0.4$, where 0.4 is the value for a spheroid of constant density. Dashed vertical arrows emphasize minimum values needed for core formation. *After Figure 2 in Hofmeister, A.M., Criss, R.E., 2012a. A thermodynamic model for formation of the solar system via 3-dimesional collapse of the Dusty Nebula. Planet. Space Sci., 62, 111–131, which has a Creative Commons Attribution 3.0 License with permissions.*

consistent with forming close to the border between the contracting and expanding parts of the nebula (now the asteroid belt). Importantly, the proposal explains why rocky bodies formed out of a nebula that has very little dust.

Our proposal implies that planetary cores are like typical iron meteorites. The similar textures and compositions of all iron meteorites (e.g. Mittlefehldt et al., 1998) provide support.

4.3.3 Heat generation

During this initial stage of accretion, which involves collapse and then contraction to a point to form the Sun, gravitational potential is converted to RE of spin and orbits, while angular momentum is conserved. The small amount of heat generated during impacts of gas, ice, and dust is radiated to space, in accord with the second law of thermodynamics. Due to the entire nebula spinning up, collisions of large fragments are unlikely. This deduction includes large iron blobs: because magnetic forces are short range, the governing force overall is gravitational.

4.4 Middle to late stages of accretion

4.4.1 Formation of satellite systems via a second collapse and contraction

Satellite systems existing for every giant planet, where the central body has a dense interior and a gassy exterior akin to the Sun, shows that a second stage of accretion occurred. These miniature "solar systems" suggest that the process for this second stage recapitulates events during the first stage. Specifically, within the rings of nebulae surrounding the Sun, pancaking is followed by densification, spinup, and roundup, as well as motion of the outer local cloud outwards (Fig. 4.5C). This process is localized about a protocore, which orbits around the forming Sun.

Another factor pertains, namely gravitational competition of the growing protocore with the large mass accumulating in the central region (Fig. 4.5B). This competition (Eq. 4.9) limited the radius of satellite formation (Fig. 4.5A). Retrograde satellites exceed this radius because these are bodies captured in the last stage (Section 4.4.2). Additionally, evidence exists in all mini-systems for a denuded region that is analogous to the asteroid belt (see Table 4 in Hofmeister et al., 2018, which is based on NASA websites). Jupiter has no satellites between r/A of $26-105$, where r is the orbital radius and A is the planet's equatorial radius. Saturn's gap is from r/A of $25-59$; Uranus' gap is from r/A $23-167$; and Neptune's gap is from r/A of $5-14$.

The inner planets lack satellite systems and gassy mantles due to gravitational competition. As these spin up, the outer clouds move out, only to be captured by the growing Sun. Gravitational competition describes the location of our Moon (Fig. 4.5), if lunar drift is accounted for (Chapter 5).

During this second stage of local collapse and contraction, gravitational potential is again converted to spin, not heat. Formation of rings is part of this stage. Formation of spinning moons requires a third stage. The large, close moons lack gassy mantles because their positioning is analogous to that of the inner Solar System planets, which are rocks. The distant moons are too small to hold on to H_2 gas; plus, the pull of the Sun is a factor.

In any of the three accretionary stages, heat production is unlikely since the PE is consumed in creating RE of orbits and spin.

4.4.2 Planetary growth by surface impacts

4.4.2.1 Impacts on formed bodies: how much heat and where?

After the initial collapse and contraction of the nebula to form a proto-Sun and its proto-planets in a disk-like configuration, the formed or forming Sun acted externally to the planets (Fig. 4.9). Due to the immense mass of the growing Sun, any material that was not in a stable orbit was drawn towards the center. Thus, the orbiting planets intercepted late arriving material drawn from great distances as it moved inward, much like a car windshield impacts flying insects during a road trip. Most planetary impacts were small and surficial so any heat released was quickly radiated to space, so that energy of collision is of no consequence to the thermal evolution of the planet. Like the car, small impactors do little damage, but damage and potential heating increase with impactor size.

Observed heavily cratered surfaces have dominated thinking regarding formation of the Solar System for 50 years (e.g., Armitage, 2011). This late stage is important for adding material, and may have given the upper mantle and Moon the high potassium and oxygen-18 concentrations they share with outer Solar system material (Chapter 1). Mercury, being close to the Sun received an excess of impacts, ejecting mantle and core material, some of which was shepherded into the asteroid belt, providing the HEDs class (Hofmeister and Criss, 2012b). Evidence is provided by Mercury's core being overly large compared to its mantle and its obit being eccentric and tilted from the shared plane of the other planets. In addition to the planet nearest the Sun being battered, Mercury and Venus lack moons, while Earth has so far been able to retain its large Moon due to its greater distance from the Sun. Lunar recession signifies that the Sun is slowly capturing the Moon (Chapter 5).

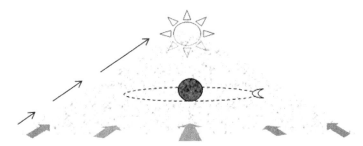

FIGURE 4.9 Sketch of the late stage bombardment in the inner Solar system. Perspective is rough, and bodies are not to scale. The debris would have been drawn towards the Sun in the central plane as indicated, but also in 3-dimensions from the nebula. Thin arrows indicate acceleration of the debris towards the Sun. The lunar orbit (dashed ellipse) presents a much larger cross section to the incoming debris than the area of the Earth.

Impacts both add material and eject material, as is well known, and have provided meteorites from Mars and the Moon, and probably from Mercury. The net gain, net loss, or mass-neutral exchange depends on whether the receiving object is well consolidated material or a conglomerate, like undifferentiated meteorites, plus the speed and size of the impactor. Fig. 4.5 suggests that the Moon accreted from the dust cloud surrounding the Early Earth in the same manner as the satellite systems accreted from the dust and gas clouds surrounding the giant planets. What is unusual is the Moon's large size relative to its companion planet, which suggests either that either material was added during the late stage, or that the Moon has recessed further than suggested by its current outward velocity. Lunar recession is accelerating with time, a consequence of the Sun applying torque to the Moon during its non-planar orbit (Hofmeister et al., in prep.; Chapter 5), and thus the former can be ruled out. Material apparently was added to the Moon, and furthermore was also added to the Earth, as gauged from comparison to Venus (Figs. 4.3 and 4.4). Progressing toward the Sun, apparent changes in the mass inventory during the final stage are:

$$\text{Earth(gain)} \rightarrow \text{Venus(neutral)} \rightarrow \text{Mercury(loss)} \tag{4.14}$$

This sequence is consistent with the focus of impactors on the Sun and their acceleration to high velocities toward the center.

Was heat added along with the late arriving material? This is almost certainly the case, but the Moon's surface indicates shallow placement of both. Equivalent behavior is expected for the Moon and Earth because the impactors are being drawn to the Sun, not to the planets, which only occasionally block and intercept their path. Any primordial heat so supplied to the planets only warmed their outer layers.

In contrast, the loss of spin and the associated heat production early on (Table 1.3) is quite large. The exponential decay suggested in Fig. 1.11 will be used in Chapter 10 to gauge heat production in the early Earth.

4.4.2.2 What giant impact?

The proposal that a giant impact occurred on Earth (e.g. Cameron and Benz, 1991) is inconsistent with Earth's orbit being near-circular, in plane, and concentric with other planetary orbits, and with Earth spinning nearly upright, like most planets, and at similar rates. The trends in Figs. 1.12, 4.3–4.5 and 4.8 could not possibly exist if a giant, destructive impact occurred, as discussed by Hofmeister and Criss (2012a). Crossing of paths and the Solar draw of impactors to the center of the Solar System requires a radial component for their trajectory. Thus, the Earth could not possibly have retained its nearly circular orbit after a large impact, had the Earth even survived.

The Moon is currently receding and thus torque exists. Torque between two spherical, attracting objects is impossible as the line of gravitational force is coincident with their radial separation, so the torque on the Moon must be an attribute of the Sun-Earth-Moon system. Thus, current models involving the transfer of spin from the Earth to the Moon's orbit are impossible. Instead the Sun is capturing the Moon, as it likely captured any former moons of Venus and Mercury. See Chapter 5 for further discussion.

Irrespective of the source of this torque, non-conservation of angular momentum of the Earth-Moon system does not prohibit their co-accretion. Instead, the Moon's current recessional velocity provides an early position that is compatible with co-accretion and gravitational competition (Fig. 4.5). Also, many fairly large impactors struck the Moon during the last stage, as documented by the crater record (Hofmeister and Criss, 2012a). Consideration of cross-sections (Fig. 4.9) shows that the Moon, rather than the Earth, received the brunt of the impacts, and these may have affected lunar angular momentum.

4.5 Differentiation in growing planets

Chapter 3 covers processes occurring inside planets. This section provides additional information on processes that were important early-on, including those before the planet attained full size, which contributed to differentiation. How heat released was distributed is important.

4.5.1 Early outgassing

Water and other ices would have condensed on dust grains in the nebula, as indicated by the compositions of comets and molecular clouds. Ice on accreting dust grains implies that initial planetary temperatures are cold, and has implications for early evolution, see Chapter 10.

As long as the accreting body remains cold, the condensed phases of H_2O and other light molecules are stable in the interior. But if heated to temperatures where the vapor pressure is significant, these truely volatile phases greatly expand. The required elevation in temperature is small and might be provided by early radioactivity, friction in the high spin Virial state, and/or impact heating on the surface of the growing planet. Except for impacts, these processes take time. Another possible cause is maintenance of Earth's surface temperature at ~ 290 K by the Solar flux, once our star ignited. Impact heating and flux on the surface would have released ice but some ice would have been buried as the layers piled on. Although the rate and the amount of buried volatiles are equivocal, as are the profiles and evolution of interior temperature, evidence exists for incorporation of ice and volatiles in Earth's interior. Evidence for deep volatiles includes the mineralogy of comets, continued outgassing of the Earth, and the redox state of the mantle, as follows.

The thermal expansion coefficient of water is $1000 \times$ larger than that of minerals and iron. Steam is far more expansive. Thus, warming of buried ice and conversion to its less dense and highly mobile forms provides great force which will exceed the yield strength of solids. Moreover, the density contrast of vapor and condensed matter is huge, providing buoyancy. Ascent would have been explosive, much like emplacement of kimberlites (Chapter 1). Explosive outgassing would allow gravitational segregation of the mixture laid on top of the protocore (see below). Otherwise, a mixture would be stable, since self-compression increases density with depth.

Segregation would have involved exothermic reactions, such as:

$$2Fe^0 + CO \rightarrow FeC + FeO$$
$$FeO + MgSiO_3 \text{ (enstatite)} \rightarrow MgFeSiO_4 \text{ (intermediate olivine)}$$
$$FeO + 5Mg_2SiO_4 \text{ (forsterite)} \rightarrow MgO + 5(Mg_{1.8}Fe_{0.2})SiO_4 \text{ (mantle olivine)}$$

(4.15)

(Hofmeister and Criss, 2012b; 2013; Chapter 11). Iron metal and water are highly reactive, and could form iron hydrides and hydroxides. These and similar reactions release heat and volatiles, while adding light elements to planetary cores, and adding oxidized iron to the mantle, which can explain the disparity in oxidation state between terrestrial rocks and meteorites. The roughly 10% light elements in the core and Fe/Mg ratio of the upper mantle of 0.1 support our deduction, as do inclusions in diamonds from the upper mantle (Chapter 1). The first reaction of Eq. (4.15) reduces the volume locally, which permits motion and thus aids gravitational segregation.

Regarding heat, for the reaction to progress, heat must be released to the colder surroundings. Because pressure drives the incorporation of volatiles into a solid, the reactions should progress from Earth's center outwards. Local warming would heat unreacted material, expanding the volatile components, which would then buoy heat upwards by advection. The process would add light elements to the core, oxidize iron in the mantle, segregate core material by density, but should provide heat to the mantle more than to core.

4.5.2 Growth of the core by segregation: a cool process

Most models rest on the work of Birch (1965) who followed the erroneous concept that Kelvin used to explain starlight. Prior to ~1830, the production of starlight was a mystery, as the prevailing notion was that the Sun was not changing (Kragh, 2016). Although Kelvin provided many great scientific advances, continued fealty to incorrect parts of his analysis has led to problems in modeling astrophysical processes (Section 4.1), to misunderstanding conditions in planetary cores (this section), and to incorrect analyses of planetary and plate cooling (Part II).

Instead of providing heat, core formation was a great *cooling* event in Earth history because this process segregated radionuclides upwards. This configuration shortens the distance over which thermal emissions must diffuse to reach the cold surface, so it increases the surface flux (Hofmeister and Criss, 2015). The effect on temperature evolution is huge (Criss and Hofmeister, 2016; Chapters 2 and 9).

4.5.2.1 Theoretical limits on heat produced during core formation

The equations and analysis for gravitational contraction of the nebula pertain to the idealized picture of catastrophic (instantaneous) core formation in the early Earth. Hofmeister and Criss (2015) recomputed the change (ΔU_g) as -1840 ± 200 J/g from comparing PREM (Fig. 1.1) and other density distributions to the fictive, homogeneous early Earth envisioned by Birch. However, Hofmeister and Criss (2015) used conservation principles, not the Virial theorem, which provides an additional constraint (Section 4.1.1). Because the total energy for both the homogeneous and the layered state is $\frac{1}{2}U_g$ per Eq. (4.2), the net energy change from the earlier work must also be halved: hence, $\Delta TE = -920 \pm 100$ J/g.

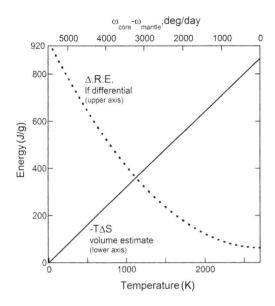

FIGURE 4.10 Energy changes in the full-sized proto-Earth, which has the maximum ΔU_g of 920 Jg^{-1} from Viral theorem constraints. Solid line and lower x-axis = -$T\Delta S$ from the ideal gas analogy (Eq. 4.5), which accounts for volumetric restriction of Fe. Dotted line and Upper x-axis = -ΔRE for the full-size Earth, assuming differential rotation. For temperatures of the present-day core (gray box), entropy-energy more than accounts for ΔU_g. *Modified after Figure 4 in Hofmeister, A.M., Criss, R.E., 2015. Evaluation of the heat, entropy, and rotational changes produced by gravitational segregation during core formation. J. Earth Sci. 26, 124−133 with permissions.*

So, what actually happened to this released inventory of energy, and what mechanisms and time frames were involved?

For an idealized conversion (i.e., like a phase transition) all of the energy released would be radiated to the surroundings, so no internal heating would exist, and the change in RE would be balanced by entropy changes. The inferred values for the entropy change (Fig. 4.10) are not affected by the Virial Theorem. Depending on the temperature, entropy changes suggested by unmixing address the change in PE, so little or no internal heating is required.

Another factor pertains in a planetary system, and that is conservation of angular momentum. Conserving both energy and angular momentum during gravitational segregation requires differential rotation between the core and the mantle (Hofmeister and Criss, 2015). Differential rotation of the core is small today (e.g., Dehand et al., 2003; Criss, 2019), which is compatible with exponential dissipation (Chapter 1). Thus, differential rotation or entropy changes can each account for the change in energy (Fig. 4.10). With a combination of both, neither a very high temperature nor large amounts of differential spin are required.

Here we note that the differential spin would eventually cause heat to be evolved in the core region. From the analysis of the crust in Chapter 1, the Virial theorem can provide the initial spin of the formed core, and heat generation can be estimated from exponential decay of spin (Chapter 10).

4.5.2.2 A more realistic transformation

Odds are against forming a full size homogeneous Earth. As discussed above, the first step of planetary accretion was probably the formation of an iron protocore. In addition, a minimum of $\sim$1200 km for core augmentation is suggested by the need for a minimum gravitational acceleration in the sorting region (Hofmeister and Criss, 2012b). The very

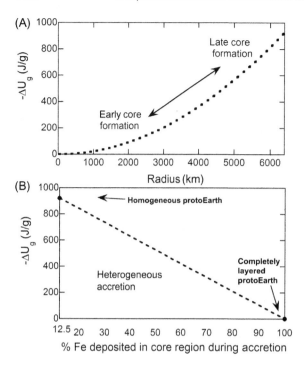

FIGURE 4.11   Trends in the negative change in gravitational potential. The Virial maximum is used here. (A) Dependence of $-\Delta U_g$ on the radius of the homogeneous Earth when core formation began. (B) Schematic of the dependence of $-\Delta U_g$ on the amount of Fe preferentially deposited in the region of the present-day core. *Modified after Figure 2 in Hofmeister, A.M., Criss, R. E., 2015. Evaluation of the heat, entropy, and rotational changes produced by gravitational segregation during core formation. J. Earth Sci. 26, 124–133 with permissions.*

central region could not sort due to high compression with low gravitational acceleration. This deduction supports our proposal of iron cores forming first, via magnetic attraction. With early core formation, some heat could be shed to the mantle even as more material is being added to that interior boundary, since this material is not part of the "bucket" where the phase transition occurs isothermally. From Fig. 4.11A, the change in energy for this continuous process is far smaller, and so this heating is also fairly small, on the order of a few hundred degrees.

Furthermore, if the initial planetary nucleus was dominated by iron, due to magnetic attraction with cold welding (Section 4.3.2), then the change in energy that would be associated with the gravitational segregation of subsequently accreted material would be much lower than would be released from the gravitational segregation of a full sized but homogeneous early Earth. (Fig. 4.11B).

4.5.3 Can buoyancy forces overcome resistance of solids to deformation?

Conventional discussions of core formation involve gravitational settling of dense particles in a medium with a certain viscosity. The model underling the analysis is Stokes' settling, which provides a terminal velocity (Eq. 3.1). If the medium is a solid, or stiff like a glass, considerable resistance impedes relative motion, and so the problem relevant to core formation is not the size of the terminal velocity but rather criteria for the onset of motion. Thus, the particle must supply enough force to overcome the yield stress (σ_{yield}) of the medium. This criterion for settling is analogous to Eq. (3.7) for convective flow of plastic solids. From the derivation of the Grashof number by Hofmeister and Criss (2018), for settling to begin requires:

$$g(r)\Delta \rho h < \sigma_{\text{yield}} \qquad (4.16)$$

where h can be particle size or the thickness of a dense layer. To estimate behavior in the Earth, we consider laboratory experiments on deformation of rocks and ceramics. For the 14 sets of experimental data on diverse solids shown in Figure 8 of Hofmeister and Criss (2018), rocks deform when the differential stress is on the order of the confining pressure or the hydrostatic pressure. Hence, we estimate σ_{yield} as the overpressure of $\rho g h$. Differential stress provided by the heavy grains is described by the LHS of Eq. (4.16), which is always less than $\rho g h$. Thus, high strength of rocks under compression prohibits core formation in a solid Earth.

Two other relevant factors exist. One is that the density in a homogenous Earth increases downwards due to self-gravitation, promoting stability. The other is that gravitational force decreases inwards. Both of these changes oppose gravitational segregation, particularly in small bodies where g is very small.

Thus, if the Earth formed as a homogeneous body, it would not sort gravitationally, unless an additional process occurred. Melting has been suggested as a requirement, but accretion does not provide high initial temperatures. Yet, isotopic studies suggest that the core formed early, within ~30 to ~125 Ma of the start at 4.5 Ga (e.g., Kleine et al., 2009). We address this riddle below.

4.5.4 Sorting above the protocore

After the protocore formed, a mixture of silicates and iron arrived (Fig. 4.6). Given sufficient gravitational force, along with some internal heat and the destabilizing effects of escaping, deeply accreted volatiles, this added material could be sorted, thereby increasing the size of the core to what we observe presently.

Mass fractions of planetary cores depend on planetary radius (Fig. 4.8A). Trends of the rocky bodies suggest that a minimum radius of ~1200 km is required to substantially augment the initially accreted protocores. Considering icy and rocky bodies for which the moment of inertia is known, differentiation is also associated with a minimum surface gravitational acceleration (g) of ~1.2 m/s (Fig. 4.8B). The trends show that an object must have a mass in excess of ~10^{22} kg to have an augmented metallic core, which is greater than the mass of the entire asteroid belt.

The observations (Fig. 4.8) are consistent with the round, self-gravitating moons having tiny protocores ($\sim 2 \times 10^{21}$ kg), whereas planets have large, augmented cores, which began with a larger nucleus of $\sim 6 \times 10^{22}$ kg. The sole exception is Io, whose spectacular volcanism evidences continued gravitational processing (Chapter 5). One implication is that the round moons have more Fe in their mantles than planets, due to lack of sorting. This inference is supported by lunar chemical compositions (see Lodders and Fegley, 1998). Bodies that are not round but were accreted from the nebula should have small protocores, suggested to be ~3% of their mass. However, for bodies derived from ejecta from previously formed bodies (e.g., Vesta: Hofmeister and Criss, 2012b), not even a miniscule protocore is expected.

4.6 Summary

Release of huge amounts of heat upon planetary formation by gravitational accretion is forbidden by thermodynamic laws and the Virial theorem. Data on the Solar System bodies provide corroboration. Gravitational accretion generates rotational energy, while conserving angular momentum. Most of this energy was sequestered in the newly formed Sun. Cold accretion is demonstrated by icy moons existing in the outer Solar System, as these bodies are too small to collect and/or retain light molecules as gas (Chapter 10).

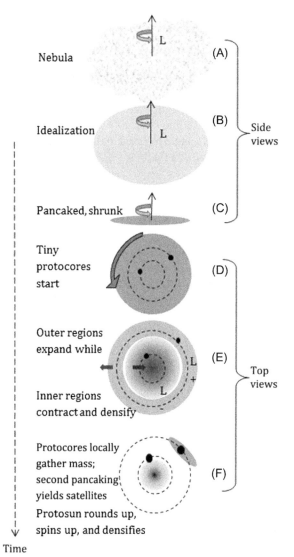

FIGURE 4.12   Schematic of formation of the Solar System from a nebula, whose slow spin maintains its stability against collapse (A). The time sequence starts with an idealized constant density oblate (B), which flattens and contracts during energy minimization (C). Planet nuclei begin forming at this time (D). Interior densification and round up is also conservative, if the outer regions expand. This process (E) is unrelated to migration of planets, which has no known cause. The final accretion stage (F) continues via energy minimization and the whole process repeats locally to form satellite systems.

Heat production by another early gravitational process (sorting) is problematic for additional reasons. One involves the forces required to overcome the resistance of stiff media to deformation. Another involves the possibility of heterogeneous accretion. We propose that the combined effects of the strong attractive force of magnetism at short range, the ductility and durability of iron particles, their propensity to weld, the high density of iron alloys, and iron's high atomic mass, all lead to iron-rich cores of planets forming first, thereby providing nuclei for additional planetary growth. Fig. 4.12 summarizes the steps of forming the planets, which were fast.

We emphasize that our model of Fig. 4.12 is conservative. Our model precludes migration of fully formed planets, which is currently popular and used to justify giant impacts, although no physical basis or cause has been proposed. Gravity pulls, but does not push. Our model explains how the co-ordinated orbits and spin of the Solar System originated in accretion of the solar nebula. The Solar System is a highly organized entity that formed and continues to evolve according to the rules of physics. But, once the planets form, these are independent entities, as demonstrated by their Keplerian orbits.

Not shown in Fig. 4.12 is the subsequent, late-stage addition of surface material by impacts, which produced surficial heat and added veneers of outer solar system material (see Fig. 4.9). This is not part of the conservative accretion process, but rather involves material not in stable orbits being pulled into the Sun.

As is observed currently and from the rock record, magmatism and explosive outgassing have largely controlled Earth's thermal and chemical evolution. Although the accretionary process was cold, permitting ices to be retained deep inside the Earth, this quiescent situation was temporary. Explosive outgassing was destabilizing was greatly destabilizing, permitting growth of the core by gravitational settling, and additional density stratification of dense oxides from silicates (Hofmeister and Criss, 2013). All processes involve loss of heat and minimization of energy. In particular, chemical segregation led to more rapid cooling of the planet, since heat-producers were carried upwards.

Spin energy is subsequently dissipated by internal friction, provided by differential rotation or other internal motions (Chapter 5), and this process would release heat, but over a protracted interval, which will be modeled in Parts II and III.

References

Armitage, P.J., 2011. Dynamics of protoplanetary disks. Ann. Rev. Astron. Astrophys. 49, 195–236. Available from: https://doi.org/10.1146/annurev-astro-081710-102521.

Bergin, E.A., Tafalla, M., 2007. Cold dark clouds: the initial conditions for star formation. Ann. Rev. Astron. Astrophys. 45, 339–396.

Bates, J.B., 1978. Infrared emission spectroscopy. Fourier Transform IR Spect. 1, 99–142.

Bernatowicz, T.J., Zinner, E.K., 1997. Astrophysical Implications of the Laboratory Study of Presolar Materials. AIP, New York.

Birch, F., 1965. Energetics of core formation. J. Geophys. Res. 70, 6217–6221.

Bowey, J.E., Hofmeister, A.M., 2005. Overtones and the 5-8 μm spectra of deeply embedded objects. Mon. Not. R. Astron. Soc. 358, 1383–1393.

Cameron, A.G.W., Benz, W., 1991. Origin of the moon and the single impact hypothesis IV. Icarus 92, 204–216.

Clark, S.P., 1957. Radiative transfer in the earth's mantle. Trans. Am. Geophys. Union 38, 931–938.

Criss, R.E., 2019. Analytics of planetary rotation: improved physics with implications for the shape and super-rotation of Earth's core. Earth Sci. Rev. 192, 471–479.

Criss, E.M., Hofmeister, A.M., 2017. Isolating lattice from electronic contributions in thermal transport measurements of metals and alloys and a new model. Int. J. Mod. Phys. B 31 (175020). Available from: https://doi.org/10.1142/S0217979217502058.

Criss, R.E., Hofmeister, A.M., 2016. Conductive cooling of spherical bodies with emphasis on the Earth. Terra Nova 28, 101–109.

Criss, R.E., Hofmeister, A.M., 2018. Galactic Density and Evolution based on the Virial Theorem, Energy Minimization, and Conservation of Angular Momentum. Galaxies 6, 115–135. Available from: https://doi.org/10.3390/galaxies6040115.

Dehand, V., Creager, K.C., Karato, S.I., Zatman, S., 2003. Earth's Core: Dynamics, Structure, Rotation. American Geophysical Union, Washington, DC, pp. 5–82.

Eddington, A.S., 1926. The Internal Constitution of Stars. Cambridge University Press, Cambridge.

Hofmeister, A.M., 2019a. Reconciling the kinetic theory of gas with gas transport data. Measurements, Mechanisms, and Models of Heat Transport. Elsevier, Amsterdam, pp. 143–179.

Hofmeister, 2019b. Modelling diffusion of heat in solids. In: Hofmeister, A.M. (Ed.), Measurements, Mechanisms, and Models of Heat Transport, 427. Elsevier, New York, pp. 359–398.

Hofmeister, A.M., 2019c. Macroscopic analysis of the flow of energy into and through matter from spectroscopic measurements and electromagnetic theory. In: Hofmeister, A.M. (Ed.), Measurements, Mechanisms, and Models of Heat Transport. Elsevier, Amsterdam, pp. 35–73.

Hofmeister, A.M., Criss, R.E., 2012a. A Thermodynamic model for formation of the solar system via 3-dimesional collapse of the dusty nebula. Planet. Space Sci. 62, 111–131.

Hofmeister, A.M., Criss, R.E., 2012b. Origin of HED meteorites from the spalling of Mercury: implications for the formation and composition of the inner planets. In: Hwee-San, L. (Ed.), New Achievements in Geoscience. InTech, Croatia, pp. 153–178. Available from: http://www.intechopen.com/articles/show/title/the-case-for-hed-meteorites-originating-in-deep-spalling-of-mercury-implications-for-composition-and.

Hofmeister, A.M., Criss, R.E., 2013. How irreversible heat transport processes drive Earth's interdependent thermal, structural, and chemical evolution. Gwondona Res. 24, 490–500.

Hofmeister, A.M., Criss, R.E., 2015. Evaluation of the heat, entropy, and rotational changes produced by gravitational segregation during core formation. J. Earth Sci. 26, 124–133.

Hofmeister, A.M., Criss, R.E., 2016. Spatial and symmetry constraints as the basis of the Virial Theorem and astrophysical implications. Can. J. Phys. 94, 380–388.

Hofmeister, A.M., Criss, E.M., 2018. How properties that distinguish solids from fluids and constraints of spherical geometry suppress lower mantle convection. J. Earth Sci. 29, 1–20. Available from: https://doi.org/10.1007/s12583-017-0819-4.

Hofmeister, A.M., Criss, R.E., 2019. The macroscopic picture of heat retained and heat emitted: thermodynamics and its historical development. In: Hofmeister, A.M. (Ed.), Measurements, Mechanisms, and Models of Heat Transport. Elsevier, Amsterdam, pp. 1–34.

Hofmeister, A.M., Wopenka, B., Locock, A., 2004. Spectroscopy and structure of hibonite, grossite, and $CaAl_2O_4$: implications for astronomical environments. Geochim. Cosmochim. Acta 68, 4485–4503.

Hofmeister, A.M., Criss, R.E., Criss, E.M., 2018. Verified solutions for the gravitational attraction to an oblate spheroid: implications for planet mass and satellite orbits. Planet. Space Sci. 152, 68–81. Available from: https://doi.org/10.1016/j.pss.2018.01.005.

Hofmeister, A.M., Criss, R.E., Criss, E.M., 2019a. How planetary tectonics relates to moon size, orbit, and body spin. Geology.

Hofmeister, A.M., Criss, R.E., Criss, E.M., 2019b. A perturbation model for evolution of the terrestrial satellite system with implications. Math. Geosci.

Kellett, B.S., 1952. Transmission of radiation through glass in tank furnaces. J. Soc. Glass Tech. 36, 15–123.

Kemper, F., Jaeger, C., Waters, L.B.F.M., 2002. Detection of carbonates in dust shells around evolved stars. Nature 415, 295–297.

Kleine, T., Touboul, M., Bourdon, B., Nimmo, F., Mezger, K., Palme, H., et al., 2009. Hf–W chronology of the accretion and early evolution of asteroids and terrestrial planets. Geochim. Cosmochim. Acta 73, 5150–5188.

Kragh, H., 2016. The source of solar energy, ca. 1840–1910: From meteoric hypothesis to radioactive speculations. Eur. Phys. J. H 365–394.

Lodders, K., 2003. Solar System abundances and condensation temperatures of the elements. Astrophys. J. 591, 1220–1247.

Lodders, K., Fegley Jr., B.J., 1998. The Planetary Scientist's Companion. Oxford University Press, Oxford.

Merstallinger, A., Sales, M., Semerad, E., Dunn, B.D., 2009. Assessment of cold welding between separable contact surfaces due to impact and fretting under vacuum. European Space Agency, Leiden, the Netherlands. Available from: http://esmat.esa.int/Publications/Published_papers/STM-279.pdf.

Mittlefehldt, D.W., McCoy, T.J., Goodrich, C.A., Kracher, A., 1998. Non-chondritic meteorites from asteroidal bodies. Rev. Miner. 36, 1529–6466.

Molster, F.J., Lim, T.L., Sylvester, R.J., Waters, L.B.F.M., Barlow, M.J., Beintema, D.A., et al., 2001. The complete ISO spectrum of NGC 6302. Astron. Astrophys. 372, 165–172.

Müller, I., 2003. Entropy: a subtle concept in thermodynamics. In: Greven, A., Keller, G., Warnecke, G. (Eds.), Entropy. Princeton University Press, Princeton, NJ, pp. 17–36.

Safronov, V.S., 1969. Evolution of the Protoplanetary Cloud and Formation of the Earth and Planet. Nauka, Moscow (Translated by the Israel Program for Scientific Translations, 1972. NASA TT-677).

Schaefer, J., 2007. Spectroscopic evidence of interstellar solid hydrogen. Chem. Phys. 332, 211–224.

Sherwood, B.A., Bernard, W.H., 1984. Work and heat transfer in the presence of sliding friction. Am. J. Phys. 52, 1001–1007.

Thompson, G.D., Corman, A.B., Speck, A.K., Dijkstra, C., 2006. Challenging the carbon star dust condensation sequence: anarchist C stars. Astrophys. J. 652, 1654–1673.

Truesdell, C., 1980. The Tragicomical History of Thermodynamics. Springer-Verlag, New York.

van Breemen, J.M., Min, M., Chiar, J.E., Waters, L.B.F.M., Kemper, F., Boogert, A.C.A., et al., 2011. The 9.7 and 18 μm silicate absorption profiles towards diffuse and molecular cloudlines-of-sight. Astron. Astrophys. 526, A152–A165.

Wetherill, G.W., 1988. In: Chapman, C., Vilas, F. (Eds.), Accumulation of Mercury from Planetesimals. Univ. Ariz. Press, Mercury. Tucson, pp. 670–691.

Yun, J.L., Clemens, D.P., 1990. Star formation in small globules – Bart Bok was correct. Astrophys. J. Lett. 365, L73. Available from: https://doi.org/10.1086/185891. Bibcode:1990ApJ...365L..73Y.

Zolensky, M.E., et al., 2006. Mineralogy and petrology of Comet 81P/Wild 2 Nucleus samples. Science 314, 1735–1739.

Websites

Friction stir welding.
https://en.wikipedia.org/wiki/Friction_stir_welding (accessed 05.30.18.).
Images of mass transfer between stars.
https://www.aavso.org/jets-bubbles-and-bursts-light-taurus (accessed 02.08.19.).
https://apod.nasa.gov/apod/ap180311.html (accessed 02.08.19.).
https://apod.nasa.gov/apod/ap170629.html (accessed 02.08.19.).
NASA Factsheets on the planets and their satellites.
https://nssdc.gsfc.nasa.gov/planetary/factsheet/ (accessed 02.09.19.).

Large-scale gravitational processes affecting planetary heat transfer

Anne M. Hofmeister, Robert E. Criss and Everett M. Criss[#]

"Two half-nothing's is nothing. It's mathematics, son. You can argue with me, but you can't argue with figures. Two half-nothings is a whole nothing." Foghorn Leghorn in "A fractured leghorn" *1950 Warner Brothers cartoon directed by Robert McKimson.*

[#] E. M. Criss is an employee of Panasonic Avionics Corporation, but prepared this article independent of his employment and without use of information, resources, or other support from Panasonic Avionics Corporation.

The popular view of the mantle being in perpetual vigorous convection that is driven by an eternally hot core is inconsistent with the low energy of Earth's internal heat sources. In particular, the core has negligible radioactivity, based on data on all types of iron meteorites (see tables in Mittlefehldt et al., 1998). A small heat source, derived from freezing of the inner core, has been postulated (e.g., Stacey and Stacey, 1998), but this concept requires a fundamental thermodynamic error. Rather, the insulating nature of the rocky mantle with its greater concentrations of the radioactive, heat-producing, lithophilic elements strongly suggests the core is melting (Criss and Hofmeister, 2016). Because the above sources of heat are insufficient to cause convection, large amounts of primordial heat have been invoked, but this is inconsistent with vigorous convection, which efficiently removes heat. The internal heating suggested by plate models (e.g., Stein and Stein, 1992) is too small anyways. Several additional theoretical problems exist with the convection hypothesis (Section 3.3). However, lateral movements of the continents and subduction of the plates cannot be denied. Currently, mantle convection is the only hypothesis for plate tectonics under investigation, despite the fact that the mechanism linking surface motions to interior circulation is unknown. Rational empiricism, in general, and the scientific method, in particular, each require that viable alternatives models for lithospheric plate motion and subduction be considered, as current models contain numerous shortcomings and logical inconsistencies.

As proven by Newton, macroscopic motions are the direct result of forces, rather than being a direct result of heat. Gravitational forces not only cause most motion in the Universe, but are intimately related to the largest energy reservoir of any self-gravitating (round) planetary body. Gravitational forces have been overlooked as the direct cause of lateral plate motions. This omission is rooted in history: Kelvin's hypothesis of gravitational contraction forming starlight, although discredited, remains entrenched in the views that accretion directly converts this immense energy source into heat. Chapter 4 summarizes why gravitational potential is instead converted to rotational kinetic energy during accretion, after Hofmeister and Criss (2012) and Criss and Hofmeister (2018). Chapter 4 focused on the early stages of planetary formation, which began with formation of a protocore by magnetic and gravitational attraction. This step is followed by the conservative contraction of the nearby nebula to form a protomantle. Our concern here regards the subsequent, non-conservative processes, including gravitational sorting and late stage impacts which add both material and heat to the outer layers. The various processes overlap in time and involve changes in the size of the growing protoplanet. Unfortunately, we have little or no direct information regarding these processes, because we have little to no information on Earth's early history, particularly on the stages of growth and differentiation. Hence, testing of formation models is limited to evaluating Solar System trends. The present chapter focuses on gravitational processes that are currently active on the full-size, differentiated Earth, for which we have considerable information and also considers the scant evidence provided by Earth's oldest rocks. Crucially and importantly, we have samples of upper mantle materials only. This includes xenoliths originating above 250 km and diamond inclusions from above 300 km (Chapter 1). Early claims of lower mantle origins were based on genetically unrelated minerals in separate inclusions. Thus, geochemical data on rock samples provide information on the chemical evolution of the upper mantle, not of the core or whole Earth.

A key gravitational process is spin, which is the end-product of the gravitational forces driving accretion of an individual body, as is evident in star data (Fig. 1.12). Spin is known to regulate the current shape of the self-gravitating Earth (Todhunter, 1873; Criss, 2019). Its importance to other present-day phenomena is underestimated, even though spin playing a role in plate tectonics is long-known (e.g., Stehli et al., 1969). Recent quantification (Riguzzi et al., 2010; Doglioni and Panza, 2015; Fig. 1.9) provides important clues to Earth's interior workings.

A problem of perceptions exists: Earth was once considered to be *unique* since our planet is currently active, unlike the rest of the inner Solar System. Subsequent to discovery of the active volcanoes of Io, the notion that other bodies should behave like the Earth has overshadowed its uniqueness. Consequently, the large volcanoes on Mars and the unique topography on Venus are now popularly interpreted as arising from planetary tectonics similar to Earth. Hamilton (2015, 2019) presents many arguments that this interpretation is in error. Crucially, no other body exhibits magnetic striping, which was a key discovery that led to acceptance of plate motions on Earth. Without magnetic striping, it will likely be difficult or impossible to demonstrate any past or present lateral motions on other bodies which may be associated with plate tectonics. In contrast, the buoyancy driven volcanism observed elsewhere is driven by radial, not lateral motions.

Gravity is the source of planetary-scale motions, and operates independently of any heat source. Importantly, motions of matter not only carry heat (advection) but can create heat via friction (viscous dissipation). Chapter 3 covers how advection and other processes of heat transport operate in planets, whereas the present chapter focuses on gravitationally induced motions.

Vertical and lateral motions on planets differ in symmetry, and are derived from gravitational forces in completely different ways, due to the spherical symmetry of Newton's law. Section 5.1 therefore begins with the familiar vertical phenomenon of buoyancy. Section 5.2 focuses on internally derived gravitational forces associated with spin, mechanisms of spin dissipation, and links of this process with plate tectonics. Section 5.3 covers polar jets, which are well-known in astronomical environments. Section 5.4 covers externally derived gravitational forces acting on the rocky planets, which can produce torque, and proposes a gravitational mechanism for plate tectonics. We argue that plate tectonics with subduction is a modern phenomenon. Section 5.5 covers the much different conditions in the early Earth, when late-arriving impacts added heat and mass. Section 5.6 summarizes, explaining how the "hot" impacts following "cold" accretion fits the observations and reconciles competing views of the early Earth.

5.1 Radial internal forces in self-gravitating bodies

5.1.1 Isolated volcanism

Currently, most volcanism on Earth is associated with the subduction zones, or with the spreading centers of the mid-ocean ridges. However, volcanoes also exist at considerable distance from plate boundaries, as exemplified by kimberlite pipes in continental interiors and by the chains of volcanoes in the middle of the Pacific Ocean plate (Fig. 5.1).

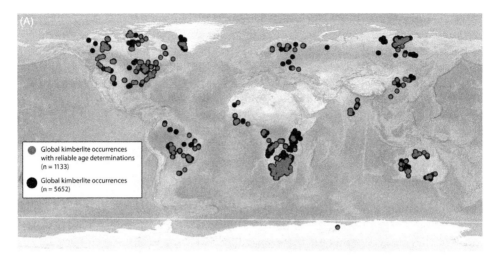

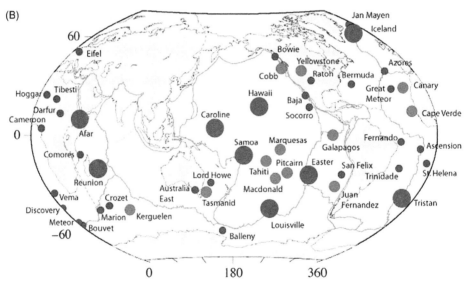

FIGURE 5.1 Locations of magmatic activity compared to plate boundaries. (A) Kimberlites are located on thick and old parts of the continents, which are inside plates. (B) Location of major volcanic centers. Large dots are those characterized as "primary," i.e. originating in the deep mantle, as ascertained by Courtillot et al. (2003). Only three primary hot spots (Hawaii, Caroline, and Reunion) are distinct from plate boundaries. *(A) Reproduced with permissions from Figure 1 in Tappe, S., Smart, K., Torsvik, T., Massuyeu, M., de Wit, M., 2018. Geodynamics of kimberlites on a cooling Earth: clues to plate tectonic evolution and deep volatile cycles. Earth Planet. Sci. Lett. 484, 1–14: https://doi.org/10.1016/j.epsl.2017.12.013. Part (B) was adapted by Foulger, G.R., 2010. Plates vs Plumes: A Geological Controversy. Wiley-Blackwell, Chichester, UK, and is open access on creative commons (https://commons.wikimedia.org/wiki/File:CourtHotspots.png)*

The diamond-bearing, brecciated kimberlites differ greatly from the intraplate basaltic lavas and the basalts that erupt quiescently on Hawaii. However, all are associated with vertical ascent of material and show that volcanism on Earth can occur independently of plate tectonics. Courtillot et al. (2003) lists 49 recently active volcanic regions considered to be independent hot-spots.

Vertical kimberlite pipes are formed during gas-backed explosions that are inferred to be quite rapid (McGetchin et al., 1973). Most of these eruptions occurred from 50 to 200 Ma ago, but some Archean examples exist (Stern et al., 2016; Faure, 2006; Tappe et al., 2018). The depth of origin for kimberlites is well below 140 km, and probably somewhat below 200 km. These eruptions convey diamond upwards through the thick continents. This depth and their explosive emplacement are consistent with metasomatism in the upper mantle (see Fig. 12.8). Stern et al. (2016) argue that plate tectonics has recycled volatiles into the upper mantle over time. These volatiles being essential to form and erupt kimberlites explains their dominantly young (<200 Ma) ages. Because diamonds form in carbon-rich areas, the ancient kimberlites (which are rare even if the sparsity of the rock record is considered) could result from the chemical and thermal evolution of the upper mantle. The vertical profiles of these pipes (Mitchell, 1986) point to the emplacement and uplift process being driven by gravitational buoyancy.

Many consider deep mantle plumes as the source of basaltic hotspots, but this concept is hotly debated (Foulger, 2010; also see Chapters 1 and 3). Importantly, all samples have shallow origins. Recent studies have recognized that H_2O can play an important role in the generation of large amounts of intercontinental (flood) basalts (e.g., Liu et al., 2017). Geochemical data are most consistent with an upper mantle origin that is abetted by volatiles (Choi et al., 2019).

Great interest exists in tracking hot spots, such as the Hawaii-Emperor chain, to provide a fixed frame of reference for analyses of plate motions (see discussion by Courtillot et al., 2003). Regardless of the depth of origin, the existence of the tracks on the ocean floor demonstrates that Hawaiian volcanism is independent of plate tectonics, and that the physical cause of intraplate volcanism is simple, vertical ascent of buoyant material.

5.1.2 Variation of properties with depth

Buoyancy requires a gravitational field. Earth's internal gravity field is nearly radial, due to very low planetary ellipticity. The value of the gravitational acceleration "g" in the mantle determined from seismic density profiles is nearly constant (Fig. 5.2), which seems counterintuitive. The cause is density stratification: Earth's core is rather dense and large whereas the outer rocky regions become progressively less dense towards the surface.

The mantle is stable due the strength of its highly compressed rocks, which require large differential stress to induce motions (Hofmeister and Criss, 2018; Chapter 3). In a high temperature gradient, the solid rock will not convect, but instead melts, producing magma with lower density and higher fluidity. But, the magma so produced can only rise if certain conditions are met: either the thermal buoyancy must be sufficiently high to overcome viscous resistance, or a path of weakness (faults/fracture zones) must be available, or the magma dissolves overlying less refractory rocks and stopes what remains.

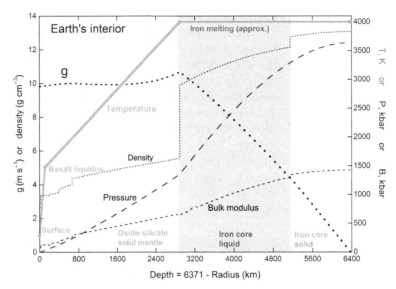

FIGURE 5.2 Interior of the Earth from the PREM seismological model, as tabulated by Anderson (2007). The kbar (= 0.1 GPa) scale was used so temperatures could be easily depicted. The temperature curve "connects the dots," i.e., is very approximate but is based on melting points of layers with definitive chemical compositions and established phase relations. The core melting temperature is overestimated, as earlier work focused on pure iron and did not reach core pressures (Aitta, 2006; Chudinovskikh and Boehler, 2007; see Chapter 8), but is shown because such data likely affected PREM.

In any case, the rate of magma ascent must be sufficiently rapid that its high temperature is retained against thermal diffusion.

5.2 Processes in spinning, self-gravitating objects

Spin is axially symmetric. Any behavior linked to spin will share this symmetry, as deduced by Newton and as exemplified by the multitude of oblate astronomical bodies. Vertical motions along the polar axis are also permissible (Section 5.3), and different internal layers in an object can spin at different rates. This section explores the latter possibility, with the aim of better understanding the link of spin to plate tectonics revealed by seismological studies (Fig. 1.9).

5.2.1 Differential rotation vs solid body spin

Spin is all organized rotation about a central axis. For coherently spinning bodies, all internal layers have a common angular velocity, which requires that the tangential velocity increases linearly with cylindrical (or equatorial) radius (Fig. 5.3A,B,E). When internal layers spin at different angular velocities, the dependence of u_{tan} on r becomes non-linear. The term "rotation curves" covers all such behavior. Differential rotation is exemplified by spinning hurricanes (Fig. 5.3C) whose rotation curves resemble those of spinning solid

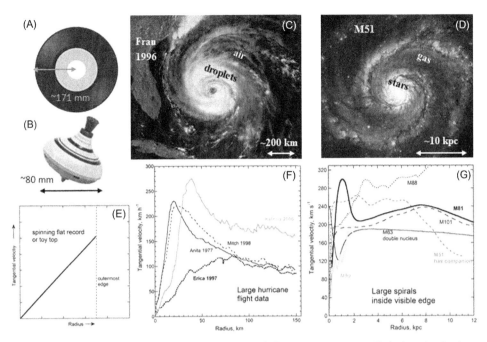

FIGURE 5.3 Comparison of familar spinning objects and their rotation curves. Scale bars in the images are approximate. (A) Cartoon of phonograph record. (B) Photo of spinning toy top. (C) Image of category 3 hurricane Fran off the Florida coast. For time-lapse images and maps of wind speed, see the NOAA website. (D) Image of the Whirlpool galaxy in visible light from the NASA/IPAC website. (E) Schematic of rotation curves, i.e., (tangential velocity vs radius) of solid bodies. (F) Rotation curves of four large hurricanes, similar to Fran, from flight data of Willoughby et al. (2006) and Dima et al. (2010), which focus on the central regions. Images of these hurricanes can be found on the NOAA website. (G) Rotation curves of complex spiral galaxies, based on data from Sofue et al. (1999; 2003). Except for M83, which has two central dense spots, the rotation curves increase steeply from the center. Images are on the NASA/IPAC website. *(B) Modified after Freestockphoto3565 - freeimageslive. Panels (C), (D), (F), (G) from Figure 1 in Hofmeister, A.M. Criss, R.E., 2017. Implications of geometry and the theorem of gauss on Newtonian gravitational systems and a caveat regarding Poisson's equation. Galaxies 5, 89–100. Available from: http://www.mdpi.com/2075-4434/5/4/89, which is open access.*

bodies near their center. Specifically, hurricane wind velocities are null at the eye, increase outward to reach a maximum value in u_{tan}, and then progressively decline at large radial distance r (Fig. 5.3F). Hurricanes resemble toy pinwheels, which are run by wind as well. Even though hurricanes are made of gas (with suspended water droplets), the rotation curves of their central zones are like those of solid objects.

Very, very large and self-gravitating bodies also spin, as is evident qualitatively in the resemblance of spiral galaxies to hurricanes (Fig. 5.3C and D), whirlpools, and pinwheels, and quantitatively in galactic rotation curves (Fig. 5.3F and G). Spin exists over an incredible range of distance and time scales. Mathematical formulae of spin for self-gravitating bodies with homogeneous density are given by Hofmeister and Criss (2017), whereas the effect of density varying with radius is addressed by Criss and Hofmeister (2018).

5.2.1.1 Evolution without friction between components

Departures of the rotation curves from the "solid-body," linear increase of velocity with radius are termed "differential rotation." A more precise term would be "differential spin" when referring to large spinning bodies.

Both galactic spin and Solar System orbits formed during the earliest stages of nebular contraction, in a manner that conserved angular momentum and energy (Criss and Hofmeister, 2018; Section 4.4; Fig. 4.3). Because no friction exists either between the 8 planets in our Solar System or amongst the billions of stars in a galaxy, the conditions of their formation are well-preserved. Currently, the evolution of the Solar System is limited to slow changes resulting from tiny gravitational interactions between bodies. Dynamics of the Earth-Moon pair are summarized in Section 5.4.

Importantly, accretionary processes densified the nebulae to form a central body (Hofmeister and Criss, 2012). During this process, the nebula had to 'spin up' in order to conserve angular momentum. As the central body coalesced, the outer parts of the nebula simultaneously moved outward and lost angular momentum (Criss and Hofmeister, 2018). The total energy and momentum of the nebula was in this way conserved. Conservative accretion of galaxies is supported by with data on their varied shapes and sizes. Their relationship between tangential velocity and density shows that outer galactic regions are much less dense than their central zones. Hence, the rotation curves of Fig. 5.3G confirm that the outer parts of spiral galaxies have moved away, losing their angular momentum and rarifying. This process is still going on in these immense entities.

Density in the archetypical spiral galaxy, our neighbor Andromeda, follows a power law with an inverse square radius ($\sim r^{-2}$) density distribution, valid up to ~ 100 kpc where Andromeda merges with our own Milky Way galaxy (Criss and Hofmeister, 2018). The fast spinning interior of Andromeda is a lot denser! However, for this immense and slowly evolving spiral galaxy, more mass ($\sim 10^{12}$ suns) exists beyond $\sim \frac{1}{2}$ the object radius than inside, regardless of whether none, one, or several black holes ($\sim 10^{6}$ suns) are present near the center. Hence, motions of stars in large spirals cannot be similar to the pattern in our Solar System, which is dictated by a single central mass. This very large Andromeda galaxy has neither finished contracting to the center, nor finished expanding toward the outside.

5.2.1.2 Initial spin of stars and planets

Nebular contraction in our "small" Solar System fully progressed to the development of its round central body. This process in miniature also proceeded simultaneously in outlying locations, to eventually form the planets and their moons. In forming the planets, the outer parts of these tiny "sub-nebulae" each moved away from their iron nucleus, which then yielded satellite systems (Section 4.4). The Sun and each planet thus compose the inner parts of their respective nebula and sub-nebulae. Thus, planetary bodies and stars should be spinning with rotation curves like solid bodies (Fig. 5.3E and G), even if not fully consolidated. Initial spin would have been fast in a conservative process.

5.2.1.3 Spin loss in stars vs planets

In contrast to orbital energy, loss of star spin with time is evident in diverse astronomical data (e.g., Epstein and Pinsonneault, 2014; Fossat et al., 2017). Stars are soft, fluid or gassy bodies, particularly towards their exteriors. Without rigidity, differential spin is possible, as also observed for hurricanes and Earth's atmosphere. Adjacent layers spinning at different rates generate heat via friction, and slowly reduce the high initial spin.

Our soft and deformable Sun has lost much of its initial spin, whereas most planets have spin rates that are relatively close to the initial value predicted (e.g., Fig. 1.12). But, the rocky planets have strong, compressed interiors, so how can these lose spin?

Crucially, the interior of Earth is not completely solid. The many, many orders of magnitude difference in viscosity between deforming rock, basaltic liquid, and molten metals (Table 5.1) permits differential rotation of interior layers.

Observations confirm spin loss. Differential spin of Earth's lithosphere is demonstrated by westward drift of the continents (e.g., Stehli et al., 1969) and by axial symmetry of plate motions, which have an *average* plate velocity of $\sim 6 \, \text{cm} \, \text{yr}^{-1}$ (Riguzzi et al., 2010; Doglioni and Panza, 2015). The underlying shear zone must be global in extent, and must occur beneath the deep continental roots. These geometric requirements plus seismological constraints on plate behavior (Figs. 1.9 and 1.2) suggest that this weak zone is $\sim 300-400 \, \text{km}$ deep. The lower symmetry behavior that is also observed of the plates, including subduction, is an added complication, and is discussed further below and in Chapter 7.

Super-rotation of the solid inner core with respect to the mantle is debated and is an active research area in seismology (reviewed by Mohazzabi and Skalbeck, 2015). Possible differential spinning of the Earth deep inside is discussed further in Chapters 8 and 9.

TABLE 5.1 Radial structure of the Earth[a] with viscosity estimates from laboratory data on analogue materials[b].

Layer	Outer radius	Material	η[c]	Yield strength	References for lab data[d]
	km		Pa s	GPa	
Lithosphere	6371	basaltic glass	10^{15}	10^{-4} [e]	Sehlke et al. (2014)
Low velocity zone	6270	basaltic melt	10^{9}	~ 0	Hofmeister et al. (2016)
Upper mantle	6170	ultramafic rock	[f]	$\sim 0.5-2$	Moghadam et al. (2010)
Transition zone	5960	wadsleite	[f]	~ 2	Nishihara et al. (2008)
Lower mantle	5700	MgO ceramic	[f]	0.01[g]	Domínguez-Rodríguez et al. (2007)
Barycenter[h]	4340–4940[h]	n/a			
Outer core	3480	molten Fe-S	10^{-2}	~ 0	Funakoshi (2010)
Inner core	1220	iron alloy	10^{12}	?	Frost and Ashby (1982)

[a]*Data from PREM, as tabulated by Anderson (2007).*
[b]*The materials do not exactly correspond to the layer, but represent selected phases that are likely present.*
[c]*Dynamic viscosity at high temperature and pressure, if available.*
[d]*Other studies give similar results: see Table 3 and Figures 6 and 8 in Hofmeister and Criss (2018), which focus on insulators. Mohazzabi and Skalbeck (2015) summarize studies on the core and core materials.*
[e]*Yield strength for glass with $\sim 30\%$ crystals.*
[f]*The experiments involve deformation under differential stress. Equations for creep can be cast in terms of viscosity. However, a more realistic number may be represented by glacial rebound studies, which occur over appropriate time frames, and provide 10^{20} to 10^{22} Pa s for the mantle, as summarized by Lambeck et al. (2017).*
[g]*MgO data is on nanocrystals at ambient pressure, providing low yield strengths.*
[h]*Monthly variation in radius: contained in lower mantle.*

Returning to stars, polar jets are a ubiquitous feature and offer another mechanism for spin loss besides friction between layers. Jets are discussed below.

5.2.1.4 *What causes differential spin?*

Differential spin involves the non-conservative force of friction, and torque exists because the drag between any two shells is perpendicular to the radius. The presence of torque indicates that friction is capable of changing the angular momentum of the spinning layers. Over time, the outer layers will progressively slow, as observed for the Earth and Sun. Energy attenuation and heat generation are the hallmarks of non-conservative forces. Crucially, because friction is produced during differential spin, it follows that this non-conservative force cannot *cause* differential spin, but is a *consequence* of differential spin. Assigning friction as the cause of differential rotation is closely related to the faulty logic underlying designs for perpetual motion machines.

For differential spin to exist in planets, forces in addition to friction are required. Some possibilities are discussed below.

5.2.2 Stress field generated by a spinning oblate spheroid

The cracks in Earth's lithosphere define the plate boundaries. The primary feature is the network of mid-ocean ridge spreading centers. Because these ridges are steadily producing oceanic crust, the plates are growing, but these must collide since Earth's body radius is fixed. Consequently, the colliding plates slide under each other, creating subduction zones, which are secondary boundaries. These primary and secondary features define the 14 large plates. A substantial number of small plates also exist on the surface of the globe (38, according to Bird, 2003), whose boundaries are defined by various types of faults of different scales. For example, the San Andreas fault is a transform fault, where plates slide past each other. In part, the complexities of plate tectonics are a geometrical consequence of covering a spheroid with many, nearly flat plates.

The primary boundaries were interpreted as conjugate zones of weakness by Hofmeister and Criss (2005) based on comparing global mid-ocean ridge patterns to fault geometry (Figs. 1.10 and Fig. 5.4). Such fault patterns are expected during shear failure while under uniaxial compression (Fig. 5.4B). Similar patterns are observed in laboratory experiments (Fig. 5.4C and D).

The oblate symmetry of the gravity field is a consequence of spin. Since the lithosphere represents only ~1 to 4% of Earth's radius, this thin layer effectively constitutes the surface, and can be viewed as a shell being subject to the axially symmetric field provided by the core and lower mantle. The dependences of Earth's properties on depth (Fig. 5.2) quantitatively confirm the control exerted by the interior, much of which arises from the gravitational pull of the dense and large central core, as discussed above.

The oceanic lithosphere is a very thin, brittle chill zone that is immediately underlain by melt or partial melt. The Southern Hemisphere, which is mainly oceanic, is fully consistent with axial symmetry (Fig. 1.10), which is the main perturbation of Earth's gravity field from a spherical distribution. The Northern Hemisphere, which has most of the continental landmass, is less symmetric (Fig. 5.4A). The six continents each include ancient cratons

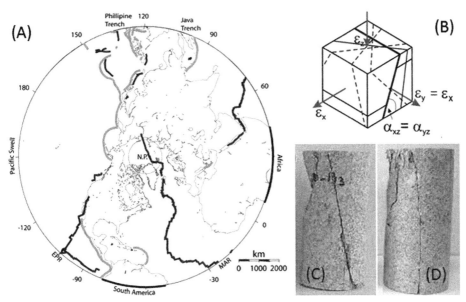

FIGURE 5.4 Failure modes. (A) Cracking in Earth's northern hemisphere. Heavy black lines show mid-ocean ridges and heavy gray lines show subduction zones. Fine lines are outlines of the continents. (B) Uniaxial stress on a homogeneous block. In this idealization, faults will form at some angle to the shortening axis, and with axial symmetry about this axis. (C) Conjugate shear failure of weathered granite vs. (D) axial splitting of fresh granite, both under monotonic uniaxial compression in the laboratory. The steep fracturing angles are consistent with the idealization. *(A) Orthographic projection from Weinelt, M., 2003. Create-a-map Web site: http://www.aquarius.geomar.de/omc. (accessed 12.17.) using the program of Wessel, P., Smith, W.H.F., 1995. The Generic Mapping Tools (GMT), version 3.0: Technical reference and cookbook (SOEST/NOOA), http://gmt.soest.hawaii.edu (accessed 12.17.). (B) Modified after Figure 11 in Healy, D., Blenkinsop, T.G., Timms, N.E., Meredith, P.G., Mitchell, T.M., Cooke, M.L., 2015. Polymodal faulting: time for a new angle on shear failure. J. Struct. Geol. 80, 57–71. Available from: https://doi.org/10.1016/j.jsg.2015.08.013 (Under a Creative Commons license), which is open access (http://creativecommons.org/licenses/by-nc-nd/4.0/). (C and D) are both reproduced with permissions from Figure 7 in Ludovico-Marques, M., Chastre, C., Vasconcelos, G., 2012. Modelling the compressive mechanical behaviour of granite and sandstone historical building stones. Construct. Building Mater. 28, 372–381: https://doi.org/10.1016/j.conbuildmat.2011.08.083.*

with deep roots to ~ 200 km. Their thickness enhances their strength. In addition, the continents have been "pressure cooked." Even so, some of the zones of weakness in the Northern hemisphere follow the longitudinal failure mode associated with a uniaxial stress field. Others are deflected (cf. Figs. 1.10 and 5.5A). This hemispheric difference is consistent with laboratory experiments, whereby strong fresh granite (analogous to continental crust) fractures longitudinally, whereas relatively weak altered granite (analogous to oceanic crust) has conjugate shears (Fig. 5.4C and D).

Thus, locations of the plate boundaries are primarily controlled by the symmetry of the gravitational field for a large spinning body. But the cracking is predicated on the existence of a thin chill zone over most of its surface. Thus, the thermal structure of the planet is also relevant regarding whether plate tectonics can exist. Yet another criterion exists: the oceanic plates must be thick and strong enough to maintain coherency while subducting.

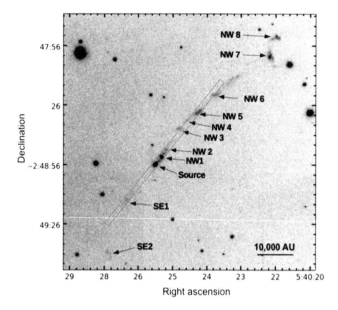

FIGURE 5.5 Image of the jets expelled from the very young brown dwarf HH 1165 in the visible. Various knots observed in the NW and SE directions. The slit orientation used for acquiring spectra is shown. *Reproduced with permissions from Figure 3 of Riaz, B., Briceño, Heathcote, S., Whelan, E.T., 2017. First large-scale Herbig—Haro jet driven by a proto-brown dwarf. Astrophys. J. 844, paper 47 (19 pp.) https://doi.org/10.3847/1538-4357/aa70e8.*

The thermal gradient needed for plate tectonics to operate as observed today is thus restricted to some range of values, depending on the material and its melting point.

No other planetary bodies have both spin and a fairly thin chill zone, nor do they have a comparatively large Moon (Section 5.4), making plate tectonics unique to the Earth. Hamilton (2015, 2019) argued for uniqueness, based on geologic evidence.

5.3 Polar jets and spin evolution of planetary bodies

One way for a self-gravitating object to become rounder, while conserving angular momentum and rotational energy, is for the object to spit out axial jets, which will reduce the kinetic energy of the star, but not its angular momentum. This phenomenon is observed for the smallest and densest types of stars, brown dwarfs. Although most jets are small, a spectacular jet with 0.26 kpc length from a ~ 36 Jupiter mass young stellar object was observed by Riaz et al. (2017) (Fig. 5.5). Jets are common for young stars and could be a part of planet formation as well. Two very different processes are sketched below.

5.3.1 Fast, low friction, gas jets

To model this phenomenon during rounding up of gassy objects, Criss and Hofmeister (2018) conserved total energy:

$$U_{g,oblate} + RE_{oblate} + U_{g,jets} + LE_{jets} = U_{g,sphere} + RE_{sphere} \tag{5.1}$$

where LE is the linear kinetic energy of the jets. To provide an approximate solution, consider a point distant from the oblate planet and its jets: together these objects act as a point mass, providing:

$$U_{g,oblate} + U_{g,jets} \cong U_{g,sphere} \tag{5.2}$$

Combining Eqs. (5.6) and (5.7) gives:

$$RE_{oblate} + LE_{jets} \cong RE_{sphere} \tag{5.3}$$

Eq. (5.3) shows that kinetic energy can be conserved during round up and spin down, because the reduction in R.E. is compensated by the kinetic energy of the jets. Total energy is conserved through the Virial theorem and the approximate relationship of Eq. (5.2). Energy balance is approximate, but provides:

$$\left(1 - \frac{m_{jets}}{M}\right) R_{final}^2 \, \omega_{final} \left(\omega_{final} - \omega_{initial}\right) \cong \frac{5}{2} \frac{m_{jets}}{M} u_{jets}^2 \tag{5.4}$$

Momentum conservation is exact when friction is not present. Linear momentum is conserved by jets departing equally in opposite directions, or, if not, by recoil of the body. Angular momentum can be conserved by requiring that changes in $I\omega$ balance during round up, where I is the moment of inertia and ω is the angular velocity. For an oblate body with homogeneous density and any ellipticity, $I = \frac{2}{5}MR^2$, where R is the equatorial radius. For balanced jets and no friction:

$$\frac{\omega_{final}}{\omega_{initial}} = \frac{R_{initial}^2}{(1 - m_{jets}/M)R_{final}^2} \tag{5.5}$$

The object spins up as it rounds up and releases jets. A detailed analysis, including mass balance, is not performed here. Note that substantial heat loss is not invoked, but some heat would be advected.

The jetting phenomenon would allow nascent rocky planets to expel volatiles during formation. In cold accretion, volatiles could be incorporated at great depths. In neglecting frictional heating, we are implicitly assuming that the phenomenon occurs before formation (compaction) is complete. Thus, the spin loss of a solid planet cannot be explained in this manner because frictionless gas jetting can only be part of the early, conservative steps of star and planet formation.

5.3.2 Slow, high friction melt "jets"

In formed planets, polar expulsions of magmatic material is possible, but this phenomenon would not conserve energy since friction is involved, producing heat which is cast off. Furthermore, these events would not cause material to leave the planet, so referring to this phenomena as "jets" may not be optimally accurate. However we retain this term to signify that the process has a gravitational origin, and to avoid use of "plumes" which connotes a different, and disputed, phenomenon of whole mantle convection.

Slow jetting involves changes in planetary gravitational potential and reduction of the moment of inertia, even though the linear kinetic energy (LE) of the volcanic jets is small.

The amount of heat produced is important to both the energy and linear momentum balance. Heat can also be advected. Angular momentum must be conserved. Therefore, jets along any direction other than polar possess angular momentum, and so can change the angular momentum of the body.

In Section 5.5.1, we discuss the potential role of polar jets in the growth of Earth's core, after Hofmeister and Criss (2015). That slow jets might have produced the giant Martian volcanoes is explored in Chapter 10.

5.4 Torque from external gravitational forces

Angular momentum changes in response to the presence of torque:

$$\vec{\tau} = \vec{r} \times \vec{F} \tag{5.6}$$

In order to create torque, the force vector must have a component perpendicular to the radius defining the angular motion. The result is an angular acceleration of some type, and thus the change in angular momentum is time dependent. Simple situations are described by:

$$\tau = I\frac{d\omega}{dt} \tag{5.7}$$

Notably, for a planet in a Keplerian orbit, changing ω is destabilizing, so the planet spirals in or out, depending on the torque. Evolutionary behavior of spinning entities and orbital systems requires torque, with one possible exception, redistribution of internal mass (Section 5.4.3).

As is well-known, the eight planets orbit almost independently of each other, in nearly Keplerian orbits. Interplanetary interactions are small, but cause tiny perturbations of the orbits. Gauss (1818) viewed the latter situation in terms of interacting rings, in pairs. Adaptations of his incomplete force model by Hill (1901) are only applied to the inner Solar System (Doolittle, 1912), because steady-state was assumed. The four outer planets interact as point masses in a time-dependent manner, as is well-known. The example considered below is the Earth-Moon-Sun system, which is demonstrably unstable.

5.4.1 Point mass interactions in a 3-body system

The non-central gravitational attraction to an oblate body differs from the Newtonian equations for spherical distributions of mass, which are gravitationally equivalent to point masses. However, these perturbing effects are extremely limited due to the near spherical shape of the celestial bodies. For the gas giants, these perturbing effects occur only within a few body radii (Hofmeister et al., 2018). Due to the far greater size of orbital radii, planet-planet or planet-moon interactions are thus described to high accuracy using the gravitational attraction of point masses, in pairs. Orbital torque cannot exist in any interaction between two spherical bodies, but must exist if a third, non-collinear body adds a significant gravitational force.

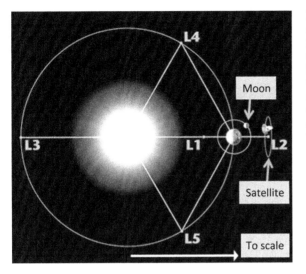

FIGURE 5.6 Schematics of idealized, planar orbits and the 5 Lagrange points (L1, etc.) of the Earth orbiting the Sun. The lunar and satellite orbits are far larger than the actual orbits. At the bottom, the white arrow has a tip which shows the maximum Earth-Moon distance, while the shaft shows the average Sun-Earth distance, to scale. *Modified after a publically available image from NASA (https://solarsystem.nasa.gov/resources/754/ what-is-a-lagrange-point/).*

The Earth-Moon-Sun system is clearly a 3-body system, because the pull of the Sun is 2.2× that of Earth on the Moon, yet the Moon seems to orbit the Earth! In actuality, the Earth and Moon orbit the Sun in a coordinated fashion. Moreover, our Moon is not located in a stable position for a 3-body planar orbit, as shown long ago by Lagrange (Fig. 5.6), so it orbit is inherently unstable.

Since the 3-body problem describes the Earth and Moon orbiting the Sun, angular momentum of the Earth-Moon subsystem cannot possibly be conserved. Thus, Earth's spin is not transferred to the Moon, which is the current model (Williams, 2000). Torque does not exist in the Earth-Moon subsystem since the point mass approximation is valid. Time-dependent torque exists in the 3-body system because the Sun-Moon-Earth plane is not only tilted with respect to the barycenter plane, but this tilt varies, with a rather long repeat pattern of ~19 years. The Sun is pulling the Moon both towards the barycenter plane and away from Earth. Due to the complicated motions, a perturbation calculation will be presented elsewhere. Our interest here is how Earth's spin is internally dissipated.

5.4.2 Three-body sub-systems

The barycenter orbits the Sun in a nearly circular orbit, while the Moon and Earth's geo-center *each* orbit their mutual barycenter in coordinated elongated elliptical paths. Hence, the system described above is actually two subsystems: the Sun-barycenter-Moon, described above, and Sun-barycenter-geocenter, discussed here. The barycenter being internal (Table 5.1) makes this complex orbital problem also a solid-body deformation problem. Earth's spin is involved, since this moves the latitude and longitude of the bary-center on a daily basis. This is augmented by the monthly radial change of the barycenter due to the lunar orbit. Two additional perturbations exist: precession of Earth's spin affects the barycenter internal location, while interplanetary interactions alter the barycenter orbit. The latter is small, and neglected here.

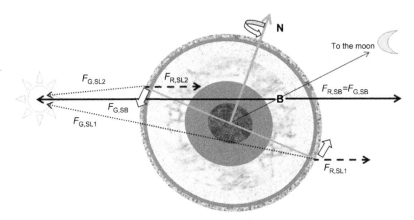

FIGURE 5.7   Schematics of torque on the Earth's lithosphere in the 3-body sub-system (Sun-barycenter-geocenter). Not to scale. Slice down the spin axis, showing the main layers. Solid gray layers are partially molten. Flattening due to spin is highly exaggerated. The barycenter B is fixed in space by the geocenters of the Earth and Moon, but these bodies both move relative to B as if it were a fulcrum. At B, the Sun's pull (denoted by subscript G) is in the same direction and has the same magnitude as the centrifugal force (shown as heavy lines). The other forces (dashed and dotted lines) refer to the lithosphere (granite pattern, subscript L), where the centrifugal forces (denoted by subscript R) are parallel to the barycenter plane but are not co-linear with the direction to the Sun. Because the Earth is a strong material, the imbalanced forces (white arrows) create internal stress. The Moon's effect is limited to tidal forces, whereas the Sun provides both tidal forces and the perturbing torques sketched here.

To understand the essential behavior of this unique, complicated system, we consider a simplified interior structure (Fig. 5.7). Crucially, the lithosphere is decoupled from the interior by melt at ~ 70 km depth (Chapter 7) and by shear in a spheroidal shell accompanying differential spin (Fig. 1.8). Since the lithosphere is furthest from the geocenter, it is subject to the maximum torque (Eq. 5.5) during any force imbalance. The perturbations discussed here are in addition to tidal forces, which have been modeled in terms of an idealized *planar* 3-body system, after Lagrange. Our concern is the out-of-plane motions which create torque. Tilt of the Moon's path around the Earth with respect to the plane of the barycenter orbit creates a non-radial force component and torque. Lunar motions also place the geocenter out of the barycenter plane, providing torque in this subsystem as well (Fig. 5.7).

5.4.2.1 Asymmetry of plate tectonics

Brittle failure of the lithosphere reasonably resembles fractures during uniaxial stress (Fig. 5.4), but is perturbed by the skewed distribution of the deeply rooted continents (Fig 5.8). The many additional fractures in an asymmetric pattern create the plates, which can then move in ways that differ from the westward drift of differential spin. But to do so requires an asymmetric force field. Fig. 5.7 shows that force imbalances caused by Earth's non-central barycenter impose a low symmetry stress field on the brittle lithosphere. Daily and monthly cycles of the barycenter, due respectively to Earth's axial spin and the Moon's orbit around the Earth, cause small periodic changes in the stress of the

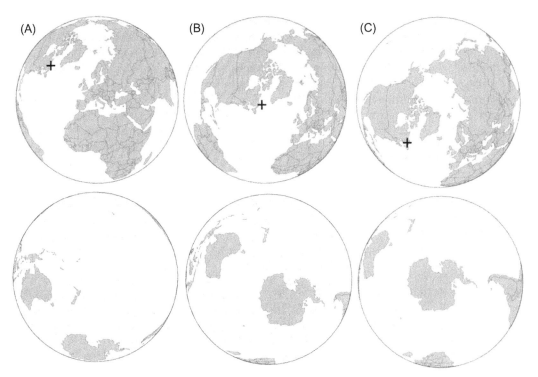

FIGURE 5.8 Comparison of the faces of the globe. The top and the bottom row show the opposite hemi-spheres, where + shows the tectonic North pole. (A) Arrangement with the greatest continental coverage on one side. (B) Arrangement with the tectonic poles of Doglioni and Panza (2015) at the center. (C) Standard views down the spin axis. *All images downloaded from https://www.jasondavies.com/maps/rotate/, with permission of Jason Davies.*

lithosphere. Such periodicity is effective at causing failure: this phenomenon is known as fatigue (Schijve, 2009). Fatigue onsets before static failure criteria are met, suggesting that a tiny perturbation of the Earth's spinning lithosphere by the Sun is sufficient to develop plate motions which are superimposed on E-W drift.

These perturbations are in fact significant, given the velocities involved: First, the distance of the barycenter from the geocenter is controlled by $B = r_m m / (m + M_E)$. Therefore, to the increase in B occurs at ~ 4 mm/y from lunar orbital data. This change is dwarfed by the monthly motion of the barycenter from 4360 to 4960 km radius, which provides a radial velocity of ~ 20 km/d. Both are miniscule compared to the tangential velocity of the barycenter over the day at $\sim \frac{1}{2}$ km/s. Stresses on the lithosphere change in response to these huge excursions of the barycenter inside Earth's mantle.

The combination of these effects (axial stress field, differential rotation, an asymmetric surface, torque in the three-body subsystem, and cyclic motions) provides plate tectonics in a co-operative manner. Both spin and the overly large Moon with its uniquely tilted elliptical orbit around the Earth are essential. The continued warmth of the Earth is another contributing factor, because the Earth retains a weak zone below the thermal chill zone at its surface. The thickness of the chill zone is important, see below. Note that heat produced by Earth's internal radioactivity is inadequate to power mantle convection and plate motions (see chapter introduction). Any heat produced by friction is a byproduct of forces causing motions, not the source of those motions. Chapter 7 discusses associated energies and temperatures.

Importantly, the tectonic and spin equators are not identical (Fig. 5.8B and C: for details see Doglioni and Panza, 2015). The tectonic equator includes highest plateau on the globe (Tibet) and the Andes Mountains. This configuration is top-heavy, with most of the continental mass in the Northern tectonic hemisphere (Fig. 5.8b), and resembles that of a child's toy top (Fig. 5.3B) which is thick around the middle. However, the lopsided mass distribution on a toy top causes interesting motions and precession. Regarding the massive Earth, the distribution of continents may affect precession: clearly, their distribution affects solar torque and plate tectonics, as follows.

5.4.2.2 Proposal for the onset of plate tectonics

The lithosphere, which receives the greatest torque from the Sun (Fig. 5.7), is grossly heterogeneous in density and thickness. The seismic character is not axisymmetric, down to nearly 400 km (Fig. 1.2), and two different types of crust cover the globe asymmetrically (Fig. 5.8A). Only about 20% of the continental landmass occupies the hemisphere of Fig. 5.8A (bottom), whereas the North Pole is well within the opposite hemisphere dominated by continents (Fig. 5.8A, top). The average continental elevation above sea level is ~ 0.6 km, whereas the average elevation of oceanic crust is ~ 3 km below sea level. Isostasy explains a difference in elevation existing between the low density, silica-rich continental crust and the high density, mafic oceanic crust, but, due to the dynamic nature of the lithosphere, this does not hold everywhere (e.g., Watts, 2001). Chapter 7 provides further discussion. Consequently, torque, as shown in Fig. 5.7, differs between the continental and the oceanic hemispheres. Moreover, the torque direction (Eq. 5.6) differs from the polar spin axis.

Not only has the barycenter moved outwards with time as the Moon recedes, but the distribution of continental mass has also changed over Earth's history. Clearly, this differential torque has changed with time. Evidence exists for continental masses clustering together, and then breaking up (Rogers and Santosh, 2004). However, tracing lateral motions becomes increasingly difficult further back in time, due to the rock record becoming increasingly sparse. Pangea certainly existed about 0.3 Ga ago. The earlier existence of Gondwana (~ 0.6 Ga) is also certain. Earlier reconstructions are ambiguous.

Meert (2014) emphasized that the configurations of Pangea, Rodina, and Columbia are very similar which may suggest bias, but could signify minimal lateral motions in the crust. Piper (2013) and Roberts (2013) argue for a different style of tectonics early on, which they term lid tectonics. Meredith et al. (2017) stated that the frequency of subduction ~ 1 Ga ago was only 10% of that in modern times. This is compatible with age distributions of rocks that had experienced subduction (high P but low T metamorphism), see Ernst (1972) or Stern (2005). Hamilton (2019) argues that plate tectonics with subduction started recently, ~ 0.65 Ga ago. Polat et al. (2016) and others argue that such phenomena began much earlier, in the Archean.

TABLE 5.2 How gravitational processes affect certain parts or aspects of the Earth.

Entity	Conditions during stages in Earth history			Change with time
	Early	Middle	Present	
Barycenter location	Deep	Lowest mantle	Mid lower mantle	Torque increases
Barycenter cycle	None	Weak monthly variations	600 km monthly excursions	Fatigue increases
Continental mass	Small, growing	Growing?	⅓ of lithosphere	Torque increases
Continental depth	Shallow, growing	Growing?	>250 km	Drag increases; isostatic imbalance increases
U, Th, K location	Not in the core	Upper mantle? Continents?	Continents	Earth's thermal profile evolves in a complex fashion
Ocean lithosphere	Impact and basal heat	Some basal heat	Little basal heat	Mafic layer thickens
Spin rate	Fast	Intermediate? Slow?	Slow	Exponential decay now; Event driven?
Uniaxial fractures	Simple "X" of shears	Continents corralled	Continents breach shear zones	Pattern increasing disrupted as continents merge
Differential spin	None initially	Onsets?	Weak, but present	Now increasing with time

Note: Details of the various processes are covered in Chapters 6 to 10.

Controversy exists for several reasons, one being the sparsity of old rocks and the exponential decline of sedimentary rock area with age. This decline is especially significant, as this rock type covers 60% of the surface (Blatt and Jones, 1975). This preservation problem is compounded by the ancient sea floor being limited to coastal regions, which causes it to be preferentially eroded (R. Gregory, personal communication). Nonetheless, several ophiolite sections with ages approaching 2 Ga have been documented (Stern, 2005) and older may exist (Kusky et al., 2001). However, formation of oceanic crust, although a prerequisite for plate tectonics, does not require organized subduction as observed today. If the oceanic crust is thin, it is too weak to subduct. The thermal model of Chapter 7 probes the issue of seafloor spreading. Here we focus on the torque required for asymmetric plate tectonics with subduction.

We propose that plate tectonics with subduction is a recent phenomenon, due to the chill zone thickness and moreover because strong lithospheric torque only developed recently. Table 5.2 lists how the Earth has changed with time due to external and internal gravitational forces. These data partially explain the motions of the continents with time, which is a key clue to the past dynamics of the Earth's outer layers. Earth's radial structure demonstrates the strong link between chemical differentiation and gravitational buoyancy. Since the important heat-producing elements are preferentially incorporated in buoyant melts and silica-rich rocks, Earth's thermal evolution must depend on gravity. Table 5.2 lists major changes in the Earth that are driven by gravity, over most of the lifetime of the planet. In recognition of incomplete knowledge and ongoing controversies re Earth's current internal processes, we represent Earth's history in terms of early, middle, and present stages.

It is irrefutable that the continents formed long ago, as evidenced by the Archean granite-greenstone terranes. Some argue for continued growth. Armstrong (1991) has argued that continental volume is steady-state since the crust-mantle recycling rate is declining with time.

Whether growth is negligible or substantial, the Moon was closer in the Archean, so the barycenter and geocenter were closer to each other as well. Torque and fatigue have thus increased with time. In contrast, the conjugate shears would have existed prominently early on, due to higher spin rates, and should have provided upwelling zones. The sinking mafic material has not been preserved in the Archean rock record, consistent with a more brittle, thin chill zone. Dome-and-keel structures unique to Archean terranes emphasize the importance of diapirs (vertical motions) early in Earth's history.

Although many questions exist regarding the details, the time dependent changes in external and internal gravitational forces, along with the time-dependence of heat transfer, point to plate tectonics with subduction being a recent development. Lateral motions in the Archean need not have been organized, because the crust was thin and Solar torque was weak. Lateral motions were likely limited to the continents drifting east-west, as indicated by the reconstructions. Chapter 6 quantifies the Wilson cycle.

5.4.2.3 Implications for the core

If differential rotation exists it is small (Criss, 2019) and is between the core and strong lower mantle. The core is close to the geocenter, so it receives less torque from the Sun with respect to the barycenter than does the lithosphere (Fig. 5.7), and is less affected by the cycles imposed by spin and the lunar orbit. Importantly, the core is strong due to compression. Fatigue is unexpected under these circumstances.

Seismic data suggest a fairly homogeneous lower mantle (Fig. 1.2). Radial symmetry is the premise of PREM (based on low ellipticity). Because the torque on the mantle is less than that on the lithosphere, the mantle would need to be more heterogeneous than the lithosphere for this mechanism to be important.

5.5 Creation of layers in formed planets

Planets have layers, with progressively lower density towards the surface (Fig. 5.2). Several key issues in current accretion models are explained by planetary growth about iron nuclei. After these iron protocores formed, the material added would have been poorly sorted, so the protoplanets would not have had an optimal density distribution. Furthermore, additional material was added in late stage impacts, along with heat. Late additions were unlikely to be of exactly the same material that formed the protoplanets. This section discusses the processes involved in creating layers.

5.5.1 Gravitational segregation in the proto Earth

Gravitational forces can sort phases with different densities, provided that motion is possible. One possibility is large-scale melting, but high temperatures are not expected for the early Earth (Chapter 2). A more likely possibility is sorting as a response to outgassing of deeply buried ices. Such processes would have changed the moment of inertia, which in turn required a change in angular velocity. Gravitational settling is opposed by friction, and so is not conservative, which makes quantification uncertain. If an iron protocore existed, subsequent sorting of phases and angular momentum changes need not be nearly as extensive as popularly envisioned. We discuss sorting above Earth's protocore in Chapter 9.

5.5.2 Late stage impacts

Copious evidence exists for the bombardment of planetary surfaces during the last stage of accretion, but this would have had little effect on the deep interior. The observations have led to the concept that accretion was a non-conservative, fractal process. However, no such process can explain the observed regularities in the Solar System (Chapter 4). Importantly, fractal accretion would have been destructive at virtually all stages, as collisions would have involved roughly equal size bodies. Late stage impacts of small bodies on large planets are substantially different, as the mass of the planet greatly exceeds the mass of the small body. These impacts can cause growth, while disrupting the locus of impact.

After planetary bodies formed and absorbed their sub-nebulae, the accretionary process changed drastically. Late stage impacts may have been frequent. The bodies responsible for these impacts are drawn to the Sun, but when intercepted (Fig. 4.9) add heat-energy and mass to the *surface* of the intercepting planet, since the force is external to the planet. This process is non-conservative, and almost certainly was heterogeneous.

Evidence for late-stage augmentation of the Earth by impacts is summarized by Hamilton (2019): Several beds of impact-melted spherules, altered from glass beads, have been documented. These beds occur in Archean upper-crustal stratigraphic sections in South Africa and Western Australia. Logically, their existence indicates nearby sizeable impacts. However, the oldest, well-documented, little-disrupted, impact-melt-lake complexes occurred in the early Proterozoic (e.g., Sudbury, Ontario, and Vredefort, South Africa). Large Hadean terranes no longer exist, so the geologic evidence of the earliest impacts on Earth has been lost. However, late impacts involving heterogeneous material is indicated by the thorium distribution on the Moon, discussed in Chapter 10.

One possibility is that the upper mantle (above 410 km; 10% of Earth's total mass) is mostly composed of late arriving impacts. If this were the case, late impacts would have provided much heat. Behavior of the inner layers is a different matter (Chapter 8). An alternative picture of chondrite formation and the final accreting stage is presented in Chapter 11.

Late-stage impacts could also have affected planetary spin. This non-conservative process is difficult to constrain. Evolution of spin and heat, which will be further explored in Parts II and III.

5.6 Summary

The Earth responds mechanically to internal and external gravitational forces. Therefore, its thermal and chemical evolution are strongly affected by these forces. Previous depictions of the Earth have inadequately addressed the role of gravity in creating vertical and lateral motions, while greatly overestimating the role of heat. These depictions erroneously attribute large-scale lateral motions to thermal processes. Likewise, the importance of heating in accretion has been grossly overestimated, resulting in consensus models that do not explain various fundamental observations (see Chapter 4). Additional errors have been introduced by overinterpreting geochemical evidence when we have no samples from the deep Earth.

This chapter proposes a new gravitational mechanism for plate tectonics that arises from gravitational forces. Our proposal is based on observational data and physical principles, with much of the former being presented in Chapter 1. The uniqueness of the Earth-Moon-Sun system is taken in to account, which explains why plate tectonics with subduction is only observed on Earth, and points to this phenomenon being a relatively new development. Divergent, evidence based interpretations regarding this observation are presented by Hamilton (2019) and Polat et al. (2016); Szilas (2018) reviews the geochemical data and argues that additional information is needed for resolution for the onset of plate tectonics. Given the lack of lower mantle samples, we are convinced that approaches other than geochemistry are essential for a better understanding of Earth's interior.

Our new mechanism for plate tectonics lacks the gross inconsistencies associated with the mantle convection hypothesis, and many specific predictions are testable. But, improved thermal profiles and evolutionary models are needed. Because heat transfer in the Earth is caused by radiative diffusion, very few parameters are needed, even if the internal chemical composition is unconstrained. Unsurprisingly, Earth's chemical and thermal evolution are intertwined, but our data only constrain the outer layers. In addition, Earth's chemical evolution is strongly affected by its internal gravitational gradient. We are greatly concerned that gravitational buoyancy stemming from volatiles incorporated during cold accretion has been misinterpreted as thermal buoyancy engendered by very high temperatures.

Part II addresses this knotty puzzle to the extent that is currently possible, using thermal models based on new data and microscopic models on heat transport of metals and insulators. These microscopic models are summarized in Chapter 3, and detailed in Hofmeister (2018).

References

Aitta, A., 2006. Iron melting curve with a tricritical point. J. Stat. Mech. Theory Exp. 12. Available from: https://doi.org/10.1088/1742-5468/2006/12/P12015Online at: stacks.iop.org/JSTAT/2006/P12015.

Anderson, D.L., 2007. New Theory of the Earth, second ed. Cambridge University Press, Cambridge.

Armstrong, R.L., 1991. The persistent myth of continental growth. Aust. J. Earth Sci. 38, 613–630.

Bird, P., 2003. An updated digital model of plate boundaries. Geochem. Geophys. Geosys. 4 (3). Available from: https://doi.org/10.1029/2001GC000252No. 1027.

Blatt, H., Jones, R.L., 1975. Proportions of exposed igneous, metamorphic, and sedimentary rocks. Geo. Soc. Am. Bull. 86, 1085–1088.

Choi, S.H., Musaka, S.B., Ravizza, G., Fleming, T.H., Marsh, B.D., Bédard, J.H.J., 2019. Fossil subduction zone origin for magmas in the Ferrar Large Igneous Province, Antarctica: evidence from PGE and Os isotope systematics in the Basement Sill of the McMurdo Dry Valleys. Earth Planet. Sci. Lett. 506, 507–519.

Chudinovskikh, L., Boehler, R., 2007. Eutectic melting in the system Fe–S to 44 GPa. Earth Planet. Sci. Lett. 257, 97–103.

Courtillot, V., Davaille, A., Besse, J., Stock, J., 2003. Three distinct types of hotspots in the Earth's mantle. Earth Planet. Sci. Lett. 205, 295–308.

Criss, R.E., 2019. Analytics of planetary rotation: improved physics with implications for the shape and super-rotation of Earth's core. Earth Sci. Rev. 192, 471–479. Available from: https://doi.org/10.1016/j.earscirev.2019.01.024.

Criss, R.E., Hofmeister, A.M., 2016. Conductive cooling of spherical bodies with emphasis on the Earth. Terra Nova 28, 101–109.

Criss, R.E., Hofmeister, A.M., 2018. Galactic density and evolution based on the Virial Theorem, energy minimization, and conservation of angular momentum. Galaxies 6, 115–135. Available from: https://doi.org/10.3390/galaxies6040115.

Dima, I., Desflots, M., 2010. Wind Profiles in Parametric Hurricane Models; Report to Air Worldwide. Available online: <www.air-worldwide.com/.../AIRCurrents-Wind-Profiles-in-Parametric-Hurricane-Models> (accessed 04.07.17.).

Doglioni, C., Panza, G., 2015. Polarized plate tectonics. Adv. Geophys. 56, 1–167.

Domínguez-Rodríguez, A., Gómez-García, D., Zapata-Solvas, E., Shen, J.Z., Chaim, R., 2007. Making ceramics ductile at low homologous temperatures. Scripta Mater. 56, 89–91.

Doolittle, E., 1912. The secular variations of the elements of the orbits of the four inner planets computed for the epoch 1850.0 G.M.T. Trans. Am. Phil. Soc. 22, 37–189.

Epstein, C.R., Pinsonneault, M.H., 2014. How good a clock is rotation? The stellar rotation–mass–age relationship for old field stars. Astrophys. J. Available from: https://doi.org/10.1088/0004-637X/780/2/159 [Online]. 780, No. 159 Available: http://iopscience.iop.org/0004-637X or.

Ernst, W.G., 1972. Occurrence and mineralogic evidence of blueschist belts with time. Am. J. Sci. 272, 657–668.

Faure, S., 2006. World Kimberlites CONSOREM Database Version 2006–2: Quebec, Canada, Consortium de Recherche en Exploration Minérale CONSOREM, Université du Québec à Montréal, https://consorem.uqac.ca/kimberlite/world kimberlites and lamproites consorem database_v2010.xls.

Fossat, E., et al., 2017. Asymptotic g modes: evidence for a rapid rotation of the solar core. Astron. Astrophys. 604, #A40. Available from: https://doi.org/10.1051/0004-6361/201730460.

Foulger, G.R., 2010. Plates vs Plumes: A Geological Controversy. Wiley-Blackwell, Chichester, UK.

Frost, H.J., Ashby, M.F., 1982. Deformation Mechanism Maps. Pergamon Press, Oxford.

Funakoshi, K.I., 2010. In situ viscosity measurements of liquid Fe–S alloys at high pressures. High Pres. High Temp. 30, 60–64.

Gauss, C.F., 1818. Determinatio Attractionis quam in punctum quodvis positionis datae ejus massa per totam orbitam ratione temporis quo singulae partes describuntur esset dispertita. Werke, vol. III, in Werke, 3, pp. 331–357. Dieterich, Gottingen. http://resolver.sub.uni-goettingen.de/purl?PPN235999628 (in Latin).

Hamilton, W.B., 2015. Terrestrial planets fractionated synchronously with accretion, but Earth progress through subsequent internally dynamic stages whereas Venus and Mars have been inert for more than 4 billion years. Geol. Soc. Am. Spec. Pap. 514, 123–156.

Hamilton, W.B., 2019. Toward a myth-free geodynamic history of earth and its neighbors. Earth Sci. Rev. (in press). Available from: https://doi.org/10.1016/j.earscirev.2019.102905.

Healy, D., Blenkinsop, T.G., Timms, N.E., Meredith, P.G., Mitchell, T.M., Cooke, M.L., 2015. Polymodal faulting: time for a new angle on shear failure. J. Struct. Geol. 80, 57–71. Available from: https://doi.org/10.1016/j.jsg.2015.08.013 (Under a Creative Commons license).

Hill, G.W., 1901. The secular perturbations of the planets. Am. J. Math. 23, 317–336.

Hofmeister, A.M., 2019. Measurements, Mechanisms, and Models of Heat Transport. Elsevier, Amsterdam.

Hofmeister, A.M., Criss, R.E., 2005. Mantle convection and heat flow in the triaxial Earth. In: Foulger, G.R., Natland, J.H., Presnall, D.C., Anderson, D.L. (Eds.), Melting Anomalies: Their Nature and Origin. Geological Society of America, Boulder CO, pp. 289–302.

Hofmeister, A.M., Criss, R.E., 2012. A thermodynamic model for formation of the solar system via 3-dimensional collapse of the Dusty Nebula. Planet. Space Sci. 62, 111–131.

Hofmeister, A.M., Criss, R.E., 2013. How irreversible heat transport processes drive Earth's interdependent thermal, structural, and chemical evolution. Gondwana Res. 24, 490–500.

Hofmeister, A.M., Criss, R.E., 2015. Evaluation of the heat, entropy, and rotational changes produced by gravitational segregation during core formation. J. Earth Sci. 26, 124–133.

Hofmeister, A.M., Criss, R.E., 2017. Implications of geometry and the theorem of gauss on Newtonian gravitational systems and a caveat regarding Poisson's equation. Galaxies 5, 89–100. Available from: http://www.mdpi.com/2075-4434/5/4/89.

Hofmeister, A.M., Criss, E.M., 2018. How properties that distinguish solids from fluids and constraints of spherical geometry suppress lower mantle convection. J. Earth Sci. 29, 1–20. Available from: https://doi.org/10.1007/s12583-017-0819-4.

Hofmeister, A.M., Sehlke, A., Avard, G., Bollasina, A.J., Robert, G., Whittington, A.G., 2016. Transport properties of glassy and molten lavas as a function of temperature and composition. J. Volcan. Geotherm. Res. 327, 380–388.

Hofmeister, A.M., Criss, R.E., Criss, E.M., 2018. Verified solutions for the gravitational attraction to an oblate spheroid: implications for planet mass and satellite orbits. Planet. Space Sci. 152, 68–81. Available from: https://doi.org/10.1016/j.pss.2018.01.005.

Kusky, T.M., Li, J.-H., Tucker, R.D., 2001. The Archean Dongwanzi ophiolite complex, North China craton: 2.505-billion-year-old oceanic crust and mantle. Science 292, 1142–1145.

Lambeck, K., Purcell, A., Zhao, S., 2017. The North American Late Wisconsin ice sheet and mantle viscosity from glacial rebound analyses. Quat. Sci. Rev. 158, 172–210.

Liu, J., Xia, Q.-K., Kuritani, T., Hanski, E., Yu, H.-R., 2017. Mantle hydration and the role of water in the generation of large igneous provinces. Nat. Commun. 8, 1824. Available from: https://doi.org/10.1038/s41467-017-01940-3.

Ludovico-Marques, M., Chastre, C., Vasconcelos, G., 2012. Modelling the compressive mechanical behaviour of granite and sandstone historical building stones. Construct. Building Mater. 28, 372–381.

McGetchin, T.R., Nikhanj, Y.S., Chodos, A.A., 1973. Carbonatite-kimberlite relations in the Cane Valley diatreme, San Juan County, Utah. J. Geophys. Res. 78, 1854–1869.

Meert, J.G., 2014. Strange attractors, spiritual interlopers and lonely wanderers: the search for pre-Pangean supercontinents. Geosci. Front. 5, 155–168.

Meredith, A.S., Collins, A.S., Williams, S.E., Pisarevsky, S., Foden, J.D., Archibald, D.B., et al., 2017. A full-plate global reconstruction of the Neoproterozoic. Gondwana Res. 50, 84–134.

Mitchell, R.H., 1986. Kimberlites. Plenum, New York.

Mittlefehldt, D.W., McCoy, T.J., Goodrich, C.A., Kracher, A., 1998. Non-chondritic meteorites from asteroidal bodies. Rev. Miner. 36, 1529–6466.

Moghadam, R.H., Trepmann, C.A., Stöckhert, B., et al., 2010. Rheology of synthetic omphacite aggregates at high pressure and high temperature. J. Petrol. 51, 921–945.

Mohazzabi, P., Skalbeck, J.D., 2015. Superrotation of Earth's inner core, extraterrestrial impacts, and the effective viscosity of outer core. Int. J. Geophys. 2015, 763716.

Nishihara, Y., Tinker, D., Kawazoe, T., et al., 2008. Plastic deformation of wadsleyite and olivine at high-pressure and high-temperature using a rotational Drickamer apparatus (RDA). Phys. Earth Planet. Inter. 170, 156–169.

Piper, J.D.A., 2013. A planetary perspective on Earth evolution: lid tectonics before plate tectonics. Tectonophysics 589, 44–56.

Polat, A., Kokfelt, T., Burke, K.C., Kusky, T., Bradley, D., Dziggel, A., et al., 2016. Lithological, structural, and geochemical characteristics of the Mesoarchean Târtoq greenstone belt, South-West Greenland, and the Chugach-Prince William accretionary complex, southern Alaska: evidence for uniformitarian plate-tectonic processes. Can. J. Earth Sci. 53 (11), 1336–1371.

Riaz, B., Briceño, Heathcote, S., Whelan, E.T., 2017. First large-scale Herbig–Haro jet driven by a proto-brown dwarf. Astrophys. J. 844. paper 47 (19pp). Available from: https://doi.org/10.3847/1538-4357/aa70e8.

Riguzzi, F., Panza, G., Varga, P., Doglioni, C., 2010. Can Earth's rotation and tidal despinning drive plate tectonics? Tectonophysics 484, 60–73.

Roberts, N.M.W., 2013. The boring billion? lid tectonics, continental growth and environmental change associated with the Columbia supercontinent. Geosci. Front. 6, 681–691. Available from: https://doi.org/10.1016/j.gsf.2013.05.004.

Rogers, J.J.W., Santosh, M., 2004. Continents and Supercontinents. Oxford University Press, Oxford.

Schijve, J., 2009. Fatigue of Structures and Materials, second ed. with Cd-Rom Springer, Berlin, ISBN 978-1-4020-6807-2.

Sehlke, A., Whittington, A., Robert, B., Harris, A., Gurioli, L., Médard, E., 2014. Pahoehoe to 'a'a transition of Hawaiian lavas: an experimental study. Bull. Volcanol. 76 (11), 876–896. Available from: https://doi.org/10.1007/s00445-014-0876-9.

Sofue, Y., Tutui, Y., Honma, M., Tomita, A., Takamiya, T., Koda, J., et al., 1999. Central rotation curves of spiral galaxies. Astrophys. J. 523, 136–146.

Sofue, Y., Koda, J., Nakanishi, H., Onodera, S., 2003. The Virgo high-resolution CO survey, II. Rotation curves and dynamical mass distributions. Publ. Astron. Soc. Jpn. 55, 59–74.

Stacey, F.D., Stacey, C.H.B., 1998. Gravitational energy of core evolution: implications for thermal history and geodynamo power. Phys. Earth Planet. Inter. 110, 83–93.

Stehli, F.G., Douglas, R.G., Newell, N.D., 1969. Generation and maintenance of gradients in taxonomic diversity. Science 164, 947–949. Available from: https://doi.org/10.1126/science.164.3882.947.

Stein, C.A., Stein, S.A., 1992. A model for the global variation in oceanic depth and heat flow with lithospheric age. Nature 359, 123–128.

Stern, R.J., 2005. Evidence from ophiolites, blueschists, and ultrahigh-pressure metamorphic terranes that the modern episode of subduction tectonics began in Neoproterozoic time. Geology 33, 557–560.

Stern, R.J., Leybourne, M.I., Tsujimori, T., 2016. Kimberlites and the start of plate tectonics. Geology 44, 799–802. Available from: https://doi.org/10.1130/G38024.1.

Szilas, K., 2018. A geochemical overview of mid-Archaean metavolcanic rocks from Southwest Greenland. Geosciences 8. Available from: https://doi.org/10.3390/geosciences8070266No. 0266.

Tappe, S., Smart, K., Torsvik, T., Massuyeu, M., de Wit, M., 2018. Geodynamics of kimberlites on a cooling earth: clues to plate tectonic evolution and deep volatile cycles. Earth Planet. Sci. Lett. 484, 1–14.

Todhunter, I., 1873. A History of the mathematical theories of attraction and thefigure of the Earth. Volume I. MacMillan and Co., London, 474 p.

Watts, A.B., 2001. Isostasy and Flexure of the Lithosphere. Cambridge University Press, ISBN 0-521-00600-7.

Weinelt, M., 2003. Create-a-map Web site: <http://www.aquarius.geomar.de/omc> (accessed 12.17.).

Wessel, P., Smith, W.H.F., 1995. The Generic Mapping Tools (GMT), version 3.0: Technical reference and cookbook (SOEST/NOAA), <http://gmt.soest.hawaii.edu> (accessed 12.17.).

Williams, G.E., 2000. Geological constraints on the Precambrian history of Earth's rotation and the Moon's orbit. Rev. Geophys. 38, 37–59.

Willoughby, H.E., Darling, R.W.R., Rahn, M.E., 2006. Parametric representation of the primary hurricane vortex. Part II: a new family of sectionally continuous profiles. Mon. Weath. Rev. 134, 1102–1120.

Websites

NASA/IPAC Extragalactic Database (NED).
https://ned.ipac.caltech.edu/ (accessed 26.06.17.).
NOAA (National Ocean and Atmostpheric Administration). Has time lasped images.
https://www.weather.gov/ilm/HurricaneFran (accessed 23.02.19.).
Maps by Jason Davies.
https://www.jasondavies.com/maps/rotate/.

The thermal state and evolution of Earth

Thermal models of the continental lithosphere

Robert E. Criss

....the continental volume steady-state model with a declining crust-mantle recycling rate is a viable replacement for the growth myth.... **R.L. Armstrong (1991).**

Heat Transport and Energetics of the Earth and Rocky Planets
DOI: https://doi.org/10.1016/B978-0-12-818430-1.00006-9

This chapter derives new analytical equations for thermal profiles and applies them to the continents using accurate thermal conductivity data and available geochemical constraints on radionuclide distributions. The results resemble measurements of continental heat flow, and suggest new explanations for several characteristics and long-term processes of continental interiors. These analytical solutions disclose functional interrelationships among the variables and material properties that cannot be easily deduced from iterative modeling. Temperature is provided in degrees centigrade for convenient reference to surface conditions and for ease of comparison with previous work.

Section 6.1 summarizes available data on the thermal gradients and heat flux of the continents and on the heat generation and thermal conductivity of representative rocks. Section 6.2 provides new, steady state solutions to Fourier's equations that accommodate variations in both conductivity and internal heat generation. Section 6.3 uses these equations to predict the surface and basal fluxes. Our method differs from the previous approaches that take surface flux as a given, and importantly demonstrates the functional interrelationships between governing parameters. Section 6.4 provides explanations for the magmatism of continental interiors and for the Wilson cycle that are grounded in thermal physics, and Section 6.5 summarizes the results of this chapter.

6.1 Review of available continental data

6.1.1 Observed temperature gradients and heat flux

More than 14,000 measurements of the near-surface geothermal gradient and heat flow have been made in continental areas, and these have been summarized in many reports and maps (e.g., Blackwell and Richards, 2004). Geothermal gradients in continental areas are commonly about $25\,°C\,km^{-1}$, but are lower for Archean cratons and much higher in areas of Cenozoic volcanism and tectonism. Histograms of continental heat flux are skewed to high values, but the mean and median of the measurements are similar at about 65 and $60\,mW\,m^{-2}$, respectively (see Fig. 1.7). The most recent average of $58\,mW\,m^{-2}$ (Vieira and Hamza, 2018) considered rock-type to address areas that are inadequately sampled. This continental flux would correspond to a power of 29 TW, if the continents covered Earth's entire surface.

Fluid flow can alter thermal gradients and fluxes. Water flowing through regional aquifers can reduce gradients and flux in zones of recharge, and increase their magnitudes in discharge areas, because heat is extracted from the rocks encountered in route. In some geothermal areas, boiling fluids emerge at Earth's surface, so the thermal gradients and fluxes of discharge areas can be very high. These effects are well known, but it is important to note that flows of water cannot alter the overall mean value of the regional heat flux, because heat robbed from one area simply emerges at another (Section 3.3.4.1; Hofmeister and Criss, 2005). Other disturbances of thermal gradients and fluxes are caused by near-surface annual temperature variations, or by the effects of mine ventilation or the circulation of drilling fluids in boreholes, etc. Many papers offer means to correct measurements for such transient or localized effects.

6.1.2 Thermal conductivity

At any position in a conductive system, the thermal conductivity of surrounding rock provides a direct link between the observed flux and the observed temperature gradient. Knowledge of thermal conductivity and its variations are therefore essential to understanding the physical process of heat flow.

Measurements are used here to constrain the thermal conductivity (K) of the continental crust, rather than theory, because the relatively low melting temperatures of felsic rocks ($\sim <1200\,°C$) allow relevant temperatures to be explored in the laboratory. Pressure is sufficiently low that its weak effect ($<4\%$/GPa: Chapter 3) need not be considered, contrary to the large effect described by Furlong and Chapman (2013). As discussed in Section 3.1, the mechanism for heat conduction is diffusion of radiation, where the frequency of light participating in the process depends on the temperature range. At the low temperatures of the continental crust, infrared light is the main carrier, providing a strong decrease in K above ambient temperature that is observed in accurate measurements of crystalline rocks.

Conventional measurements of K depend on experimental details such as sample thickness, number of thermal contacts, and the color and grain-size of the sample. Conventional data commonly provide an incorrect continual decrease in K with T, due to extrapolating data collected at low temperature to high temperature. Also, pressure derivatives derived from conventional contact measurements are often unrealistically high due to compression of pore space. Laser-flash measurements of thermal diffusivity lack the systematic experimental errors of contact losses and spurious ballistic radiative gains known to compromise direct measurement of thermal conductivity. For details, see Hofmeister et al. (2007), Hofmeister (2019), and references therein.

By definition:

$$K = \rho C_P D \tag{6.1}$$

where D = thermal diffusivity, C_P = heat capacity, and ρ = density. Density depends only weakly on temperature and pressure, so readily available ambient values suffice for the lithosphere. Heat capacity is known for common minerals (Robie and Hemingway, 1995) and can be estimated for rocks from oxides or mineral sums, if data are not available (Robertson and Hemingway, 1995). Thus, K can be determined from accurate data on thermal diffusivity. Results on ~ 200 minerals and rocks from laser flash analysis are summarized by Hofmeister (2019), most of which are posted on the website: (http://epsc.wustl.edu/~hofmeist/thermal_data/).

Granodiorites have long been considered to represent the upper continental crust, which typically extends to ~ 35 km (Mooney et al., 1998). Thermal diffusivity of 6 different granodiorites all depend similarly on temperature (Fig. 6.1). Results are limited to $\sim 1050\,°C$, above which point granodiorites increasingly outgas while partially melting. The $D(T)$ curves are shifted from one another due to slight variations in mineral proportions, particularly of quartz, because this common crustal mineral has high D compared to the feldspars, which dominate granodiorite mineralogy and have low D near 1 mm^2 s^{-1} at 298 K (Branlund and Hofmeister, 2007, 2012; Pertermann et al., 2008). Small amounts of quartz are evident in $D(T)$ curves, which take on a λ-shape at this transition. Granodiorite from Bear Butte has $<5\%$ quartz, given the weakness of the transition, and also has the

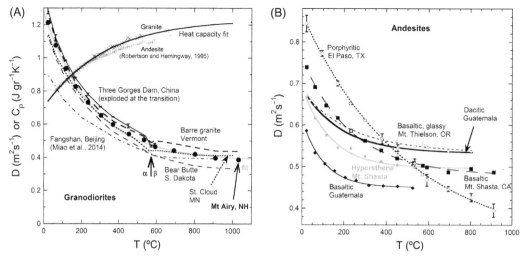

FIGURE 6.1 Thermal diffusivity as a function of temperature. Laser flash data were acquired as described by Merriman et al. (2013). Partial melting occurred at higher T. Most samples have been studied by various other techniques, with reported mineralogy and chemical compositions as designated. (A) Granodiorites. Barre granite is on the edge of this compositional field. Uncertainties are shown for the Three Gorges sample, which outgassed and exploded at the displacive transition of quartz. Observation of this well-known transition near 573 K shows that our temperature measurement is well calibrated. This Archean sample had larger grains than the others. Gray curve = the fit to Eq. 6.2, which smooths D over the α-β quartz transition. Greater proportions of quartz and alkali feldspar increase D, whereas more plagioclase and mafic minerals lower D. Granodiorite from Miao et al. (2014) which has low quartz and high porosity has similar, but lower D. Heat capacity of Westerly granite and Lake County Andesine from Robertson and Hemingway (1995) is plotted on the same scale as D, but the units are $J\ gr^{-1}K^{-1}$. (B) Andesites. Samples from Mt. Shasta and Mt. Thielson are historic USGS samples (Diller, 1898; Clarke, 1910). Samples from the Andes were remelted and studied by Hofmeister et al. (2016). The porphyritic andesite has a very fine grained plagioclase groundmass (Hoffer, 1970). Data acquisition was terminated upon melting.

lowest thermal diffusivity of the samples studied in our laboratory. Granodiorite from Fanghan, China has lower D because this has low quartz (10%), high feldspar (68%) and non-negligible porosity of 2.6% (Miao et al., 2014). The trends of low porosity granodiorites are very close to those of three Archean granulites measured by Merriman et al. (2013), which were composed mostly of plagioclase plus lesser mafic minerals.

The D data in Fig. 6.1 were fit to:

$$D(T) = b_1 + b_2 e^{-b_3 T} + b_4 T \tag{6.2}$$

where the b_i's are constants. The b_4 term is not needed for the low-temperature granodiorite data. This form reproduces other measurements of D, as well as data on $C_P(T)$. Eq. (6.2) is consistent with the model of heat transfer by radiative diffusion presented by Hofmeister (2019) and summarized in Chapter 3. The $b_4 T$ term is equivalent to the HT term from experimentally determined power-law fits (Hofmeister et al., 2014). This term describes diffusion of the overtones, whereas the other terms describe diffusion of infrared fundamentals. Because framework silicates strongly absorb light at very high frequency ($\sim 1200\ cm^{-1}$), and because temperature is proportional to frequency according to Wien's displacement law, the infrared fundamentals control heat transfer at temperatures of the continental crust, and the b_4T term is not germane.

Pores and their fillings also have an effect, which slightly reduces D from mineral sums at ambient pressure. Calculations of D from mineral data are realistic for the lower crust, where porosity is reduced or eliminated by compression and elevated temperatures (Branlund and Hofmeister, 2008).

The thermal diffusivity of andesites is reduced by porosity (Fig. 6.1B). Glassy andesites have very flat trends of D with T, whereas samples with high proportions of crystals have higher D near room temperature and steeper decreases in D with increasing T: but all andesites have rather constant values. More silicic extrusive rocks have higher D than mafic extrusives, if the crystallinity and porosity are similar (cf., the Guatemalan samples in Fig. 6.1B). Campus Andesite from Texas with a groundmass of plagioclase and large plagioclase crystals (Hoffer, 1970), provides D more like that of the intrusive rocks in Fig. 6.1A.

6.1.2.1 *Values for K used in thermal models*

The thermal conductivity $K(T)$ of the upper crust is estimated by combining D for the Mt. Airy granodiorite (Fig. 6.1) with data on ρ and $C_P(T)$, as shown in Fig. 6.2. For the lower crust and sub-continental mantle, thermal conductivity is estimated by combining available laboratory data for an assemblage of proxy minerals, namely 20% labradorite $+ 20\%$ bytownite $+ 25\%$ augite $+ 20\%$ low sanidine $+ 15\%$ quartz. Branlund and Hofmeister (2012) provide details on LFA determinations of plagioclase $D(T)$ and calculations of K using Eq. (6.1). The results are shown in Fig. 6.2, along with granodiorite thermal conductivity, which represents the upper continental crust, and is fit by:

$$K = 1.113 + 1.362e^{-0.002277T} \tag{6.3a}$$

The sub-continental mantle is fit by:

$$K = 2.157 + 1.606e^{-0.00444T} \tag{6.3b}$$

where K is in $Wm^{-1}K^{-1}$ and temperature T is in centigrade. Clearly, K is about $2.5\ Wm^{-1}K^{-1}$ under ambient surface conditions, and about $2.2\ Wm^{-1}K^{-1}$ near the melting point of upper mantle rocks. However, Eq. (6.3a) shows that the conductivity of granodiorite in the lower continental crust would be somewhat lower than that of the subjacent, sub-continental mantle (Fig. 6.2). Because K is expected to range only from 2.5 to $1.5\ W\,m^{-1}K^{-1}$ along a 200 km deep profile beneath the continents, a value of $2\ W\,m^{-1}K^{-1}$ was adopted for the initial calculations of the geotherm (below).

Andesites have low and nearly constant K because their diffusivity varies little. Fig. 6.2 shows the results for basaltic andesite from Mt. Shasta, because this had few pores and some mm size crystals, but was not porphyritic. Its thermal diffusivity lies roughly in the middle of that of the 6 other andesites studied. Its reported composition (Clarke, 1910) is similar to granodiorite, and also is reasonably close to the andesite for which heat capacity is available (Fig. 6.1A).

Constant K is appropriate given the heterogeneity in the continental crust, lack of constraints on the mineralogy in the sub-continental lithosphere, and that the following models depict *average* crust. Using constant K simplified the development of analytical

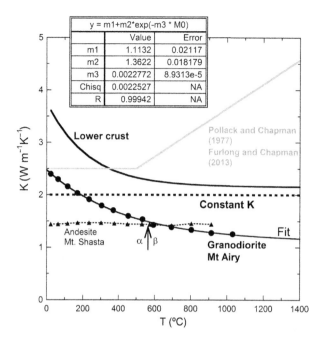

FIGURE 6.2 Thermal conductivity as a function of temperature. Black curves show experimentally determined values from laser-flash analysis, which are presented in Eq. (6.3). Constant K is a compromise value which approximates known heterogeneity of the crust and limited constraints. Solid gray curve shows the unrealistic K (T) values used by Pollack and Chapman (1977), whose thermal models are replicated by Furlong and Chapman (2013). Basaltic andesite from Mt. Shasta is shown because this lies in the middle of the range of the lavas studied here.

equations, and furthers understanding of the key processes. Furthermore, our calculations below show that the narrow range of K for the continental crust and lithosphere ($\sim$1 to 3 Wm^{-1} K^{-1}) only weakly affects the thermal models.

6.1.2.2 Values for thermal conductivity used in previous thermal models

Furlong and Chapman (2013) discuss why thermal conductivity is important, yet their review omits the corpus of laboratory studies completed over the last two decades, and instead refers to old data and formulae for K that those studies have overturned (e.g., Hofmeister et al., 2007). Oddly, they do not even use their formulae. Instead, they present old calculations made by Pollack and Chapman (1977) in their Figures 4a and 5, who erroneously assumed that thermal conductivity strongly increases with temperature above 500 °C (shown in gray in Fig. 6.2). Thermal curves in figures 4b of Furlong and Chapman (2013) differ slightly from their figures 4a and 5: this calculation is described as being based on $K(T)$ which also unrealistically increases with T at modest T but with a slightly different numerical value than that of Pollack and Chapman (1977). Below we show that these old profiles differ from those based on realistic K data by as much as several hundred degrees.

Recent use of modern measurements of heat transport properties in thermal models has shown that much behavior attributed to unusual circumstances such as huge radioactive concentrations, actually results from the behavior of K with T for crustal rocks (Whittington et al., 2009). Section 6.3 pursues a different modeling approach.

6.1.3 Internal heat generation

6.1.3.1 *Dependence of internal heating on K, U and Th concentrations*

The thermal evolution of continental rocks depends significantly on the concentration and distribution of the important heat-generating elements, specifically K, U, and Th. The rate of heat production is readily calculated from the known decay energies and half-lives of the radionuclides ^{235}U, ^{238}U, ^{232}Th and ^{40}K. However, a correction for neutrino energies is significant for ^{40}K (e.g., Van Schmus, 1995), but was omitted from earlier literature. The following equation provides the present-day, volumetric heat generation for a rock with a density of 2700 kg m^{-3}:

$$\Psi = 0.095\,(\%K) + 0.266\,(ppmU) + 0.071\,(ppmTh) \tag{6.4}$$

where Ψ is heat generation in $\mu W\ m^{-3}$, %K is the potassium concentration in percent, and U_{ppm} and Th_{ppm} are the uranium and thorium concentrations of the rock in ppm. All concentrations are for the three bulk elements as reported by conventional analysis, and specifically are not the concentrations of oxides, nor the concentrations of particular isotopes.

If the rock density differs from 2700 kg m^{-3}, the value for Ψ returned by Eq. (6.4) should be multiplied by that actual density divided by 2700. If values are needed for a rock in the geologic past, Eq. (2.25) provides the heat generation per kilogram, which is easily converted to the volumetric heat generation by multiplying by the actual rock density.

Eq. (6.4) allows heat generation to be readily computed for any rock with known concentrations of the indicated elements. For example, the average, upper continental crust is estimated to have about 2.8%K, 2.5 ppm U and 10 ppm Th, so Ψ is about 1.6 $\mu W\ m^{-3}$.

Typical rocks have a Th/U concentration ratio of about 4. Given this, good estimates of total heat production for any particular rock with known K and U concentrations can be made from a graph of K vs. U concentrations (Fig. 6.3).

6.1.3.2 *Vertical variations*

Roy et al. (1968) showed that the heat flow from several geologic provinces is linearly related to the heat generation of rocks exposed at the surface (Fig. 6.4). These strong trends project to a flux at zero heat generation that they and subsequent researchers argue represents a regionally uniform contribution from the subjacent mantle and lower crust. In many areas this intercept represents roughly half of the total heat flow. Roy et al. (1968) show that these linear trends are compatible with many geologic models, including constant heat production above a depth that is uniform from province to province. Lachenbruch (1970) showed that an exponential decline of heat production with depth is compatible with such linear trends along with a variable crustal depth. For an exponential model, available data suggest that the scale length, representing the vertical distance over which the heat generation decreases by a factor of "e," is typically about 10 km, as suggested by the slope of the trend line (Fig. 6.4). The analysis of Section 6.3 provides alternative interpretations for the slopes and *y*-intercepts of these observed trends.

Data from the 12.3 km Kola Superdeep borehole provide an enlightening example of geologic variability that differs from the standard exponential model (Fig. 6.5). Chemical analyses show that the heat productions of the rocks from the top half of the borehole are

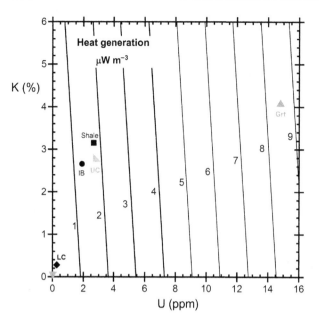

FIGURE 6.3 Graph showing how contours of heat generation, in $\mu W\ m^{-3}$, depend on K and U concentrations, assuming that the Th/U concentration ratio is 4. Plotted points depict typical rocks and estimates, from various sources including Lodders and Fegley (1998). Grt (granite), IB (Idaho batholith), Shale (North American shale composite); ultramafic rock (gray triangle) plots near the origin. UC and LC are estimates for the average composition of the upper and lower continental crust, respectively.

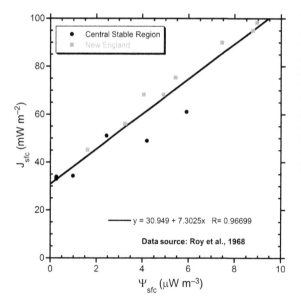

FIGURE 6.4 Relationship between heat flow (J) and the heat generation in surface rocks from the Central USA and the New England area, updated after Roy et al. (1968). Heat generation was calculated from the reported K, U and Th concentrations using Eq. (6.4); in cases where Th concentrations were not analyzed, they were estimated as $4\times$ the U concentration. The regression line represents all data and has a slope of 7.3 km; the y-intercept of 31 $mW\ m^{-2}$ has been interpreted as the "basal flux" but other interpretations are possible.

significantly *lower* than those from the bottom half, opposite to geochemical expectations (Kremenetsky and Ovchinnikov, 1986). This behavior occurs because at this particular locality, the Archean granite-gneiss basement is overlain by a ~6 km-thick section of Proterozoic metasedimentary and metavolcanic rocks that are less silicic and have

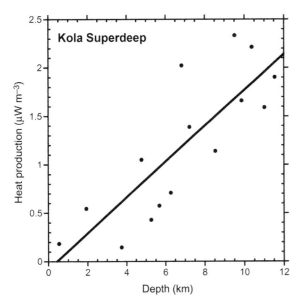

FIGURE 6.5 Heat generation of rocks as a function of depth in the 12 km-deep Kola Superdeep drill hole. Heat generation increases with depth at this site, contrary to conventional expectations. Points were calculated using Eq. (6.4) and the summary data in Table 5 of Kremenetsky and Ovchinnikov (1986).

generally lower U and Th contents. However, their regional geologic and geochemical studies suggest that the rock expected below 12.3 km would have lower heat generation than most rocks along the drill hole, in better agreement with expectations. Various models are examined below to address different radionuclide distributions.

6.1.3.3 Crustal inventory

Geochemists have argued that the amount of uranium and thorium in Earth's crust exceeds the amount that a chondritic Earth can provide (e.g., Brown and Mussett, 1984), suggesting complete sequestering of these elements in the crust. Moreover, at the present time, heat generated in the Earth by U and Th decay is much more important than that contributed by K. Combining the above argument with this constraint, the average heat production in the continental crust (Ψ_{avg}) can be readily calculated from:

$$\frac{M_{cc}\Psi_{avg}}{\rho_{cc}} = \frac{M_E\Psi_E}{\rho_E} \tag{6.5}$$

where M_{cc} and ρ_{cc} are the mass and density of the continental crust, M_E and ρ_E are the mass and density of the Earth, and Ψ_E is the heat production of the bulk Earth, that can be estimated from meteorites and other arguments. Because U and Th dominate present-day heat generation, and because the Th/U concentration ratio of typical rocks is ~ 4, uranium alone can be used to make estimates. Available estimates for the concentrations of K, U and Th in the bulk Earth and constraints provided by chondritic meteorites (e.g., Lodders and Fegley, 1998) can be used to estimate Ψ_E as roughly 0.03 μW m^{-3}. The ratio M_E/M_{cc} is about 400 for a 35 km-thick crust, or about 70 if the ~ 200 km-deep sub-continental mantle is included, as done below. The ratio ρ_E/ρ_{cc} is about 2. Given these assumptions and constraints, the average crustal heat production of the crust and sub-continental mantle would be about $1\,\mu W\,m^{-3}$. As shown below, this is too high.

If heat production declines exponentially with depth, the surface concentration is easily calculated from the average crustal concentration, the scale length $a^{\ddagger}$, and the total depth L of the continental crust:

$$\Psi_{sfc} = \Psi_{avg}\frac{L}{a^{\ddagger}}\frac{1}{\left(1 - e^{-L/a^{\ddagger}}\right)} \tag{6.6}$$

For scale lengths less than $\sim\frac{1}{3}$ of the depth L used for the continental crust, the exponential term can be neglected, and the result is approximated by:

$$\Psi_{sfc} = \Psi_E\frac{L}{a^{\ddagger}}\frac{M_E}{M_{cc}}\frac{\rho_{cc}}{\rho_E} \tag{6.7}$$

For a scale length of ~ 10 km suggested by the datasets of Roy et al. (1968), the required heat production of the topmost rocks of the continental crust would be about $20\,\mu W\,m^{-3}$. Instead, most rocks on Earth's continental surface have lower values of $1-4\,\mu W\,m^{-3}$. Thus, the above estimates for the average heat production of continental crust are much too high, proving that the geochemical arguments are oversimplified. One explanation for the discrepancy is that significant amounts of K, U and Th reside in Earth's upper mantle, albeit in very low concentrations. Nevertheless, their aggregate amount is considerable because the upper mantle is far more voluminous than the crust.

6.1.4 Continental growth

Armstrong (1991) proposed that the present-day continental volume appeared early and has been maintained in steady-state ever since. Many other researchers have suggested that the aggregate volume of Earth's continents has progressively grown, while a few others argue that continental mass grew quickly early in Earth's history, and is now slowly decreasing. Available rock and isotopic records are not sufficient to definitively resolve this question at this time, and perhaps not ever. However, geologic evidence supporting an approximation to steady-state includes the required destruction, presumably by subduction or tectonic erosion, of practically all pelagic sediments that should have been produced in prodigious volumes over geologic time; and evidence for the approximate constancy of sea-level and continental freeboard over geologic time. Given these observations and the lack of persuasive evidence for either growth or destruction of continental mass over time, the steady-state model should stand as the most reasonable description.

Models for continental mass are commonly tied to plate tectonic models, and evidence for the existence, or lack of existence, of ancient ophiolites is commonly invoked to support or deny that sea floor spreading has persisted since the Archean (cf. Hamilton, 2019; Polat et al., 2016). Robert Gregory (pers. comm., 2019) points out that the potential for ophiolite preservation is very low, stating that "to find an ophiolite that is 3 billion years old would require it to survive >20 half-lives for surface exposures which are on the order of 130 million years for a few kilometers of stratigraphic section! It is hard to prove when plate tectonics started because it is not a continental phenomenon; the old rock record is continental crust." Note that the present-day volumes of old continental materials depend on their preservation potential as well as on their initial volumes. Continuous recycling of Earth's materials is a well-known geologic reality.

6.2 Steady state models

The following sections utilize solutions to Fourier's equations for the case of steady-state. Steady state conditions are approached if layer thickness is not great compared to transport timescales. For continental crust and timescales on the order of a billion years, this assumption is reasonable.

Fourier's equations, for one dimensional (vertical) heat flow in the case of steady state, are provided in Chapter 2 and are as follows:

$$\frac{\partial^2 T}{\partial z^2} = -\frac{\Psi}{K} \tag{6.8}$$

The flux $\Im$ is:

$$\Im = -K\frac{\partial T}{\partial z} \tag{6.9}$$

The negative sign in Eq. 6.9 will disappear if z is taken as depth from the surface, as done below, rather than increasing radially upward. If z represents depth, temperature gradients and fluxes are positive quantities, as generally described by geologists.

6.2.1 Model 1: one layer, constant ψ and K

For constant Ψ and K in a continental slab of total thickness L, the solution is:

$$T - T_0 = \frac{z}{L}\left(T_L - T_0 + \frac{\Psi L(L-z)}{2K}\right) \tag{6.10}$$

where z is depth, T_0 is the constant surface temperature, and T_L is the constant basal temperature. The temperature gradient is provided by differentiation:

$$\frac{\partial T}{\partial z} = \frac{T_L - T_0}{L} + \frac{\Psi(L-2z)}{2K} \tag{6.11}$$

and the flux is provided by multiplying each side by K. Clearly, the gradient and flux at the surface are:

$$\frac{\partial T}{\partial z}\bigg|_{z=0} = \frac{T_L - T_0}{L} + \frac{\Psi L}{2K} \quad \text{and} \quad \Im_{z=0} = \frac{K(T_L - T_0)}{L} + \frac{\Psi L}{2} \tag{6.12}$$

For some cases where Ψ is large, these equations predict that a maximum temperature T_{max} is reached within the continental slab at depth z^*:

$$T_{max} = \frac{T_L - T_o}{2} + \frac{\Psi L^2}{8K} + \frac{K(T_L - T_o)^2}{2\Psi L^2} \quad \text{at} \quad z^* = \frac{L}{2} + \frac{K(T_L - T_o)}{\Psi L} \tag{6.13}$$

This computed maximum temperature is meaningful only if $z^* \le L$. If so, some of the heat generated within the crust flows downward into subjacent rocks, rather than all flowing upward as generally envisioned.

6.2.2 Model 2: two layer model

A steady-state model can be constructed for continental crust of thickness L_1, heat generation Ψ_1, and conductivity K_1 that overlies the sub-continental mantle of thickness L_2, heat generation Ψ_2, and conductivity K_2, where all parameters are constants. The boundary conditions are the temperatures at the Earth's surface ($z = 0$) and at the base of the sub-continental mantle ($z = L_1 + L_2$). The temperature and heat flux must match at the interface ($z = L_1$) between the two layers.

The temperature T_1 of the upper layer is similar to Eq. (6.10), above:

$$T_1 - T_0 = \frac{z}{L_1} \left(T_{L1} - T_0 + \frac{\Psi_1 L_1 (L_1 - z)}{2K_1} \right) \tag{6.14}$$

where T_o is the surface temperature and T_{L1} is the temperature at the layer interface. The temperature T_2 of the sub-continental mantle is:

$$T_2 - T_{L1} = \frac{z - L_1}{L_2} \left(T_{L1+L2} - T_{L1} + \frac{\Psi_2 L_2 (L_1 + L_2 - z)}{2K_2} \right) \tag{6.15}$$

where T_{L1+L2} is the fixed temperature at the base of the sub-continental mantle. These equations incorporate the interface temperature T_{L1}, which is:

$$T_{L1} = \left(\frac{K_2 T_{L1+L2}}{L_2} + \frac{K_1 T_o}{L_1} + \frac{\Psi_1 L_1 + \Psi_2 L_2}{2} \right) \bigg/ \left(\frac{K_1}{L_1} + \frac{K_2}{L_2} \right) \tag{6.16}$$

The flux at various depths in the upper and lower layers is found by differentiation and multiplying by K:

$$K_1 \frac{\partial T_1}{\partial z} = \frac{K_1 (T_{L1} - T_o)}{L_1} + \frac{\Psi_1}{2} (L_1 - 2z) \tag{6.17}$$

and

$$K_2 \frac{\partial T_2}{\partial z} = \frac{K_2}{L_2} (T_{L1+L2} - T_{L1}) + \Psi_2 \left(\frac{L_2}{2} + L_1 - z \right) \tag{6.18}$$

6.2.3 Model 3: exponential variations in Ψ and K

If Ψ and K vary inside the crust, the appropriate steady-state form of Fourier's equation in one dimension is:

$$\frac{\partial}{\partial z}\left(K\frac{\partial T}{\partial z}\right) = -\Psi \qquad (6.19)$$

Many analytical solutions are possible, depending on the functional forms of Ψ and K. Geologic and laboratory evidence make it appropriate to consider heat generation Ψ as a function of depth and K as a function of T:

$$\Psi = \Psi_m + \Psi_o e^{-z/a} \quad \text{and} \quad K = g + he^{-bT} \qquad (6.20)$$

where $\Psi_o + \Psi_m$ is the heat generation of surface rocks, Ψ_m is that of deep rock, and a, b, g and h are constants. The solution is:

$$gT + \frac{h}{b}\left(1 - e^{-bT}\right) = a^2\Psi_0\left\{1 - e^{-z/a} - \frac{z}{L}\left(1 - e^{-L/a}\right)\right\} + gz\frac{T_L}{L} + \frac{zh}{bL}\left(1 - e^{-bT_L}\right) + \Psi_m\frac{Lz - z^2}{2} \qquad (6.21)$$

where the surface temperature has been taken as zero.

Importantly, the above solution is written in the form, $f(T) = f^*(z)$, where f and f* are specified functions. This equation can therefore be solved by choosing a value for z between 0 and L, calculating a numerical value for the RHS, and then iterating T until the LHS matches that particular value.

The flux at any depth $0 \le z \le L$ can be directly computed without iteration:

$$K\frac{\partial T}{\partial z} = -\frac{a^2\Psi_0}{L}\left(1 - e^{-L/a}\right) + a\Psi_o e^{-z/a} + g\frac{T_L}{L} + \frac{h}{bL}\left(1 - e^{-bT_L}\right) + \Psi_m\left(\frac{L}{2} - z\right) \qquad (6.22)$$

The temperature gradient is found by dividing the flux at any position by K, but computing K requires determining the temperature at that position, and in general this must be found by the iteration procedure described above. However, because the surface temperature is fixed, the respective value of K is 2.45 W m^{-1} K^{-1}, so the corresponding temperature gradient is easily determined by dividing the surface flux by this value.

Computed temperature profiles must asymptote to these respective gradients at the surface and the continental base. In many cases, these profiles closely approximate two linear trends, which intersect at the position (T_{int}, z_{int}):

$$z_{int} = \frac{1250 - LG_L}{G_{sfc} - G_L} \quad \text{and} \quad T_{int} = z_{int}G_{sfc} \qquad (6.23)$$

where G_{sfc} and G_L are the temperature gradients at the surface and at the base.

6.3 Comparison of steady-state models and data

6.3.1 Surface heat flux and gradients

In what follows and for simplicity, the exponential model for heat generation was selected as the default model for calculation of temperature profiles and heat fluxes. To facilitate comparisons, $K = 2$ W m^{-1} K^{-1} is used throughout. We consider two cases for basal conditions: (1) for cold, thick continental lithosphere, a constant temperature of

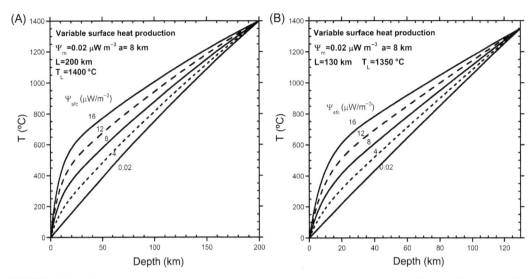

FIGURE 6.6 Effect of varying the near-surface heat production on temperature profiles calculated according to the exponential model for (A) cold and thick lithosphere and (B) hot and thin lithosphere. The reference model assumes that the heat production of deep rocks are $0.02\,\mu W\,m^{-3}$, that the exponential attenuation factor is 8 km, and that thermal conductivity is $2\,W\,m^{-1}\,K^{-1}$. The curves show the effects of varying the surface heat production, as labeled.

$1400\,°C$ is assumed for the sub-continental mantle at a depth of 200 km, but for (2) hot, thin lithosphere, a constant temperature of $1350\,°C$ is assumed for a shallower base at 130 km.

Successive figures show how temperature profiles depend on the heat generation of surface rocks (Fig. 6.6), or on that of the deep rocks constituting the sub-continental mantle (Fig. 6.7), or on the attenuation factor (scale length) of heat generation for the cold (part A) and hot (part B) cases (Fig. 6.8). We begin with a reference model which assumes that heat productions of surface rocks and deep rocks are 4 and $0.02\,\mu W\,m^{-3}$, respectively, and that the scale factor is 8 km, and then vary these parameters one at a time. The effect of changing the parameters is shown by contours on the diagrams. Details are as follows.

Fig. 6.6 shows how temperature profiles depend on variable heat generation of surface rocks, while keeping the heat production of deep rock as $0.02\,\mu W\,m^{-3}$ and the scale factor as 8 km. The indicated surface heat flux and temperature gradients approximate observed crustal values if Ψ_{sfc} is between 4 and $8\,\mu W\,m^{-3}$. Temperatures become much too high at shallow depth if $\Psi_{sfc} > 8\,\mu W\,m^{-3}$. However, the deep temperature gradient approximates that suggested by xenoliths only for the lowest values of Ψ_{sfc} of $\sim 0.02\,\mu W\,m^{-3}$.

Fig. 6.7 shows how temperature profiles depend on the heat generation of deep rock, assuming that the heat production of surface rock is $4\,\mu W\,m^{-3}$ and that the scale factor is 8 km. In all cases the indicated values for surface heat flux ($50 \pm 10\,mW\,m^{-2}$) and gradient ($25 \pm 5\,°C\,km^{-1}$) approximate observed crustal values. However, temperatures at depth become much too high if the heat generation of deep rock is $>0.1\,\mu W\,m^{-3}$.

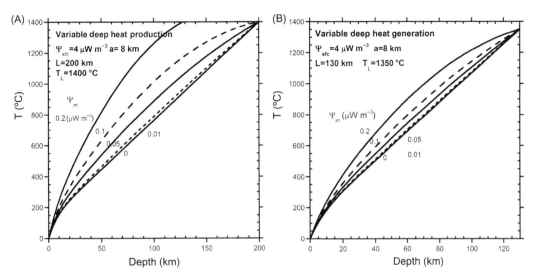

FIGURE 6.7 Effect of varying deep heat production on temperature profiles calculated according to the exponential model for (A) cold and thick lithosphere, and for (B) hot and thin lithosphere. The reference model assumes that the heat production of near surface rocks is $4\,\mu W\,m^{-3}$, that the exponential attenuation factor is 8 km, and that thermal conductivity is $2\,W\,m^{-1}\,K^{-1}$. The curves show the effects of varying the deep heat production, as labeled.

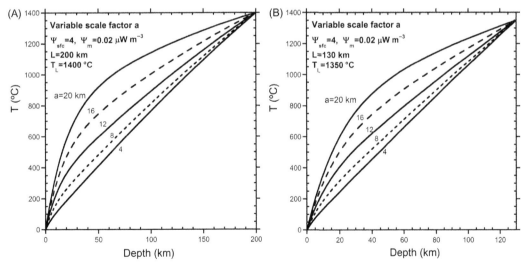

FIGURE 6.8 Effect of varying the scale length on temperature profiles calculated according to the exponential model for (A) cold and thick lithosphere and (B) hot and thin lithosphere. The reference model assumes that the heat production of crustal rocks and deep rocks are 4 and $0.02\,\mu W\,m^{-3}$, respectively, and that thermal conductivity is $2\,W\,m^{-1}\,K^{-1}$. The curves show the effects of varying the attenuation factor, as labeled.

Fig. 6.8 shows how temperature profiles depend on the scale factor for the attenuation of heat generation, assuming that the heat production of surface and deep rocks are 4 and $0.02 \, \mu W \, m^{-3}$, respectively. The indicated surface heat flux and temperature gradients approximate observed crustal values if the scale factor is between 5 and 12 km, but are either too low or too high for values outside that range.

Finally, Fig. 6.9 shows how the surface heat flux varies with surface heat production and its vertical attenuation factor, for two different estimates of the depth to the partially-melted low velocity zone, each for two different estimates of the heat production of deep rock. Taken together, Figs. 6.6–6.9 show that the exponential model is compatible with typical surface fluxes and near-surface temperature gradients in continental areas only if the heat generation of surface rock is $3 \pm 1 \, \mu W \, m^{-3}$, that of deep rock is $<0.1 \, \mu W \, m^{-3}$, and the attenuation factor is 10 ± 5 km.

6.3.2 Basal heat flux

The basal heat flux $\mathfrak{I}_L$ is readily calculated for the various models, simply by taking $z = L$ in the appropriate equations, e.g., (6.22). The largest term is $K(T_L\text{-}To)/L$, so the basal flux primarily reflects the depth L to the partially melted low velocity zone (LVZ). To very good accuracy, the basal flux is lower than the surface flux by this depth times the average radioactivity of overlying rocks. Thus, for a given depth to the LVZ, the basal flux in a particular area becomes lower as the crustal heat generation becomes larger. Stated another way, $\partial \mathfrak{I}_L/\partial \Psi_{sfc} \sim \text{-}L/2$ for the simple model, or $\sim \text{-}a^2/L$ for the exponential model. This quantitative result is opposite that of conventional thinking, and requires reconsidering the meaning of the y-intercept in plots of flux vs. heat generation (Fig. 6.4). Specifically, the contours of basal flux in Fig. 6.10 vary in a manner opposite to those of surface flux in Fig. 6.9; i.e., flux magnitudes decrease towards the upper right of the graph.

6.3.3 Relationship between flux and internal heat generation

The following sections use the equations provided above to determine and interpret the slopes and y-intercepts of the compelling, observed trends of continental heat flow vs. the heat production of surface rocks. Hereafter this is called a "$\mathfrak{I}$ vs. Ψ plot", as exemplified by Fig. 6.4.

6.3.3.1 $\mathfrak{I}$ vs. Ψ plot: data vs. model 1

For the simple, one-layer model for the continental crust, the slope and y-intercept of the predicted $\mathfrak{I}$ vs. Ψ plot is evident from Eq. (6.12):

$$\text{Slope} = \frac{L}{2} \quad \text{and} \quad \text{y-intercept} = K\frac{T_L - T_0}{L} \tag{6.24}$$

where L is the thickness of the crust, K is its thermal conductivity, and T_L and T_0 are the temperatures at the surface and at the base, which is at depth L. Note that the y-intercept is approximated by K times the average geothermal gradient of the crust. Given the simplicity of this model, estimates of the resultant $\mathfrak{I}$ vs. Ψ trends are easily calculated. For typical

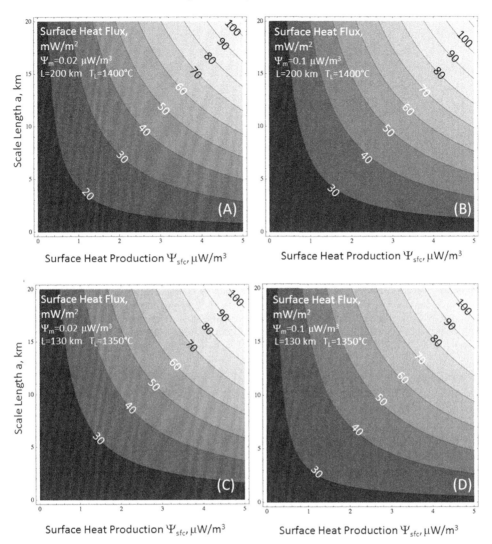

FIGURE 6.9 Contours of surface heat flow as a function of the heat generation of surface rocks and the scale length of its assumed exponential vertical attenuation (Eq. 6.22), using $K = 2$ W m^{-1} K^{-1}. The surface temperature gradient in deg km^{-1} is strictly equal to one half of the numbers on the contours, but would approximate that number divided by 2.45 if the surface rock is granodiorite. See text. (A) Assuming that the heat production of deep rock is 0.02 μW m^{-3} and that the temperature at a depth of 200 km is 1400°C. (B) Assuming that the heat production of deep rock is 0.1 μW m^{-3}, and that the temperature at 200 km is 1400°C. (C) Assuming that the heat production of deep rock is 0.02 μW m^{-3} and that the temperature at 130 km is 1350°C. (D) Assuming that the heat production of deep rock is 0.1 μW m^{-3} and that the temperature at 130 km is 1350°C.

continental crust with $K \sim 2$ W m^{-1}-K, $L \sim 35$ km, and a thermal gradient of ~ 15 deg km^{-1}, the slope would be ~ 18 km and the intercept would be ~ 30 mW m^{-2}. The slope is much too high but this rough estimate for the y-intercept resembles that of Fig. 6.4, which used J for flux.

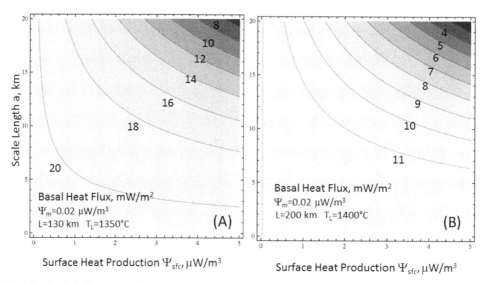

FIGURE 6.10 (A) Contours of basal heat flow as a function of the heat generation of surface rocks and the scale length of its assumed exponential vertical attenuation (Eq. 6.22), assuming that the heat production of deep rock is $0.02\,\mu W\,m^{-3}$, the thermal conductivity is $2\,W\,m^{-1}\,K^{-1}$, and that the temperature is $1350\,°C$ at a depth of $130\,km$. (B) As above, except that the temperature is $1400\,°C$ at a depth of $200\,km$.

This particular model assumes that the heat generation Ψ of crustal rock is constant. Suppose instead that Ψ varies linearly with depth, decreasing to some small value at the continental base. Then Ψ for surface rock would be twofold higher than this average value of Ψ, and the slope of the trend would be reduced by about a factor of two, which would conform much better with observation (cf. Fig. 6.4). More detailed models and calculations support this first-order argument, and like the above, also show that the "y-intercept" does not represent the "mantle flux".

6.3.3.2 ℑ vs. Ψ plot: data vs. model 2

For the two-layer model for continental crust of thickness L_1 overlying sub-continental mantle of thickness L_2, the slope and y-intercept of the predicted ℑ vs. Ψ plot can be obtained from Eqs. (6.16) and (6.17):

$$\text{Slope} = \frac{L}{2}\left\{1 + \frac{K_1 L_2}{K_1 L_2 + K_2 L_1}\right\} \quad \text{and}$$

$$y\text{-intercept} = -\frac{K_1 T_0}{L_1} + \frac{K_1 K_2 T_{L1+L2} + K_1 K_2 T_o L_2 / L_1 + \Psi_2 K_1 L_2^2 / 2}{K_1 L_2 + K_2 L_1}$$

(6.25)

The two layer model produces reasonable temperatures, temperature gradients and surface fluxes, but the slope of the predicted ℑ vs. Ψ trend is much too high. In fact, if $L_2 \gg L_1$, this predicted slope is $\sim L_1$ km, which is nearly twice as high as it was for Model 1, which was already unacceptably high. The predicted y-intercept corresponds with observed values if the heat generation in the sub-continental mantle is about $0.2\,\mu W/m^3$. Once again, the y-intercept does not correspond to a regional "mantle flux".

6.3.3.3 ℑ vs. Ψ plot: data vs. model 3

For the model that presumes that heat generation of the rocks decreases exponentially with depth, the slope and y-intercept of the predicted ℑ vs. Ψ plot can be obtained from Eq. (6.22). Thus:

$$\text{slope} = a - \frac{a^2}{L}\left(1 - e^{-\frac{L}{a}}\right)$$

$$\text{y-intercept} = \left(\frac{a^2}{L}\left(1 - e^{-\frac{L}{a}}\right) - a + \frac{L}{2}\right)\Psi_m + \frac{gT_L}{L} + \frac{h\left(1 - e^{-bT_L}\right)}{bL} \qquad (6.26)$$

For realistic parameter choices, calculated quantities (Fig. 6.11) provide reasonable matches to observations (Fig. 6.4). However, the y-intercept does not correspond to the "mantle flux", nor need it be laterally uniform in a given region, as suggested by the contours in Fig. 6.10.

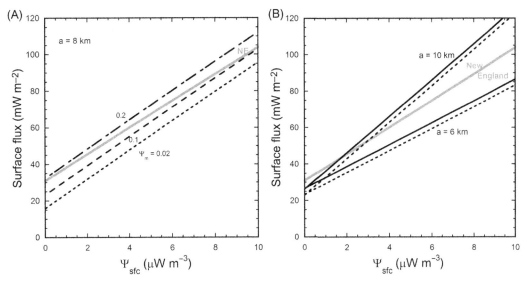

FIGURE 6.11　Lines showing the dependence of surface heat flow varies on the heat generation of surface rocks, comparing the data in Fig. 6.4 (gray line "New England") with theoretical lines calculated with Eq. 6.24. (A) Dependence of flux lines on the heat generation Ψ_m of deep rock (dashed lines), for values of 0.02, 0.1 and $0.2\,\mu W\,m^{-3}$, assuming $a = 8$ km, $L = 200$ km, $T_L = 1400\,°C$, and $K = 2$. (B) Data from Fig. 6.4 (gray line) compared to calculated lines for $a = 6$ and $a = 8$ km, for the cases of $L = 200$ km and $T_L = 1400\,°C$ (dashed lines) and for $L = 130$ km and $T_L = 1350\,°C$ (solid black lines).

6.3.3.4 Previous calculations of thermal profiles

Calculations presented in Furlong and Chapman (2013) were constructed assuming high heat generation in the lower crust in order to provide high intercepts on the flux vs Ψ plot, previously considered to stem from significant deep crustal or mantle flux. Feasible temperature profiles were obtained in this and other studies because K was assumed to strongly increase with temperature, unlike measurements. Overly large and increasing K with depth compensates for the incorrect deep placement of radioactive elements. As discussed by Transtrum et al. (2015), lumped parameters in models allow calculations to agree with data, even if the underlying physics is wrong.

6.4 Relevance of thermal processes to continental behavior

The following sections discuss possible connections between the thermal characteristics of crustal rocks and the tectonic and magmatic behavior of continents. These conclusions are partly based on the time-dependent equations presented in Chapter 2, because evolutionary processes are not steady-state.

Many equations presented in this chapter and in Chapter 2 appear to be based on a large number of parameters. For example, the exponential model of Eq. (6.22) presumes a total thickness, fixed basal and surface temperatures, differing heat generations of surface rocks and deep rocks, a scale length for the attenuation of heat-production, and relevant values for thermal conductivity and its associated functional coefficients. However, these quantities have not been treated as free parameters, but rather are closely constrained by field, geochemical, and laboratory data. Only by allowing these parameters to vary over large ranges, as in Figs. 6.6—6.9, can we infer *how* these parameters individually control calculated temperature profiles and fluxes.

6.4.1 Relevance to volcanic plateaus, stratovolcanoes, and subvolcanic batholiths

Smith et al. (1961) identified an evolutionary process that dominates Tertiary volcanism and plutonism in the continental interior of the western USA. This process typically begins with the eruption of basalts and basaltic andesites, followed by the eruption of lavas that tend to be progressively more silicic, such as andesites and dacites. A volcanic plateau with a thickness of one kilometer or more that extends over thousands of square kilometers can be produced over a period that typically spans several million years. Concomitantly, giant magma chambers commonly evolve in the subsurface, and then gradually stope upward. Ultimately, a circular block of the superjacent volcanic rocks can founder into the molten mass below, producing a huge caldera sometimes reaching many tens of kilometers in diameter, with this down-dropped block displacing the magma below. This displacement produces a colossal pyroclastic eruption of ash flow tuff, whose volume can be 100s or even 1000s of cubic kilometers. Continued evolution of the pile can feature the eruption of inter-caldera rhyolites, resurgent doming, meteoric-hydrothermal alteration, and most importantly the emplacement and crystallization of huge subvolcanic batholiths (e.g., Criss and Taylor, 1983).

Thermal physics provides helpful insight into how this remarkable process works. Because the thermal diffusivity and thermal conductivity of volcanic rocks is low (Figs. 6.1B and 6.2), eruption of a volcanic cap will provide an insulating blanket, trapping heat below and fostering more melting. Progressive heating will lower the thermal conductivity of the subjacent crust (Fig. 6.2), creating a positive feedback that will amplify the retention of heat (Whittington et al., 2009). Meteoric-hydrothermal systems will advect heat toward the center of the pile, producing a thermal dome (Norton and Knight, 1977; Criss and Taylor, 1983), while simultaneously producing hydrous assemblages with low melting temperatures that will also foster the continuation of magmatic activity, as well as its evolution to more silicic compositions. Ultimately, much of the trapped heat is lost via a catastrophic eruption that initiates the cooling of the terrane and the crystallization of a subvolcanic batholith.

In short, low-conductivity volcanic rocks can trap heat in localized areas, and the positive feedback of increased temperature and lower conductivity will sustain melting and magmatism in the same area. This phenomenon can also explain repeated eruptions from a single vent, producing huge stratovolcanoes if the erupted magmas are too viscous to spread laterally. The thermal properties of surficial volcanic rocks can likewise explain observations that are commonly attributed to mantle plumes.

6.4.2 Relevance of thermal processes to the Wilson and supercontinent cycles

The Wilson Cycle is a long term tectonic process thought to operate over an interval of 300–500 million years. This conceptual cycle includes the thinning and rifting of a continent, followed by the progressive growth of an intervening ocean basin, followed by a period of convergence that culminates in a continental collision. The Wilson Cycle is a simplification of the more complex supercontinent cycle that begins with most of Earth's continents being assembled into a single large super-continental mass. Subsequently, this immense conglomeration is disrupted into many continental and smaller fragments with intervening ocean basins, only to ultimately converge again.

So, what initiates the breakup process? Conventional thinking attributes breakup to the formation of a mantle plume that just happens to form beneath the center of the continent, or supercontinent. As is usual for such "explanations", the hypothetical plume exists wherever and whenever it is needed, and performs whatever action is required (Hamilton, 2019). In contrast, the thermal analysis provided above suggests two possible explanations for continental breakup that are grounded in physical principles.

The first explanation is based on the low thermal conductivity of the continental mass compared to the underlying sub-continental mantle, which in turn is lower than the conductivity of subjacent peridotite under oceanic lithosphere. Due to this configuration, any heat from deep sources, and any heat generated within the continental and sub-continental masses by radioactive decay, will cause the continental interior to heat up. According to the equations in Chapter 2, the temperature will be greatest in the central part of the mass; moreover, a time-scale of ~ 0.5 Ga is sufficient to generate the required temperature increase.

The second explanation is based on the combination of high heat generation and low thermal conductivity of the continental crust itself. Rock that generates $1 \, \mu W \, m^{-3}$ will heat up by 1000 °C in 100 Ma, if conductive losses are negligible. While the latter condition cannot be strictly attained, the positive feedback between high temperatures and low conductivity fosters heat retention. Granitic rocks will therefore become hot in an environment of low conductivity, given time intervals commensurate with the Wilson Cycle, and this heating will foster the thinning and breakup of continental interiors. In addition, any burial of granitoids beneath other rock, as appears to have happened on the Kola peninsula (Fig. 6.5), or any other deep emplacement of radionuclides, will foster such interior heating (see Figs. 6.6—6.8).

6.5 Conclusions

Analytical solutions explain the physical causes of Earth's temperature profiles and fluxes, and reveal the tradeoffs and ambiguities that plague multi-parameter models. Steady state equations provide a reasonable approximation of the thermal character of the continental lithosphere. Thermal profiles are controlled by the fixed temperatures of the surface, the temperature and depth of its partially melted base, the thermal conductivity of the rock column, and the distribution of heat-generating radionuclides. A nominal value of $2 \, W \, m^{-1} \, K^{-1}$ for thermal conductivity, and typical values of a few $\mu W \, m^{-3}$ for the heat generation of shallow rock, are adequate to calculate generalized profiles, as we have pursued here. Calculations that assume that the heat generation of rocks decreases exponentially with depth and use reasonable values for thermal conductivity resemble typical temperature gradients and thermal fluxes observed in continental areas. Interestingly, the surface flux increases as the concentration of radionuclides in the crust becomes higher, yet the basal flux becomes progressively lower, other things being equal. The basal flux is primarily controlled by the depth to the LVZ, less this smaller correction. The sub-continental lithosphere must have a small, but not negligible, amount of radioactivity.

Small amounts of internal heat production are needed in the sub-continental lithosphere to reproduce the observed flux. Thus, Earth's inventory of radionuclides cannot all reside in continental rocks above 35 km.

More data on thermal conductivity and heat production of rocks representing the deep crust and the sub-continental mantle are needed. Data from the Kola Superdeep drill hole depart from standard assumptions. Simple thermal models provide insight into processes that occur in areas of continental magmatism, and into processes that occur over far longer time scales relevant to the production and destruction of supercontinents.

References

Armstrong, R.L., 1991. The persistent myth of continental growth. Aust. J. Earth Sci. 38, 613—630.

Blackwell, D.D., Richards, M., 2004. Geothermal Map of North America, AAPG, scale 1:6,500,000.

Brown, G.C., Mussett, A.E., 1984. The Inaccessible Earth. George Allen & Unwin, London.

Branlund, J.M., Hofmeister, A.M., 2007. Thermal diffusivity of quartz to 1000 degrees C: effects of impurities and the α-β phase transition. Phys Chem. Miner. 34, 581—595.

Branlund, J.M., Hofmeister, A.M., 2008. Factors affecting heat transfer in SiO_2 solids. Am. Miner. 93, 1620–1629.

Branlund, J.M., Hofmeister, A.M., 2012. Heat transfer in plagioclase feldspars. Am. Miner. 97, 1145–1154.

Clarke, F.W., 1910. Analyses of Rocks and Minerals From the Laboratory of the United States Geological Survey, 1880 to 1908. Government Printing Office, Washington D.C., USGS Bulletin 411.

Criss, R.E., Taylor Jr., H.P., 1983. An $^{18}O/^{16}O$ and D/H study of Tertiary hydrothermal systems in the southern half of the Idaho batholith. Geol. Soc. Am. Bull. 94, 640–663.

Diller, J.S., 1898. The Educational Series of Rock Specimens Collected and Distributed by the United States Geological Survey. Government Printing Office, Washington D.C., USGS Bulletin 150.

Furlong, K.P., Chapman, D.S., 2013. Heat flow, heat generation, and the thermal state of the lithosphere. Annu. Rev. Earth Planet. Sci. 41, 385–410. 2013.

Hamilton, W.B., 2019. Toward a myth-free geodynamic history of Earth and its neighbors. Earth Sci. Rev. (in press).

Hoffer, J.M., 1970. Petrology and mineralogy of the Campus andesite pluton, El Paso, Texas. Geol. Soc. Am. Bull. 81, 2129–2136.

Hofmeister, A.M., 2019. Measurements, Mechanisms, and Models of Heat Transport. Elsevier, Amsterdam, pp. 201–250 (see Chapter 4 on methods; Chapter 7 on mineral and rock data, and Chapter 11 on a radiative diffusion description of heat conduction).

Hofmeister, A.M., Criss, R.E., 2005. Earth's heat flux revised and linked to chemistry. Tectonophysics 395, 159–177.

Hofmeister, A.M., Sehlke, Alexander, Avard, Geoffroy, Bollasina, Anthony J., Robert, Geneviève, Whittington, Alan G., 2016. Transport properties of glassy and molten lavas as a function of temperature and composition. Journal of Volcanology and Geothermal Research 327, 380–388.

Hofmeister, A.M., Dong, J.J., Branlund, J.M., 2014. Thermal diffusivity of electrical insulators at high temperatures: evidence for diffusion of phonon-polaritons at infrared frequencies augmenting phonon heat conduction. Journal of Applied Physics 115, 163517. Available from: https://doi.org/10.1063/1.4873295.

Hofmeister, A.M., Pertermann, M., Branlund, J.M., 2007. Thermal conductivity of the Earth. In: Price, G.D., (Ed.), V. 2 Mineral Physics. In: Schubert, G. (Ed. In Chief), Treatise in Geophysics. Elsevier, The Netherlands, pp. 543–578.

Kremenetsky, A.A., Ovchinnikov, L.N., 1986. The Precambrian continental crust: its structure, composition and evolution as revealed by deep drilling in the U.S.S.R. Precambrian Res. 33, 11–43.

Lachenbruch, A.H., 1970. Crustal temperature and heat production: implications for the linear heat-flow relation. J. Geophys. Res. 75, 3291–3300.

Lodders, K., Fegley, B.J., 1998. The Planetary Scientist's Companion. Oxford University Press, Oxford.

Merriman, J.D., Alan, G.W., Hofmeister, A.M., Nabelek, P.I., Benn, K., 2013. Thermal transport properties of major Archean rock types to high temperature and implications for cratonic geotherms. Precambrian Res. 233, 358–372.

Miao, S.Q., Li, H.P., Chen, G., 2014. Temperature dependence of thermal diffusivity, specific heat capacity, and thermal conductivity for several types of rocks. J. Therm. Anal. Calorim. 115, 1057–1063.

Mooney, W.D., Laske, G., Masters, T.G., 1998. CRUST 5.1: a global crustal model as $5° \times 5°$. J. Geophys. Res. 103, 727–747.

Norton, D., Knight, J., 1977. Transport phenomena in hydrothermal systems: cooling plutons. Am. J. Sci. 277, 937–981.

Pertermann, M., Whittington, A.G., Hofmeister, A.M., Spera, F.J., Zayak, J., 2008. Thermal diffusivity of low-sanidine single-crystals, glasses and melts at high temperatures. Contrib. Miner. Petrol. 155, 689–702. Available from: https://doi.org/10.1007/s00410-007-0265-x.

Pollack, H.N., Chapman, D.S., 1977. On the regional variation of heat flow, geotherms, and lithospheric thickness. Tectonophysics 38, 279–296.

Polat, A., Kokfelt, T., Burke, K.C., Kusky, T., Bradley, D., Dziggel, A., et al., 2016. Lithological, structural, and geochemical characteristics of the Mesoarchean Târtoq greenstone belt, South-West Greenland, and the Chugach-Prince William accretionary complex, southern Alaska: evidence for uniformitarian plate-tectonic processes. Can. J. Earth Sci. 53 (11), 1336–1371.

Robertson, E.C., Hemingway, B.S., 1995. Estimating Heat Capacity and Heat Content of Rocks. U.S. Geological Survey, Reston VA, Open-file Report 95–622.

Robie, R.A., Hemingway, B.S., 1995. Thermodynamic Properties of Minerals and Related Substances at 298.15 K and 1 Bar (10^5 Pascals) Pressure and at Higher Temperatures. U.S. Government Printing Office, Washington D.C., U.S. Geological Survey Bulletin 2131.

Roy, R.F., Blackwell, D.D., Birch, F., 1968. Heat generation of plutonic rocks and continental heat flow provinces. Earth Planet. Sci. Lett. 5, 1−12.

Transtrum, M.K., Machta, B.B., Brown, K.S., Daniels, B.C., Myers, C.R., Sethna, J.P., 2015. Perspective: sloppiness and emergent theories in physics, biology, and beyond. J. Chem. Phys 143. Available from: https://doi.org/10.1063/1.4923066.

Smith, R.L., Bailey, R.A. and Ross, C.S., 1961. Structural evolution of the Valles caldera, New Mexico, and its bearing on the emplacement of ring dikes. U.S. Geological Survey Professional Paper 424-D, 145−149.

Van Schmus, W.R., 1995. Natural radioactivity of the crust and mantle. In: Ahrens, T.J. (Ed.), Global Earth Physics. American Geophysical Union, Washington, D. C., pp. 283−291.

Vieira, F., Hamza, V., 2018. Global heat flow: new estimates using digital maps and GIS techniques. Int. J. Terrestrial Heat Flow Appl. Geotherm. 1, 6−13. Available from: https://doi.org/10.31214/ijthfa.v1i1.6.

Whittington, A.G., Hofmeister, A.M., Nabelek, P.I., 2009. Temperature-dependent thermal diffusivity of Earth's crust and the implications for magmatism. Nature 458, 319−321.

Thermal models of the oceanic lithosphere and upper mantle

"There are two ways of doing calculations in theoretical physics. One way, and this is the way I prefer, is to have a clear physical picture of the process that you are calculating. The other way is to have a precise and self-consistent mathematical formalism. You have neither." *Enrico Fermi, as quoted by Freeman Dyson (2004).*

All existing cooling models of the moving oceanic lithosphere have serious flaws. To avoid repeating previous mistakes, we re-examine key models that are used, for example, to interpret seismologic results (Schmerr, 2012). Section 7.1 focusses on the most serious problems: specifically, that the plate and half-space formulations do not actually solve Fourier's heat equation for the conditions on Earth's surface, while mass is not conserved in modeling subsidence of plates. Some details are given in the Appendix to this chapter. Section 7.2 covers relevant data on the lithosphere that are needed to correctly posit a valid thermo-mechanical model. Section 7.3 provides a steady-state model and a geotherm for the upper mantle. Section 7.4 summarizes.

7.1 Fundamental errors in cooling models of the oceanic lithosphere

Plate and half-space models poorly match heat flux measurements, particularly at young ages (Fig. 7.1). Attribution of this mismatch to hydrothermal circulation is fallacious, as discussed by Hofmeister and Criss (2005a) and in Section 3.3.4.1. Acceptance of these models, and their use to calculate the global power after 1992, rests on an apparent match with ocean floor depths. However, reproducing depths does not validate the heat

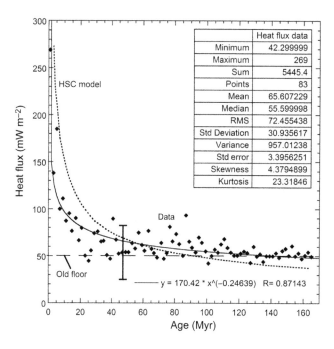

FIGURE 7.1 Data from the North Atlantic and North Pacific Oceans that were modeled by Stein and Stein (1992). Diamonds = 2 Ma bin averages. Error bar = typical standard deviation. Box shows the statistical evaluation of the data. Dotted line = the half-space cooling model of Stein and Stein (1992) which is nearly indistinguishable from the plate model of Parsons and Sclater (1977). Solid line = the power fit listed at the bottom, which is greatly influenced by the two points closest to 0 age. A similar power fit was obtained by Hofmeister and Criss (2005b), who fitted the binned data of Pollack et al. (1993), which averaged many regions using wider intervals. Dashed line = the asymptotic value at old ages.

	Heat flux data
Minimum	42.299999
Maximum	269
Sum	5445.4
Points	83
Mean	65.607229
Median	55.599998
RMS	72.455438
Std Deviation	30.935617
Variance	957.01238
Std error	3.3956251
Skewness	4.3794899
Kurtosis	23.31846

y = 170.42 * x^(−0.24639) R= 0.87143

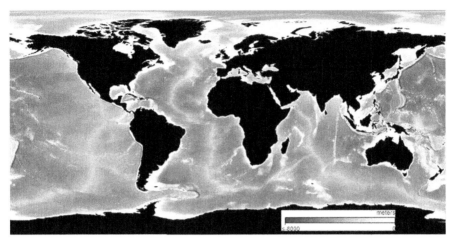

FIGURE 7.2   Ocean floor depths. *Imagery by Jesse Allen, NASA's Earth Observatory, using data from the General Bathymetric Chart of the Oceans (GEBCO) produced by the British Oceanographic Data Centre. Publically available from https://visibleearth.nasa.gov/view.php?id = 73963 (accessed 17.03.19.).*

flux calculations, and vice versa, because these calculations are based on different formulae for temperature vs time: see Sections 7.1.1.3 and 7.1.2.4.

The match of the model with ocean depths is unpersuasive. Overall, plate models provide better fits for progressively older ocean floor (e.g., Parsons and Sclater, 1977; Honda and Yuen, 2004), but the Southern Argentine Basin cannot be fit well (Hohertz and Carlson, 1998). Systematic misfits indicate that plate models fail to explain the variation of mean ocean depth with age over all ages (Carlson and Johnson, 1994). The underlying problem is that measured bathymetry of the ocean floor is not due to subsidence, but instead depends on slab dips. Mid-ocean ridges perturb depths in various ways (Fig. 7.2) because how these volcanic ridges rise above their surroundings depends on the depth and lateral extent of magma feeding the ridge, its density contrast with the surroundings, and the rate that the magma arrives (Section 7.2).

Introduction of additional free parameters gives the appearance of a better match with data, but the fundamental errors discussed below remain. For example, Hofmeister and Criss (2006) examined the plate model of Wei and Sandwell (2006) and showed that 7 of their 8 equations contained an error, including division by zero. The appendix examines Hasterok's (2013) recent model.

7.1.1 Mathematical and conceptual flaws in previous descriptions of temperature

The key equation of plate models allegedly describes the variation of temperature (T, in °C) with depth (z) at a certain time (t). Vertical heat flux is determined by taking the derivative with z and multiplying it by thermal conductivity (K). We elucidate the physics by using constant values for the material properties, as assumed in most models. This section uses typical values: thermal diffusivity $D = 1 \text{ mm}^2\text{s}^{-1}$; speed $u = 6 \text{ cm y}^{-1}$; plate thickness $L = 100$ km; and $K = 3$ W m^{-1}K^{-1} = $\rho C_P D$, where density $\rho = 3$ g cm^{-3} and heat capacity $C_P = 1$ J g^{-1}K^{-1}.

7.1.1.1 Geometry and history

Fourier's 1-dimensional (1-d) equation with constant D:

$$\frac{\partial T}{\partial t} = D\frac{\partial^2 T}{\partial z^2} \tag{7.1}$$

leads to the temperature profile of a cooling half-space (Fig. 7.3):

$$T(t, z) = T_{mtl}\,\text{erf}\left(\frac{z}{2\sqrt{Dt}}\right) \tag{7.2}$$

where erf is the error function and t is time. Eq. (7.2) applies to a surface that is held at $0\,°C$ for all time and T_m is the temperature in $°C$ of the half-space initially and at $Z = \infty$ always (Carslaw and Jaeger, 1959).

Equation 7.2 was inappropriately used by Kelvin to make his now famous, gross underestimation of Earth's age as 94 Ma. Obviously, Earth is a sphere, not a huge Cartesian block extending to infinity below some surface (the $+z$ direction), and laterally to infinity. Oceanic lithosphere plates are instead finite in all directions. A model with infinite y is reasonable, since slices of lithosphere perpendicular to y behave similarly. Plate models, being finite in z and x, are considered to be a better approach than the half space.

7.1.1.2 Time is wrongly equated to distance

All plate and half-space cooling models use the substitution:

$$x = ut \tag{7.3}$$

not to analyze motions, but instead to determine temperatures. However, t and any spatial variables are independent in Fourier's equation for any coordinate system (Chapter 2).

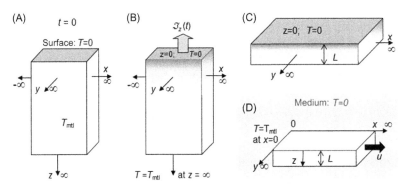

FIGURE 7.3　Model geometry. (A) Initial conditions in the cooling half-space. Gray color = upper surface at $z = 0$ fixed at $0\,°C$. The solid extends to infinity in all directions away from that surface. (B) The half-space at time t, where the surface has cooled according to Eq. (7.2), such that the base at $z = \infty$ is held at the initial T_{mtl}. (C) The infinite plate after a small time. Temperatures are nearly the same as in the cooling half-space. (D) A plate moving in the x direction, which has a constant temperature at $x = 0$. This situation is discussed in the Appendix.

PRESENT DAY   CRETACEOUS 65 million years ago   JURASSIC 150 million years ago   TRIASSIC 200 million years ago

FIGURE 7.4 Reconstruction of the motions of continents over the time frame considered in plate and half-space cooling models. *Public domain from Kious, J., Tilling, R.I., Kiger, M., Russel, J., 1995. This Dynamic Earth: The Story of Plate Tectonics. (online ed.). United States Geological Survey, Reston, Virginia, USA. ISBN 0-16-048220-8. <http://pubs.usgs.gov/gip/dynamic/historical.html> (accessed 08.03.19.).*

Eq. (7.3) being inapplicable is obviated by considering what is being modeled. The heat flux (Figs. 1.6 and 1.7) and the subsidence depth data have been collected over the last ~50 years. Because plates move at a few cm per year, and the Earth is cooling slowly, the database represents a snapshot of the current spatial variation. The data being modeled are a function of x, not of t.

If two spatial variables are relevant, then Fourier's 2-d heat equation describes time-dependent changes, but only if advection is not important. The plates do move (Fig. 7.4). Over the time frame being modeled, the supercontinent Pangea broke up. If a model is to address the effect of motion of the plates on the surface flux, then an advection term must be considered. The equation for heat transfer with no generation and motion in x and subsidence in z is:

$$\frac{\partial T}{\partial t} = D\frac{\partial^2 T}{\partial x^2} + D\frac{\partial^2 T}{\partial z^2} - u_x\frac{\partial T}{\partial x} - u_z\frac{\partial T}{\partial z} \tag{7.4}$$

Clearly, the substitution of Eq. (7.3) into solutions for Eq. (7.1) does not solve Eq. (7.4) and thus does not describe behavior of moving, cooling plates. McKenzie's (2018) approach is invalid because time and position are independent variables.

7.1.1.3 Plate model formulae do not satisfy the stated initial condition

The temperature equation underlying many plate models comes from McKenzie (1967), and is identical to the formula presented by Parsons and Sclater (1977):

$$T(t, Z) = T_{mtl}\frac{Z}{L} + \frac{2T_{mtl}}{\pi}\sum_{j=1}^{\infty}\frac{(-1)^j}{j}\exp\left\{-\left(\left[\left(\frac{uL}{2D}\right)^2 + j^2\pi^2\right]^{1/2} - \frac{uL}{2D}\right)\frac{X}{L}\right\}\sin\left(\frac{j\pi Z}{L}\right) \tag{7.5}$$

Eq. (7.5) provides the required boundary values of $T = 0$ at $z = 0$, and $T = T_{mtl}$ at $z = L$. However, at $t = 0$ (i.e., $x = 0$ from Eq. (7.3)), Eq. (7.5) reduces to:

$$T(0, Z) = T_{mtl}\frac{Z}{L} + \frac{2T_{mtl}}{\pi}\sum_{j=1}^{\infty}\frac{(-1)^j}{j}\sin\left(\frac{j\pi Z}{L}\right) = 0 \tag{7.6}$$

The RHS was evaluated using the auxiliary formulae for Fourier series (e.g., Selby, 1973, p. 477):

$$\frac{Z}{L} = -\frac{2}{\pi}\sum_j \frac{(-1)^j}{j}\sin\left(\frac{j\pi Z}{L}\right) \quad \text{for} -L < Z < L \tag{7.7}$$

Thus, McKenzie's formula yields $T = 0$ at $t = 0$. The plate is already cold, so how can it cool further?

First, we take Eq. (7.5) at face value. Computations performed using Mathematica confirm that $T = 0$ at $t = 0$ and at short times. At long times, T_{mtl} is attained at $z = L$, and so their basic formula describes a plate that warms up, rather than cools. Next, we look to subsequent literature. Then, we evaluate the mathematical representations.

7.1.1.4 Plate model formulae presented subsequently

Stein and Stein (1992) provide a slightly different formula than Eq. (7.5):

$$T(t, Z) = T_{mtl}\frac{Z}{L} + \frac{2T_{mtl}}{\pi}\sum_{j=1}^{\infty}\frac{1}{j}\exp\left\{-\left(\left[\left(\frac{uL}{2D}\right)^2 + j^2\pi^2\right]^{1/2} - \frac{uL}{2D}\right)\frac{X}{L}\right\}\sin\left(\frac{j\pi Z}{L}\right) \tag{7.8}$$

which they attribute to McKenzie (1967) and Parsons and Sclater (1977). Our literature search unearthed neither a correction nor an explanation for substituting 1 for $(-1)^j$. Stein and Stein's (1992) formula satisfies the boundary and initial conditions, which we ascertained using Mathematica. Eq. (7.8) is an improvement over Eq. 7.5, which we now abandon, but is nevertheless not correct, as discussed below.

McKenzie (2018) states that his earlier isotherms and heat flux were calculated not from Eq. (7.8), but by solving Eq. (7.1). Carslaw and Jaeger (1959, p. 104) and some algebra lead to the formula for cooling of a stationary slab from T_{mtl}:

$$T(t, Z) = T_{mtl}\frac{Z}{L} + \frac{2T_{mtl}}{\pi}\sum_{n=1}^{\infty}\frac{1}{n}\exp\left(-Dt\frac{n^2\pi^2}{L^2}\right)\sin\left(\frac{n\pi Z}{L}\right) \tag{7.9}$$

McKenzie (2018) states that McKenzie (1967) substituted Eq. (7.3) into (7.9) to provide T vs distance from the ridge, which is invalid (Section 7.1.1.2).

How Parsons and Sclater (1977) obtained their results is not clearly stated. From the match with the half-space model shown by Stein and Stein (1992), they used Eq. (7.9). McKenzie (2018) represents the results of Parsons and Sclater (1977) using Eqs. (7.9) and (7.3), combined.

7.1.1.5 Mathematics of plate model formulae

At long times, the exponential term in Eq. (7.9) approaches 0 and $T \to T_{mtl}Z/L$. This steady-state limit describes conditions of old plates far from the ridges. The model is valid at this limit because x and time are irrelevant.

This steady-state limit is also reached in (7.8) because the sum goes to 0. Otherwise, Eq. (7.9) for a stationary plate cannot be reconciled with Eq. (7.8). In particular, for $u = 0$

(i.e., a stationary plate), Eq. (7.8) does not depend on time or if $x = ut$ is substituted, it reduces to steady state, whereby $T = T_{mlt}z/L$. The appendix provides further discussion.

Convergence of all of the above series is slow, requiring several thousand terms as demonstrated by using Mathematica. Errors can be significant if fewer than a few hundred terms are used. The cited papers provide insufficient details about their computations.

7.1.1.6 Summary of the thermal component of plate models with implications

The flaws described above have not previously been recognized, and obviously have not been remedied in any later model. We have shown that plate models reduce to a linear thermal gradient at the conditions where these models apply, namely large distances from the ridges and steady-state conditions. Plate models do not reproduce the measured heat flux at short distances from the ridge, as shown in Fig. 7.1, because they do not describe a 2-dimensional system.

The cooling plate model is poorly reasoned. In actuality, the lower boundary condition is changing. Existence of sheeted dikes show that the thermal gradient near the ridge is predominantly horizontal, not vertical as assumed. A wholly different model is needed to address thermal behavior near the ridges.

7.1.2 Mathematical and conceptual flaws in previous descriptions of ocean floor depth

Because the equations for temperature are flawed, the subsidence equations, which are based on contraction of the cooling plates, cannot be correct. Matching the measured subsidence required making additional compensating errors.

7.1.2.1 An infinite half-space cannot shrink

Contraction or expansion of matter is described by one of:

$$\alpha_{vol} = \frac{1}{V}\frac{\partial V}{\partial T} \quad \text{or} \quad \alpha_{lng} = \frac{1}{L}\frac{\partial L}{\partial T} \tag{7.10}$$

where vol denotes volume (V) changes and lng denotes length (L) changes. The choice of α depends on the situation. Of interest is motion of heat along a certain direction, e.g., along a bar that is heated at one end and cooled at the other. If the bar is infinite, dividing by $L = \infty$ gives thermal expansivity $= 0$. The half-space cannot contract as it cools because no matter how much the half-space cools, its length is still infinite. Davis and Lister (1974) purportedly describe half-space expansion in their second equation. They did not divide by length and therefore their approach is not consistent with either definition of α.

7.1.2.2 Isostasy does not apply to an infinite half-space

An infinite half-space extends to infinity vertically (Fig. 7.3). The concept of isostasy is based on two boats floating in a sea displacing different amounts of water in accord with their different masses and areas. Each boat extends above and below the water line. A half space has infinite mass, and no bottom, so this concept cannot be applied. In describing isostasy and subsidence of the half-space, Parker and Oldenburg (1973), Davis and Lister

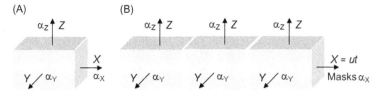

FIGURE 7.5 Schematics of directional heating or cooling. (A) Expansion of a stationary anisotropic solid, such as olivine. (B) Contraction/expansion of an anisotropic solid that is moving in the x-direction. If the solid contracts, then the velocity is slower than it would have been under isothermal conditions, and conversely. Thus, the thermal changes along x are not relevant: the measured velocity combines (lumps) the true velocity with expansion or contraction along x.

(1974), and Stein and Stein (1992) avoid discussing infinities, and base their equations on their flawed notion of shrinking an infinite bar (see above).

7.1.2.3 Improper use of thermal expansivity

All plate models use the volumetric definition of thermal expansivity, but as Hofmeister and Criss (2006) showed, cooling of a 1-d plate instead requires using linear thermal expansivity. For an isotopic solid, expansion in any one direction is governed by:

$$\alpha_{lng} = \frac{\alpha_{vol}}{3} \tag{7.11}$$

Linear thermal expansion (Fig. 7.5) is a real phenomenon that must be considered when designing bridges, buildings, and aircraft (e.g., Incropera and DeWitt, 2002), and is also used in temperature control. Linear expansion is measured via dilatometry and interferometry.

A general mathematical description of density changes during plate contraction, based on the commonly used linearization for thermal expansion (Incropera and DeWitt, 2002) is:

$$\rho(t) = \rho_i \left\{ 1 - \alpha_z [T_z(t) - T_{zi}] - \alpha_y [T_y(t) - T_{yi}] - \alpha_x [T_x(t) - T_{xi}] \right\} \tag{7.12}$$

Because temperature only depends on z, only expansion along z is relevant, so the parameter is $\alpha_Z = \alpha_{lng}$ for an isotropic solid medium.

For lithosphere moving in the x direction, expansion in x is undetectable, since any such changes are indecipherable from the plate motions (Fig. 7.4B). McKenzie (2018) agreed that α_{lng} is the correct parameter, but then postulated that the 3-fold error in existing models poses no problem, because of subsidence. His claim is invalid, as follows:

7.1.2.4 Improper subsidence calculations in plate models

Subsidence (water depth) was calculated from:

$$d_{wat} = \frac{\alpha \rho_{mtl} T_{mtl} L}{2 \left(\rho_{mtl} - \rho_{wat} \right)} \left[1 - \frac{8}{\pi^2} \sum_{j=1}^{\infty} \frac{1}{j^2} \exp \left\{ -\left(\left[\left(\frac{uL}{2D} \right)^2 + j^2 \pi^2 \right]^{1/2} - \frac{uL}{2D} \right) \frac{X}{L} \right\} \right] \tag{7.13}$$

where index j is odd only (Stein and Stein, 1992). Because Eq. (7.13) is unrelated to Eq. (7.9), which was used to calculate heat flux, the calculations of heat flux and subsidence in plate models are independent.

For $x = 0$, $d_{wat} = 0$. For large x, the ocean thickness trends to:

$$d_{wat}(X \to \infty) = \frac{\alpha \rho_{mtl} T_{mtl} L}{2(\rho_{mtl} - \rho_{wat})} = \frac{\alpha T_{mtl} L}{2(1 - \rho_{wat}/\rho_{mtl})} \tag{7.14}$$

If the correct parameter α_{lng} is used, while keeping other parameters the same as in previous models, then the water depth is three times smaller than that previously reported. For plate models to reasonably depict subsidence data requires either that L be 3-fold larger (i.e., ~ 300 km) or likewise for T_{mtl} (i.e., $\sim 3750\,°C$). Or, both could be multiplied by 1.73 (~ 173 km and $\sim 2160\,°C$).

Thus, subsidence is greatly affected by the erroneous use of α_{vol} in 1-d cooling. Furthermore, because material properties are well established, the only parameter with leeway is plate thickness. Seismic determinations give an average 65 km for the Gutenberg discontinuity, which marks the transition from the brittle lithosphere to the asthenosphere (e.g., Schmerr, 2012), and has been connected with melting (see Section 7.2). Using appropriate α_{lng} and any reasonable value for L in existing plate models does not describe ocean depths.

Subsequent plate models are more complicated, with more free parameters. The interplay of K and α in plate models was recognized by Honda and Yuen (2004). Better fits exist in multi-parameter models because multiplied (lumped) parameters provide ambiguities which allow data to be fit with incorrect physics (Transtrum et al., 2015).

7.1.2.5 Violations of mass conservation

Mid-ocean ridges add ~ 36 km^3 of mafic crust per year, which imperceptibly displaces ocean water, since the ocean volume is $\sim 10^9$ km^3. Furthermore, old crust is being subducted at an equal rate. In plate models, "subsidence" is obtained by mathematically converting rock to water, as was first pointed out by Hofmeister and Criss (2006).

In detail, isostasy is viewed as the depth of the ocean increasing to compensate for the changes in the rock column. Comparing the initial state and old lithosphere (Fig. 7.6) leads to:

$$\rho_i L + \rho_{wat} d_{wat} = \rho(t)L(t) + \rho_{wat} d_{wat}(t) \tag{7.15}$$

Although the lithosphere may solidify and cool, the mass of the rock column should not change as it is carried across the mantle. Conservation of rock mass requires:

$$\rho_i L = \rho(t)L(t) \tag{7.16}$$

Combining the above:

$$d_{wat}(t = 0) = d_{wat}(t) \tag{7.17}$$

Seafloor subsidence cannot be obtained from this model if rock mass is conserved.

Depth dependent density has no effect on this conclusion. Nor does including a small thickness for the asthenosphere wedge make any difference. This deduction can be confirmed by computing mass from the results of McKenzie et al. (2005), who utilized more parameters in his recent plate models. Including the wedge is dubious in any case because isostasy is based on a flat reference surface (Fig. 7.6).

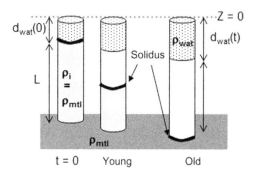

FIGURE 7.6 Schematics of subsidence in plate models. Dotted pattern = ocean; white = rock; light gray = asthenosphere/melt; dark gray = wedge below lithosphere. Thick black arc = solidus. Physical characteristics of the left-most column at $t = 0$ (which has no lithosphere) contrasts with the right-most column at old ages (which has no wedge) and the middle columns at intermediate ages (which have all three layers).

The underlying problem is that subsidence involves the mass of a block, not its height, for constant area. Cooling a body cannot change its mass.

7.1.3 Summary of existing cooling models

Different formulations for temperature vs depth were used to calculate the flux and subsidence, so these calculations are independent. The thermal model does not match the flux near the ridge, because a 1-d solution to Fourier's equation was used where a 2-d solution is required. Far from the ridge, the model reduces to steady-state. But, a cooling model is not needed when time is not germane.

Regarding subsidence, the half-space cannot contract. Plates can, but due to thermal expansivity being low and the boundary temperatures being fixed, contraction is negligible. Subsidence calculated using plate models is an artifact that corresponds with mathematical disappearance of rock mass. Notably, the formulae for temperature that were used to model subsidence do not reduce to that of a stationary plate, and therefore are not germane. The appendix provides a correct solution for a moving plate. Comparing this result with plate models provides a clue as to the error made in the original derivation of McKenzie.

7.2 Behavior of the oceanic lithosphere

7.2.1 Heat flux measurements provide a snapshot in time

The measured heat flux describes the geometry of Earth's internal emissions at the present time. Histograms of heat flux from the oceans peak at 50 mW m^{-2} whereas the median or mean values are near 60–62 mW m^{-2} (Fig. 1.7). Interesting high flux MOR areas have been preferentially sampled, whereas cold, difficult to reach deep trenches and the Artic region are under-sampled (Fig. 1.6A). Recent assessments (Vieira and Hamza, 2018) suggest a range of 50–70 mW m^{-2} for the average oceanic flux (Table 1.2), in view of sampling uncertainties. Away from the ridges, flux is fairly uniform and near 50 mW m^{-2} (Fig. 1.6B; Fig. 7.1).

7.2.2 Temperatures are roughly steady-state

Steady-state conditions are indicated by several lines of evidence. (1) Flux over most of the ocean floor is fairly constant (Fig. 1.6; 1.7; Table 1.2). Note that volcanic ridges have only about twice the heat flux of the flat, old ocean floor (Fig. 7.1) and cover only $\sim 12\%$. (2) Lava production is actually tiny. The total volume of upwelling magma is ~ 30 km^3 per year, deduced from the average spreading rate of 6 cm y^{-1}, the average thickness of 7 km of oceanic crust (Section 1.1.2.2), and the total length of the ridges which is about 80,000 km (from various sources). For comparison, the volume of Pacific Ocean water is 10^9 km^3. Individual eruptions create sheeted dikes of ~ 1 m wide, which are miniscule compared to the lateral extent of seafloor, $\sim 10^6$ m. (3) Oxygen isotopes from ophiolites (oceanic crust uplifted and preserved on the continents) show that the heated subsurface ocean water is buffered by the rocks (Gregory and Taylor, 1981). The large mass of the rock and the continual circulation of water is the key reason for chemical buffering, whereas the slow circulation rate prevents disturbance of the ocean temperature from that maintained by the immense Solar flux. The hydrothermal equilibration time of ~ 0.1 Ga revealed by ophiolite studies (Gregory and Taylor, 1981) indicates that seafloor spreading has occurred over several such intervals. (4) The oldest magnetically-striped basaltic sea-floor being ~ 0.34 Ga (Granot, 2016) corroborates that the present day seafloor production has gone on for a substantial length of time, and sets a minimum. (5) A maximum of 1.8–1.9 Ga is set by several occurrences of ophiolites (e.g., Stern, 2005). Preservation of older oceanic lithosphere is difficult due to their coastal location and erosion (R.T. Gregory, personal communication). Moreover, the generally low-lying ocean crust must be somehow uplifted unto continents, so special tectonic circumstances are needed to provide this important window into the past (e.g., Gray and Gregory, 2003).

Mafic crust similar to that existing today has been produced over a substantial fraction of Earth's recent history. This requires steady-state conditions.

7.2.3 Ocean floor depths are mechanically controlled

The key feature of the ocean floors is the broad expanses of uniform depths (Fig. 7.2). The ocean floor is highest near the continents, and lowest near subduction zones, but its variations are rather subdued. Statistical assessment shows that ocean floor is on average at -3440 m with respect to sea level (Fig. 7.7, after Weatherall et al., 2015). This average is only 190 m lower than the ocean ridge mean at -3250 m (Vérard, 2017, who used the earlier GEBCO dataset of 2009: see websites). The curve for the mid-ocean ridge depth (not shown: see Vérard, 2017) is a broad and skewed Gaussian, with a full width at half maximum of ~ 900 m. Thus, mean ridge elevation is indistinguishable from the average depth of -3200 m for Earth's entire rocky surface (Weatherall et al., 2015). These three statistical parameters differ little (Fig. 7.7) due to the flatness of the ocean floor (Fig. 7.2).

Gaussian fits to the ocean floor have a peak at -4350 m. The peak position varies little for any reasonable division between land and ocean (Fig. 7.7). The arrow at -200 m represents the flat lying continental shelves, but the continental slope extends further outboard. The Gaussian distribution represents the deep ocean plains, which are the

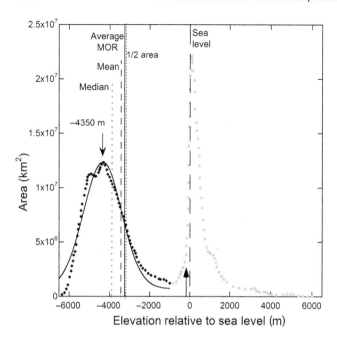

FIGURE 7.7   Area covered as a function of elevation. The curves are from Weatherall et al. (2015) analysis of 2014 GEBCO data. The average depth of the mid-ocean ridges is from (Vérard, 2017). The 2008 and 2014 databases yield similar statistics, even though the more recent data base covers considerably more area. Arrow = depth of −200 m for the continental shelves. The Gaussian fit has a peak at −4350 m for the −700 m division shown and a baseline for the Y axis of 0. The peak position varies from −4365 to −4330 m, for the ocean depth starting point of −200 to −2000 m, and whether a horizontal baseline is used for the fit or not.

most common type of ocean floor (Fig. 7.2). The statistical mean of −3440 m is less negative because this includes the continental slopes, which we confirmed by integrating the curve in Fig. 7.7. Thus, the mafic lithosphere is better represented by a depth of about −4340 m.

Deviations of bathymetric depths from the average depth at −3440 m, or the depth of the plains at −4340 m, are controlled by two main factors. One is attachment to the continents, which are less dense and pull the edges of the denser ocean floor up when abutted (e.g., the Atlantic Ocean). The second factor is the tilt of the slabs, since subduction has the opposite effect of pulling ocean floor down. Slabs dipping eastward subduct at a shallow average angle of 27° whereas westward dipping slopes are very steep, almost vertical in their middles, due to differential rotation (Fig. 1.9; Riguzzi et al., 2010; Doglioni and Panza, 2015). The overall flatness of the ocean floor stems from the sum of these competing effects.

Because Earth's outermost layer is laterally dichotomous, isostatic balance concerns the distribution of the two lithosphere types (Fig. 7.4). The oceans lie on top of the mafic crust and lap on top of 17% of the continents for a starting depth of −700 m, or 10% for a starting depth of −200 m (from integrating the data in Fig. 7.7). Water depth contributing to isostasy, as assumed in plate and half-space models, is a red herring because oceans overly both types. From Table 7.1, the isostatic balance between felsic and mafic crust occurs in the lithosphere, between roughly 50 and 100 km depth, depending on the shape and distribution of the continental roots, and their density. A reasonable depth is the average lithosphere thickness, but the balance could extend to 320 km, where the eastward dipping slabs break and differential rotation occurs (Fig 1.9), which is similar to the depth where upper mantle seismic profiles converge (Fig. 1.2).

TABLE 7.1 Characteristics of Earth's outer layers.

Layer	Depth range, km	Density, g cm^{-3}	Mass, kg[a]
Oceans	0–3.44	1	0.125×10^{22}
Oceanic crust	3.44–11	2.9[b]	0.78×10^{22}
Continental crust	0–40[c]	2.83[d]	1.67×10^{22}
Continental lithosphere	0–100[e]	2.83	4.3×10^{22}
Oceanic lithosphere	3.5–63.5[f]	3.0	6.3×10^{22}
Upper mantle[g]	~0–410	3–3.4	62.3×10^{22}
Transition zone[g]	410–660		44.8×10^{22}

[a]Computed from the volume of spherical shells, considering areal coverages.
[b]Carlson and Raskin (1984). The deeper lithosphere is assumed to be slightly denser.
[c]Average thickness from Mooney et al. (1998).
[d]Christensen and Mooney (1995). The sub-continental mantle is likely denser.
[e]Because continental roots extend to 200–250 km, and the oceanic slabs partially underplate continents, the rounded sum of 40 + 60 km is used to represent the thickness of the continental lithosphere.
[f]The average depth of the lithosphere-astenosphere boundary is ~65 km (Schmerr, 2012), which we round to 60 km, given the focus of his study on the Pacific, which is thicker than the Atlantic.
[g]From percentages Anderson (2007). The mass of the Earth is 5.972×10^{24} kg.
Notes: the oceans cover 70% of the globe, and the land 30%.

7.2.4 Seismic data on plate thickness and its changes with subduction

Studies of earthquake foci provide information on the thickness of subducting plates. Double zones of earthquake foci at intermediate depths have been well documented since the 1970s. Similar behavior occurs at great depths (Fig. 7.8).

Most intermediate depth earthquakes show down-dip compressional focal mechanisms in their upper planes and down-dip tensional focal mechanisms in their lower planes (Fujita and Kanamori, 1981). Tsukahara (1980) argued that the difference is related to unbending of the slab below the hinge. For the well-studied slabs near Japan, these zones are ~10 km thick each and merge with depth, disappearing at ~200 km (e.g., Tsukahara, 1980; also see Fig. 1.7, from Doglioni and Panza, 2015).

In certain slabs, like the intermediate depth earthquakes, two merging planes are observed for deep events (e.g., Fig. 7.8). The tension and compression of the planes are mostly opposite those of the intermediate depth earthquakes, which is consistent with the Tonga slab bending when it strikes the top of the strong lower mantle, rather than unbending, as occurs near the surface. Thus, the double planes of earthquakes have the same mechanisms regardless of the depth. Wiens et al. (1993) assumed that the lower plane of the deep earthquakes was internal to the slab. This hypothesis is based on their thickness at depth equaling the parameter L of 100 km from plate models advocated by his thesis advisor (Stein and Stein, 1992). The problematic nature of this model (Section 7.1) was not then recognized, but logically, plates warm as they descend. Thinning is required since the lower boundary is defined by temperature, is evident in the decreasing separation of the zones with depth, and is quantified below.

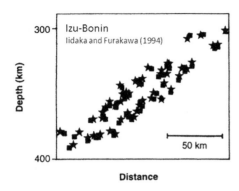

FIGURE 7.8 Example of a double seismic zone. Symbols distinguish earthquake foci from two datasets. Vertical and horizontal axes are on the same scale. *Reproduction of Figure 2 from T. Iidaka and Y. Furukawa, 1994, Double seismic zone for deep earthquakes in the Izu-Bonin subduction zone. Science 263, 1116–1118, with permissions.*

Fig. 7.9A A compares zone separations under Japan to three additional geographic locations, which represent quite different conditions. The slabs are categorized in accord with their spreading velocities and plate ages just before subduction onsets, using data from Müller et al. (2008) and the cited references. "Fast" indicates >8 cm y^{-1}, whereas "slow" indicates 3–6.5 cm y^{-1}. Ages for the young to very old categories are: $\sim$10 Ma, $\sim$40 Ma, $\sim$80 Ma, and >110 Ma.

From Fig. 7.9A, the slower slabs thin more rapidly with depth than the faster, which is consistent with temperatures increasing gradually with depth and with thermal equilibration of the exterior preceding that of the interior. This finding is supported by zones of thick (old) and thin (young) slabs merging at the same rate, if their speeds are similar. The parallel trends exist because all slabs have the same composition, the same K, and thus warm at the same rate.

Thinning with depth is supported by the data in Fig. 7.9B, which was constructed from the consistent analysis of zone separation for intermediate depth earthquakes of Brudzinski et al. (2007). We did not include their dataset in parts (A) and (C) of Fig. 7.9 because their method yields a slab width that is narrower than those reported in other studies and our graphical determinations. Fig. 7.9B shows that slabs which are old and thick upon subduction remain thicker at $\sim$120 km depth than slabs which start out young and thin. Fig. 7.9B further shows that the thickness of the slab upon production at the ridge is near 7.5 km, which matches the thickness of the solidified ocean crust.

Slab thickness at the surface is not determined by earthquakes, but instead is defined by study of seismic reflections. The Moho depth, which defines the oceanic crust, is so determined. The next deeper boundary is that between the brittle lithosphere and the much less rigid asthenosphere, the lithosphere-astenosphere boundary (LAB), which is ascertained from a decrease in shear velocities. Results from Schmerr (2012) in Fig. 7.9C show that depth to LAB increases with the age of the ocean floor, but considerable scatter exists. Less scatter exists in the trend of LAB depth with distance from the ridge, which is the geographic variable (see figure S2 in Schmerr, 2012).

Our intercepts at the surface from the double zones of earthquakes for old slabs, determined by fitting in Fig. 7.9A, coincide with LAB depths. In contrast, our intercepts for young slabs agree with crustal thickness (Fig. 7.9C). The trend is smooth. Independent

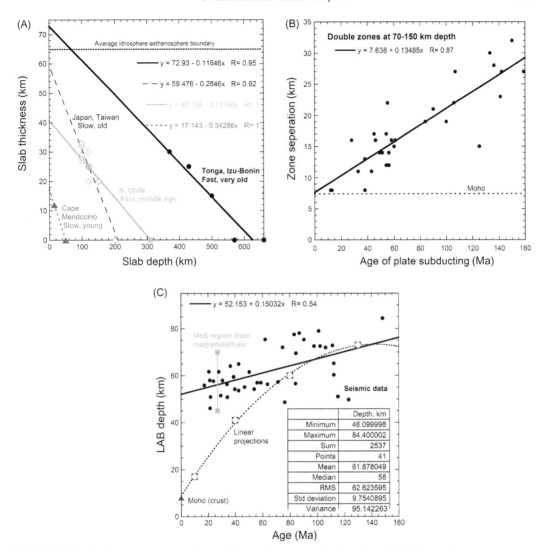

FIGURE 7.9 Various measures of plate thickness from seismologic studies. (A) Distances between planes of earthquakes, determined graphically. Data from Zhao et al. (1994) and Kao and Rau (1999) were combined in fitting. Otherwise, individual studies (Smith et al., 1993; Wiens et al., 1993; Comte et al., 1999) were fit. The depth where slabs disappear were also taken from Doglioni and Panza (2015): see Fig. 1.7. Average LAB from Schmerr (2012). (B) Data on the separation of double Benioff zones re-analyzed by Brudzinski et al. (2007). (C) Distance from the surface to the lithosphere-asthenosphere boundary. Dots and statistics = the compilation of Schmerr (2012) with shear velocity difference of 5% or greater. A few points above 160 Ma were excluded as these may be affected by locations at the edge of continents. Gray = the depth range of melt offshore of Nicaragera of Naif et al. (2013), which indicates that seismic determinations provide depths near the middle of the melt zone. Squares = the linear projections of the trends in part A to the surface. The fit here is a second order polynomial. Triangle = the average crustal thickness, which was not used in the polynomial fit.

electrical measurements of the melted region (Naif et al., 2013) indicate that the average LAB describes the middle of the melted region. The difference between projected thickness and LAB depth at the ridge is attributable to magma upwelling at the ridge. Many have argued for a magma chamber under MOR, but the geometry is debated. In view of the above, the intercepts are the best indication of the thickness of the solid plate, which asymptotes to ~ 70 km for the old Pacific plates.

7.2.5 Mid-ocean ridge characteristics

Mid-ocean ridges are volcanic fissures, and like the Hawaiian Island volcanic chain in the middle of the Pacific Ocean and Mt. Kilimanjaro on the African plains, are higher than the surrounding terrain. An obvious explanation is that magma is hotter and less dense than adjacent rock, and rises through the crust via some zone of weakness, and so the mid-ocean ridges are elongated hot bumps on the oceanic plates. This inference is quantifiable:

Heat flux correlates with spreading rate (Fig. 7.10A). The y-intercept is consistent with the average flux over all ocean floor (Table 1.2), although this average is affected by over-sampling the hot ridge areas. Because larger spreading rates mean proportionately more magma erupting, the linear trend is consistent with ridge heat flux being a magmatic per-turbation. With weak spreading, the heat flux is little elevated over the global average. This behavior is like Hawaii and other "hot spots" (e.g., Courtillot et al., 2003; Foulger, 2010; Chapter 1). The volume of magma annually building the Hawaiian chain is tiny com-pared to that of the ridges, so extra flux is not resolved within the uncertainty of the mea-surements. However, Hawaiian flux, being similar to the oceanic average, is slightly higher than that from the old crust at plate edges.

Ridge elevations likewise correlate with spreading rates (Fig. 7.10B), despite the individ-ual data points having considerable scatter, and some ridges being near trenches. This con-nection results from more magma upwelling at faster spreading centers. Ridge elevation involves not only buoyancy of the magma, but the time that the path is utilized. Eruptions following the same vertical path for a substantial time are observed on land (Mt. Kilamanjaro) and under oceans (Hawaii).

The MOR system is a zone of weakness that we attribute to longitudinal fractures imposed on the brittle lithospheric shell by the combined forces of Earth's spin and self-gravitation (see figures and discussion in Chapters 1 and 5). Overall similar spreading rates and heat flux for all ridges is consistent with this assignment, as are slightly slower spreading rates (figure 3 in Müller et al., 2008) and lower heat flux (Fig. 1.6B) near the poles. The consistency of the values, along with the longevity of the system and other observations (Section 7.2.2), points to heat evolution at the ridges being nearly steady-state.

Heat released at the ridges is a combination of heat advected by magma, latent heat released upon magma solidification, and heat evolved during cooling of the formed lavas. These processes cause the flux near the MOR to be higher than that under the old, flat middle of the oceans. From Chapters 1 and 5, plate cracking and motions originated in dif-ferential rotation and Solar torque. The energy involved is evaluated below.

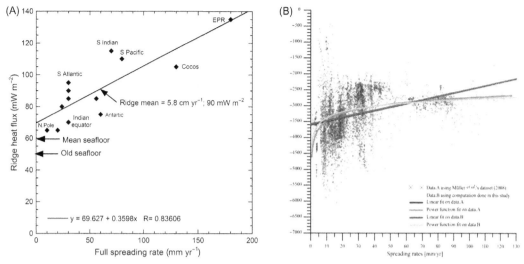

FIGURE 7.10   Dependence of ridge characteristics on spreading rate. (A) Heat Flux. The points here represent averages over segment length, not peak values which are commonly reported in the literature. The averages in this panel were estimated from the maps of heat flux in Vieira and Hamza (2018) and Müller et al. (2008), which show binned data. Uncertainties are about ± 10 on either axis. (B) Ridge elevation, showing the data of Müller et al. (2008), and linear and power law fits of Vérard (2017) to his own processing of the data and to the Müller et al. (2008) dataset. Vérard's (2017) data extend closer to $X = 0$. *Part (B) is a reproduction of Figure 6b of Vérard, C. (2017) Statistics of the Earth's Topography. Open Access Library J. 4, E3398. Available from: https://doi.org/10.4236/ oalib.1103398, which is licensed under the Creative Commons Attribution International License (CC BY 4.0).*

7.2.6 Sub-oceanic processes and their associated energy and power

The measured heat flux is unlikely to contain significant contributions from secular cooling of the upper mantle or its radioactive emissions. As argued by many, the continental crust contains most of the uranium and thorium that would be expected from a chondritic mantle. If heat sources deeper than ~ 660 km are present, these emissions do not reach the surface over the age of the Earth (Criss and Hofmeister, 2016; Chapters 2 and 8). The source of heat emitted currently from ocean floor originates in friction, as follows:

Heat is produced by friction at the base of the plates, whose motions vary in direction and magnitude, with a range of ~ 0 to almost 11 cm y^{-1} (see the figures in Doglioni and Panza, 2015). However, the thermal base of the lithosphere is defined by the solidus, so any frictional heat generated at the base of the lithosphere will cause melting. The magma so produced will uptake and store heat generated by friction as latent heat (Fig. 7.11). However, due to tilting of the plates away from the ridges, magma made at the base will flow towards the ridges. Hence, some of the heat generated below the lithosphere is advected to the ridges. Melting at the base provides lubrication, which reduces friction and heat production.

Differential rotation exists between the surface and the interior. This is a global phenomenon so it must occur beneath the continents. The frictional heat rises through the low velocity zone (Fig. 7.11). For certain compositions and depths, melting can occur, which

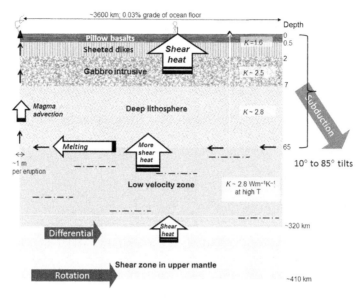

FIGURE 7.11   Schematic of heat generation and transport beneath Earth's oceans. Layers are shown with different fill patterns, with depths listed on the RHS. High temperature values of thermal conductivity (K) from Section 7.3 are indicated in the light colored boxes. Single arrows indicate motions of various entities. Subduction (medium gray) involves the crust and the lithosphere, which cross the LVZ and either break off by ~320 km, if eastward-dipping, or bend at 660 km, if westward-dipping. Differential rotation (dark gray) is independent of subduction. Location in the solid upper mantle, from about 410–320 km, generates substantial heat (large white arrow). This heat should weaken the low velocity zone, and provide some melt. The lithosphere base is defined by the solidus, which is a mechanical and thermal boundary. The magma migrates and rises at zones of weakness, which leads to a slight tilt of the ocean surface, if long lived, and positive feedback. Flowing magma advects heat (medium-sized white arrows).

uptakes latent heat, but then releases it as the material solidifies at cooler, higher levels. In this manner, the flux generated by differential rotation diffuses upwards, and seismic velocity in the LVZ is low due to partial melting (Anderson and Sammis, 1970), and temperatures in the LVZ are buffered to the solidus.

The plates also generate heat by friction as they descend from the surface. Some melt is produced on their top surfaces by water from sediments lowering the melting point: this melt rises vertically and does not provide lubrication or lateral redistribution. The plates cross the LVZ, terminating in the upper mantle at various depths (Fig. 1.8). Heat is released below the surface during plate descent to as much as 660 km, but this cannot be distinguished from heat from differential rotation.

The heat flux of 50 Wm^{-2} far from the ridges represents the basal shear without a magmatic contribution plus deeper heat due to differential rotation. Near the ridges, all three processes contribute to the heat flux. Multiple processes occurring, which are interrelated, creates uncertainties in estimating the heat produced. A few observations, in addition to those discussed above, are helpful.

The symmetry of the surface heat emissions is roughly spherical, and is analyzed as such, due to the sampling limitations and to widespread application of spherical harmonics to the planets (Fig. 1.6A; Vieira and Hamza, 2018). However, both the heat flux and spreading rates are low towards the poles and high near the equator. This gradation is consistent with dissipation of Earth's spin being an important heat source. However, the heat flux is unlikely to be zero even at the poles because the tectonic polar axis is tilted from the spin axis (Fig. 5.8), and heat flux measurements have considerable scatter (Fig. 7.1). Uncertainties in the data are significant (e.g., Figs. 7.1 and 7.10).

As sketched in Fig. 7.11, buoyant melt will reach the surface if some path is available (e.g., Hawaii, in the middle of the Pacific). The mid-ocean ridge system is slightly elevated with respect to the ocean basins, which promotes flow towards these features. This is a positive feedback loop. However, the tilt of the subducting plates far exceeds the gentle slope observed near the ridges, and therefore likely controls magma flow. This deduction is demonstrated by the East Pacific rise being the fastest spreading center, where it is parallel to the Peru-Chili trench. Slightly to the north, on the Cocos plate where the East Pacific rise is not parallel to the Mexican trench, the plate velocity is a bit slower, but remains faster than most mid-ocean ridges. Mid-ocean ridges other than the East Pacific rise are not near subducting plates: all spread about the same rate, where their full spreading rate (Fig. 7.10A) is similar to the average plate velocity.

Back arc basins have different geometries than MOR, where the spreading centers are short in length, slabs are close and dip westward and steeply towards the centers, and hinge rollback occurs. Spreading rates of the Manus, Lau, Marianas, and E. Scotia basins vary from 3 to 14 cm yr^{-1} (e.g., Taylor and Martinez, 2003), so are overall faster than the average MOR, and their heat flux is also slightly higher (70–80 mW m^{-2}: Fig. 1.6). These data are consistent with magmatic advection elevating emissions at very active oceanic volcanic centers.

7.2.6.1 Power from the measurements

The average magmatic flux at MORs (Table 7.2) was calculated in two ways: one is from the yearly addition of lava ($\sim$30 km^3; Section 7.2.2) which releases latent heat at 400 J g^{-1} and additionally advects heat at 1 J g^{-1}K^{-1}. The second is the product of the area of the high flux ridge regions (12% of the ocean floor from Fig. 7.1) and their typical flux (70 mW m^{-2} from Fig. 7.9A). These estimates of 3.5 and 3 TW, respectively, agree within uncertainties.

The total heat flux (power) from the oceans is that of the old floor (the basal value of 50 mW m^{-2}) and magmatic advection. The sum (21–22 TW for the oceanic crust) is slightly lower than the previous estimates, which are affected by over-sampling of high flux areas. (Note: most reports provide power for the globe, which sums continental plus oceanic areas.)

As a cross-check, we integrated the power fit to the heat flux data shown in Fig. 7.1. The result of 23 TW (Table 7.2) is consistent with the summed basal and magmatic values.

All data are in agreement, which strongly supports a near-surface, magmatic origin of heat flow at the ridges. Also supported is that the heat derived from the mantle is rather uniform (the basal value). Note that the currently popular value for global heat flux of 40–44 TW is based on erroneous plate models, see Section 7.1.

TABLE 7.2 Comparison of velocity, kinetic energy, radioactive energy, and power.

Process	Velocity	Energy	Power	Proposed origin; notes
	km s^{-1}	J	TW	
Ridge magma flux[a]	$\sim 10^{-12}$		3[b]; 3.4[c]	Advection below oceanic lithosphere
Old ocean floor flux[a]			18[d]	Differential rotation under oceans
Average ocean floor flux[a]			21	Both (all friction from spin down)
Measured total ocean flux[a]			23	From fitting data (Fig. 7.1)
Continental surface flux[a]			8.9[e]	Average radioactivity plus spin down under land
Continental basal flux[a]			0.2[e]	Spin down under land from Fig. 1.3
Plate tectonics	$\sim 10^{-12}$	220000[f]	0–10[g]	Solar torque: released at MOR
Total westward drift	$\sim 10^{-12}$	10^6	~ 28	Differential rotation of globe; see text
Car on freeway	0.03	500000	10^{-7}	Human activity
Earth's expected spin[h]	?	4.89×10^{29}		Formational[h]
Earth's current spin	0.46[i]	2.14×10^{29}		Gravitational, value as measured
Spin dissipation			Fig. 7.10	Assume loss proportional to energy
Spin dissipation			~ 4[j]	Unspecified, estimated from coral rings

[a]*Power is calculated from the flux and the area over which the flux is emitted, so the total ocean and land values must be added for comparison with the global power, described in the bottom half of the table. Earth's area is 5.101×10^8 km^2, and oceans cover 70%.*
[b]*Calculated from Fig. 7.1, which shows that $\sim 12\%$ of the ocean surface is affected by ridge magmatism, and from Fig. 7.9A, which shows that the average ridges have a flux of 70 mW m^{-2}.*
[c]*From 33.5 km^3 y^{-1} mid-ocean ridge production, using latent heat of 400 Jg^{-1}, density of 2.9 g cm^{-3}, cooling from 1450 K to 750 K (liquidus to black smoker temperatures), and heat capacity of 1 Jg^{-1}K^{-1}. Considering cooling to 950 K gives 3.1 TW. Lava production is uncertain by $\sim 20\%$, giving a spread of 2.2–4.2 TW.*
[d]*From 50 mW/m^2 (Fig. 7.1; Vieira and Hamza, 2018) and the oceans covering 70% of the globe.*
[e]*Our surface estimate is based on the surface average of 58 mW m^{-2} (Vieira and Hamza, 2018) and that the continents cover 30% of the globe. The continental flux derived from radioactivity in felsic rock does not pertain to the mantle, due to its high placement (see Ch. 6). Basal flux is set at 12 mW m^{-2} from xenolith data (Fig. 1.3) and using K = 2 W m^{-1}K^{-1} in the thermal models of Ch. 6.*
[f]*From average drift = 6 cm y^{-1} (Doglioni and Panza, 2015) and total lithosphere mass.*
[g]*Calculated from power = $\varepsilon M_L g v_{plate}$, assuming that the coefficient of friction ε is 0.01, but this could be lower due to lubrication by melt, or higher, if melt is near solidus.*
[h]*Currently, Earth's spin is lower than the line defined by the planets in Fig. 1.12, indicating an additional mechanism for dissipation, which is consistent with Earth being the only body with plate tectonics. We therefore consider the planetary line as formational, and this difference as powering spin dissipation and plate tectonics.*
[i]*Surface tangential velocity from the NASA factsheets.*
[j]*Change in spin rate computed from data on Devonian coral growth (Mazzullo, 1971). The tangential velocity in the Devonian (0.36–0.42 Ga ago) was 0.53 km s^{-1} from these measurements, spin was about 15% faster than today.*

7.2.6.2 Energy and power estimates for plate tectonics

Evaluating the frictional heat *energy* produced by plate motions requires no assumptions regarding the cause. The kinetic energy associated with plate tectonics is very small. For example, a car traveling on a freeway possesses greater kinetic energy than a drifting continent (Table 7.2).

However, estimating the power is uncertain, because the coefficient of friction (ε) is involved:

$$\Pi = \varepsilon M_{lithos}\, g u_{ave} \tag{7.18}$$

Drag can be close to zero with lubrication and/or melting, as is well-known. The 0–10 TW range in Table 7.2, considers values of ε from 0 to 0.01. This estimate uses the mass of the mafic lithosphere under the oceans (Table 7.1) where its base is defined by the solidus (Fig. 7.11). With frictional heating between blocks of similar materials, each block is heated equally. For the lithosphere with its base at the solidus, we assume all the frictional heat goes into the melted interface. Although our estimate is rough, it is compatible with the 3–4 TW of heat advected at the mid-ocean ridges. If we require these two independent estimates to balance, then the effective coefficient of friction is 0.003 at the lithosphere base, which is consistent with lubrication.

7.2.6.3 *Energy and power for the current spin loss*

Evaluating the frictional heat *energy* produced by differential rotation likewise requires no assumptions regarding the cause. The 6 cm yr^{-1} tangential velocity of westward drift is associated with the layers above 320 km (Fig. 1.9). The mass involved is 8.6% of Earth's total mass, which consists of the dicotomous crust, the lithosphere, and most of the upper mantle (Table 7.1). The energy is not large (Table 7.2).

Power is roughly estimated from Eq. (7.18), using 9.8 m s^{-2} (Fig. 5.8). For the coefficient of friction, we use 0.003, because this value for ε produces a power for motions of the surface plates that equals the extra heat emitted at the ridges. The result is a power of 28 TW for the current westward drift. This rough estimate is based on the mass of the lithosphere and therefore represents the sum of the power out of the ocean floor (21–22 TW + 3–4 TW advection) plus the power emitted inside the upper mantle. The difference of $\sim$3 TW, which is equivalent to $\sim$20 mW m^{-2} under the continents, is a rough estimate. Nonetheless, this value is compatible with the calculated basal flux of $\sim$18 mW m^{-2} for a 130 km thick lithosphere and our reference conditions (Fig. 6.7B). Thus, the frictional heat in the upper mantle is generated throughout, and increases as the surface is approached.

7.2.6.4 *Power estimated from spin dissipation with time*

The 6 cm yr^{-1} tangential velocity of differential rotation is negligible compared to 0.5 km s^{-1} of the spinning globe. Thus, the kinetic energy associated with plate tectonics and differential rotation is a tiny fraction of the energy in Earth's spin (Table 7.2). Dissipation is estimated as follows:

First, we assume that the loss is evenly distributed over time, back to the time of onset. Since the loss driving plate tectonics is apparently unique to the Earth, its rate can be quantified from comparing Earth to the planetary trend of spin energy against gravitational energy in Fig. 1.11. On this basis, Fig. 7.12 shows that linear spin down over all of Earth's history produced very little heat, 2 TW. If today's ridge magmatism were the *only* process occurring, this could be explained by spin down over most of Earth's history. Accounting for the higher basal flux under old ocean floor requires dissipation over a much shorter time. The power required for plate tectonics, which involves ridge

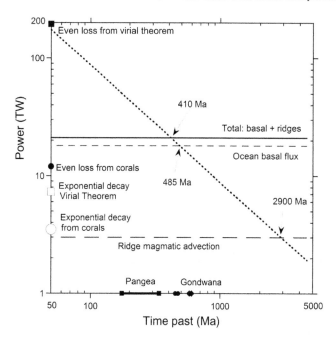

FIGURE 7.12 Power produced by loss of Earth's spin energy, calculated in various ways. Dotted line = dissipation of spin energy linearly over time, assuming that the energy loss is that relative to the trend of the other spinning planets (Fig. 1.11). Horizontal lines with symbols = time span of supercontinents, which is shown on the x-axis for convenience. Horizontal lines, as labeled = power associated with the measured heat flux on the ocean floor and the global projection of all sources. Arrows with ages point to the intersections, which suggest the onset of the process. Open symbols = estimates for the current power from the Viral theorem and data on corals, presuming exponential decay and plotted on the y-axis for convenience. Filled symbols = estimates presuming even loss of spin over time. Table 1.3 and Chapter 1 describe these calculations.

magmatism and differential rotation, requires dissipation over 260–410 Ma. The trends suggest that plate tectonics with subduction began about the time of Pangea, ∼340 Ma, which is also the age of the oldest ocean floor.

More energy may have been available over Earth's history than calculated, if the other planets are spinning down, or lost spin energy at some point in their history: this is likely for Mars (Part III). Therefore, the ages suggested above, and in Fig. 7.12 are minimal.

Second, we estimate maximum powers (Table 1.3) from the possible high spin rates developed during accretion. The Viral theorem provides abundant energy, if the loss occurred evenly over history. Loss at the beginning could have been large (Chapter 9). But, addition of some mass to the Earth after it accreted from the local nebula and spun-up would have resulted in less spin than the Virial estimate (Chapter 5). Thus, spin down provides much of the energy, but apparently not quite enough today, and so a second mechanism is needed. This finding supports torque from the imbalance solar forces participating in plate tectonics: this mechanism was invoked to explain the symmetry of plate motions being lower than uniaxial, which would be provided by spin down alone (Chapter 5).

7.2.7 Thermal conductivity data on Earth's mafic layers

Laser-flash measurements of thermal diffusivity on ∼200 crystals and rocks as a function of temperature are summarized by Hofmeister et al. (2014), Hofmeister and Branlund (2015), and Hofmeister (2019). Thermal conductivity is accurately calculated from the definition ($K = \rho C_P D$) because heat capacity C_P is well-known, or can be accurately determined from oxide or mineral summations (Robie and Hemingway, 1995; Robertson and Hemingway, 1995). Likewise, density ρ is well-known plus thermal expansivity is low and varies little among geologic materials.

Results above 273 K for the minerals and layers of the oceanic lithosphere (Fig. 7.13) are fit to:

$$K = k_0 + k_1 e^{-bT} + HT \qquad (7.19)$$

(Table 7.3). This form merges the universal equation ($D = FT^{-G} + H^* T$, for $T > \sim 250$ K, where the coefficients H and H* differ, but are closely related), with well-known behavior of C_P and ρ at high T. Heat capacity increasing with T, whereas ρ decreases, largely offset each other, and so the linear increase stems from $D(T)$ and is generally small to negligible. Eq. (7.19) is also compatible with radiative diffusion in the infrared (the first two terms) and in the near-infrared (the HT term). These terms add, due to weighting of K by heat capacity C_P for multiple mechanisms (Criss and Hofmeister, 2017; Chapter 3).

Also included are data on a fresh interior of a pillow basalt from the East Pacific rise (Galenas 2008: sample 3963-3 from Perfit et al., 2003), using C_P data from Robertson and Hemingway (1995) on similar material. Its thermal diffusivity is similar to remelted MORB glass (Hofmeister et al., 2016), due to the groundmass of pillow lavas consisting of glassy material. Highly disordered material has low and flat K above ambient temperature due to compensating behavior of D, C_P, and ρ (e.g., Branlund and Hofmeister, 2012). For this reason, we fit K of basalt and the feldspars to a polynomial (Table 7.3). Eq. (7.19) is equivalent to a linear or quadratic equation for small b. Results for two plagioclase are averaged, due to this mineral being affected by phase transitions.

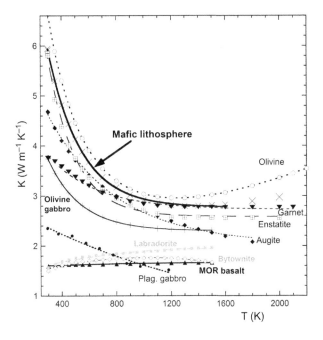

FIGURE 7.13 Thermal conductivity obtained from thermal diffusivity. Symbols and lines as labeled. Data on basalt (Galenas, 2008) and the two plagioclases were fit to a second order polynomial: all other fits were to Eq. (7.19) with $H = 0$, except for olivine (circles). Two mineral sums are shown: $\times$ = sum of 65% Fo90 + 15% En89 + 15% augite + 10% garnet. + = sum of 50% olivine + 25% labradorite + 25% bytownite. For comparison, measurements of a plagioclase gabbro by Miao et al. (2014) are included.

TABLE 7.3 Fits of thermal conductivity to Eq. (7.19) in W m^{-1}K^{-1} for temperature in Kelvins.

Material	k_0	k_1	b	H	χ^2	T_{max}
Olivine (Fo90)	1.72626 ± 0.06	12.622 ± 0.2	0.003381 ± 0.00007	0.00080561	0.027	1800
Orthopyroxene (En89)	2.5817 ± 0.025	13.745 ± 0.8	0.0049078 ± 0.0002	–	0.09	1800
Augite	1.8255 ± 0.04	4.5441 ± 0.10	0.001675 ± 0.000077	–	0.09	1800
Garnet	2.7238 ± 0.02	2.4939 ± 0.13	0.0027854 ± 0.0001	–	0.02	1600
Olivine gabbro[a]	2.2943 ± 0.003	4.944 ± 0.055	0.0041241 ± 0.000036		0.00006	1600
Lithosphere/UM[a]	2.7689 ± 0.01	10.992 ± 0.018	0.0041777 ± 0.00005	<0.00009	0.007	1800
		Polynomial fits				
Labradorite (An65)	1.4482	0.00068057 T	$-2.3392 \times 10^{-7}T^2$	–	0.009	1500
Bytownite (An59)	1.3282	0.00086025 T	$-4.228 \times 10^{-7}T^2$	–	0.004	1500
MOR basalt[b]	1.5738	0.00012085 T	$-4.2717 \times 10^{-8}T^2$	–	0.0006	1500
MOR magma	1.30 ± 0.15	–	–	–	–	<1450

[a]Calculated by summing the mineral data; gabbro = 50% olivine + 25% bytownite + 25% labradorite whereas ultramafic mantle = 60% olivine + 15% opx + 15% augite + 15% garnet.
[b]Assumed to represent the sheeted dikes. Due to the small size of the derivative, 1.60 W m^{-1}K^{-1} represents both materials over crustal temperatures.
Notes: all digits need to be used in the fitting parameters, due to trade-offs. See text for data sources. Abbreviations are:
Fo = forsterite, Mg$_2$SiO$_4$. En = enstatite, MgSiO$_3$. An = anorthite, CaAlSi$_2$O$_8$.

MOR basalt is accurately described as constant $K = 1.63 \pm 0.02$ W m^{-1}K^{-1} up to melting, or as 1.60 W m^{-1}K^{-1} near room T. We approximate the sheeted dikes, which lavas have fine-grained groundmass with glass, as having K similar to MOR basalt.

MOR magma has even lower $K = 1.34$ W m^{-1}K^{-1} (Hofmeister et al., 2016). Other basaltic melts, with silica contents from 51 to 39 wt%, yielded similar results. Transport properties of the silica rich magmas are also nearly independent of temperature. Lower thermal conductivity of melt than glass, which in turn is lower than its crystalline counterpart, contributes to the steeper geothermal gradient at the ridges (see below).

7.3 Thermal models of oceanic lithosphere

The oceanic crust and lithosphere are reasonably well-described by steady-state conditions. The Earth is radiating heat to space, but heat is continuously supplied to the uppermost mantle and plates from spin down, as indicated by slow plate velocities and the relatively tiny amounts of magma produced. The current state of plates subducting in an organized fashion has existed for 350 Ma or so. Although the geometry of the continents has changed during this interval (Fig. 7.4), the configuration ~65 Ma ago is similar to that today. Radioactive heat production of mafic lithosphere and the underlying peridotite mantle is small, and indistinguishable from the frictional sources (Table 7.3). Shear heating (Fig. 1.8) accompanying differential rotation and relative plate motions (Fig. 1.9) is recorded as the basal value of $50\,mW\,m^{-2}$ over most of the oceanic surface. From Figs. 7.1 and 7.10, the ridges have an additional $20{-}40\,mW\,m^{-2}$ due to advection. This excess is distributed over an area that is $\sim 12\%$ of the total ocean floor, yielding an average oceanic flux of $55\,mW\,m^{-2}$. This result matches the peak in the oceanic histogram (Fig. 1.7). The peak is an important measure because the mean is affected by the low flux Artic being under-sampled. Importantly, heat advected near the ridges is disconnected from the thermal gradient across the solidified lithosphere covering most of the ocean floor.

To describe steady-state, we integrate Fourier's equation:

$$\Im = -K(T)\frac{\partial T}{\partial Z} \tag{7.20}$$

using an initial temperature of $0\,°C$ for $z=0\,km$, and the basal value for the flux. Thermal conductivity is used as appropriate for the layers (Fig. 7.12; Table 7.3). The solution is:

$$-z\Im = k_0 T + \frac{H}{2}T^2 - \frac{k_1}{b}\exp(-bT) + \text{constant} \tag{7.21}$$

where the constant of integration depends on the surface condition (e.g., $0\,°C$ at $0\,km$). For constant K, the solution is trivial. Otherwise, temperature was calculated using Mathematica by iterating, constructing a table, and requiring that the RHZ converge to the LHS for T not varying more than 0.02 degrees among the three T-dependent parts. Achieving this degree of accuracy with an incremental calculation with variable K would require steps of $\Delta z \sim mm$, which was not recognized earlier.

7.3.1 Steady-state thermal profiles for mafic lithosphere older

Our model geotherm (Fig. 7.14, heavy black line) describes the solid crust (the thermal boundary layer), and is parameterized in Table 7.4. This curve is bracketed by results for lithospheres consisting entirely of basalt and dunite, which respectively have lower and higher K at any given T.

The highly insulating basalt layer provides a steep thermal gradient initial. The gradient is lower in the underlying gabbro layer. Below this layered crust, temperatures in the peridotite lithosphere climb even more slowly, but then steepen towards the base of the lithosphere, due to K decreasing with T up to melting near $\sim 1600{-}1800\,K$ (Fig. 7.14). Chapter 1 discussed direct measurements of the ocean floor and uncertainties.

7.3.1.1 Thermal boundary layer thickness under old ocean floor

Pressure experiments on peridotite phase equilibria constitute *closed* systems. The water content does not change during these experiments, which differs from the Earth, which is an open system. Measurements of oxygen isotopes, except for highly confined and chemically unusual ore deposits, provide confirmation that volatiles escape and that open system behavior is ubiquitous (Gregory and Criss, 1986).

The solidus of KLB-1 (Kilbourn Hole peridotite) from Takahashi (1986) is used to determine thermal boundary layer thickness because (1) this material appears to have a small amount of hydrous impurities, based on subsequent studies of dewatered KLB-1 (Zhang and Herzberg, 1994), and (2) the sub-lithospheric mantle should have a small amount of

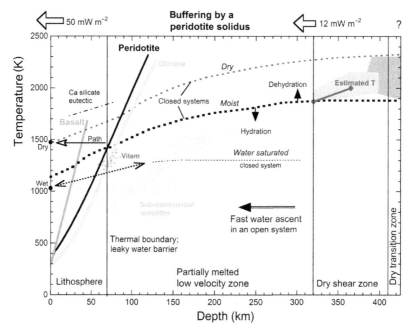

FIGURE 7.14 Steady-state thermal profiles calculated from a layered oceanic crust and lithosphere. Heavy lines = geotherms calculated for 50 mW m^{-2} and $K(T)$ from Table 7.3, as labeled. Fits in Table 7.4. Gray dots = sub-continental xenolith data (Pearson et al., 2014; see Chapter 1). Dotted curves = various peridotite solidi, all of which are collected in capsules and therefore constitute closed systems. Heavy dots = the solidus of Takahashi (1986), which involves tiny amounts of water; medium = the dry solidus (Zhang and Herzberg, 1994); and fine dots = the water-saturated solidus (Kawamoto and Holloway, 1997), which is extrapolated to melting of wet MOR basalt glass under argon (Galenas, 2008). Dot-dashed line = solidus of calcium silicates (Huang et al., 1980). Open arrow = ascent of basalt from the lithosphere base. Vertical lines separate the seismically determined boundaries connected with the lithosphere and low velocity zone. Single black arrows indicate the effects of water and its upward buoyancy in the Earth, which is an open system. Large white arrows = flux values. Gray boxes show estimated temperatures for the deeper regions, which are not molten, and have provided no samples. Dark gray line, shows a conductive geotherm for the shear region, presuming 12 mW m^{-2} is generated at the center.

TABLE 7.4 Polynomial fits to geotherms for the oceanic lithosphere distant from ridges, T in K and z in km.

Material	T_0	$T_1 z$	$T_2 z^2$	$T_3 z^3$	z range
MOR basalt	273.15	31.25	–	–	0-melting
Lithosphere, lavas	273.15	31.25	–	–	0–2
Lithosphere, gabbro	295.65	20.0	–	–	2–7
Lithosphere, peridotite	354.71	10.59	0.090787	– 0.00035572	7–130
Dunite (Fo90)	274.3	5.9495	0.16078	– 0.00068889	0–150

Notes: all digits need to be used in the fitting parameters, due to trade-offs. See text for data sources.

water, given measurements of mantle minerals (Bell and Rossman, 1992). The water saturated solidus (Kawamoto and Holloway, 1997) was determined above 50 GPa, so Fig. 7.14 includes the melting point of MOR wet basalt glass at 1 atm (Galenas, 2008). Laser flash experiments showed that glasses with 0.004−0.8 wt% H_2O melted near 1000 K. Linking our LFA data with the a water-saturated trend provides a solidus that parallels the dryer peridotite trends. Bubbles were observed on the upper surfaces of LFA samples recovered after melting. Low viscosity of basaltic melts (Hofmeister et al., 2016), combined with the high thermal expansivity of water ($\alpha = 1000 \times$ silicate melt), which increases with pressure (e.g., Chaplin, 2018), causes rapid ascent of water and water vapor in laboratory samples that are not enclosed in a small capsules or in the mantle, which are both subject to $g \sim 10$ m s^{-2}.

LFA measurements constitute an *open* system, since the vapor escapes from the top surface of the disk. The rapidity of outgassing prevented quantifying the liquidus temperature as a function of water content for the thin (~ 1 mm) samples examined in LFA: only for partially dehydrated glasses with bubbles did we observe melting at an intermediate temperature of ~ 1200 K. In contrast to the glasses, fine-grained lavas, even if hydrated, melt at the liquidus of 1475 K in LFA experiments (Galenas, 2008). The behaviors differ because heating the lavas above ~ 1000 K releases water from the tiny grains of glass, after which the water is loosely bound to the grain boundaries and so can easily ascend in the gravity field of the laboratory. Due to low magma viscosity, coupled with the kinetics, mobility, and high expansivity of water, basaltic melt in experiments on open systems is effectively produced under anhydrous conditions.

The concept of a glass transition, which is not a phase transition, has led to an oversimplified view of melting in the presence of water. This notion originated with Gibson and Giauque (1923), who studied glycerine with the purpose of understanding the 3rd law. Experimental procedures of that era would not have dehydrated glycerine, which is highly hydroscopic. Peridotite is not hydroscopic, yet dehydrating this material requires a concerted effort (see above).

For less polymerized silicate compositions, the rate of ascent is constrained in LFA studies of glass melting (Galenas, 2008; Hofmeister et al., 2016). In outgassing basaltic melts, the speed of ascent of water exceeds ~ 900 cm yr^{-1}, which is far faster than plate velocities. Hence, water at the bottom of the LVZ reaches the surface in ~ 0.4 Ma. Resurfacing the ocean is far slower, since ocean floor commonly has ages that exceed 100 Ma. From this speed and the above information, the nearly dry solidus of Takahashi (1986), reasonably represents behavior in the low velocity zone, where melting occurs in the presence of small amounts of water. Our choice is corroborated by data on sub-continental xenoliths (Fig. 1.3), which indicate that temperatures are below the "moist" peridotite solidus in the sub-continental mantle (Fig. 7.14). The Vitem xenoliths are Cenozoic (Ionov et al., 1993), whereas the other xenoliths are ~ 1 to 3 Ga old (see Pearson et al., 2014). Because flux-melting is associated with the kimberlites carrying xenoliths upwards, producing these magmas requires deep fluid storage. In contrast, the shearing motions in the oceanic lithosphere are conducive to water ascent, so the sub-continental mantle may be wetter than sub-oceanic. Xenolith data are thus compatible with Takahashi's (1986) solidus.

Eruptions of basalt at 1475 K at the ridges is also compatible with Takahashi's phase equilibria. First, temperatures should be slightly above the solidus because a significant volume of melt is observed. Second, the P, T path should be at nearly constant T: if the material cools, it solidifies and is no longer magma. Third, and in support of the above, if the magma releases significant water, it will also cool and solidify. Low viscosity magma cannot hold its water, per the open system LFA experiments and physical properties of water and steam (Chaplin, 2018).

Based on the "moist" peridotite solids, sub-oceanic lithosphere is $\sim$69 km thick under most of the oceans where the flux is 50 mW m^{-2}. The uncertainty is $\pm$7 km, accounting for leeway in various parameters (see below). Our result is consistent with seismic determinations of the lithosphere-asthenosphere boundary for the older lithosphere, and with projections of the thickness determined from earthquakes to the surface (Fig. 7.9C).

The currently popular plate model thickness of 100 km is significantly outside the uncertainty of our calculation, which is in accord with the problematic nature of previous models (Section 7.1). Such a large thickness requires an anhydrous upper mantle with a lithosphere of pure dunite (Fig. 7.13), which is unrealistic.

7.3.1.2 Effect of flux, mineralogy, and hydration on geotherms and boundary layer thickness

Near the ridges, the flux is higher than the basal value. Flux of 90 mW m^{-2} for many locations (Fig. 7.10) suggests that the gradient dT/dz is about twice as steep and that the thickness is $\sim$36 km near an average MOR: this thickness is consistent with independent measurements (Fig. 7.8C).

Xenoliths and material dredged from below the ridges have varying proportions of mafic minerals (see Chapter 1). To estimate the effect of composition, we consider two end-members. The much lower K for MOR basalt suggests a minimum limit on thickness of $\sim$30 km, whereas the high and T dependent K of 100% olivine yields an upper limit of 90$-$100 km. Both thicknesses are unrealistic.

Plagioclase, which was not included in the peridotite model, will lower K and thus would decrease thermal boundary layer thickness. Roughly, uncertainties in mineralogy provide an uncertainty in thickness of $\sim$5%. The solidus temperature decreases with hydration. Variation from water-saturated to anhydrous conditions would alter the thermal boundary layer thickness by <20%. Neither of these cases is expected, and so the location of the peridotite solidus provides an uncertainty of <5% to our determination of 69 km.

Pressure should increase K at the base of the lithosphere by 5% (Chapter 3). Although this is a systematic error, another systematic error exists that would change the boundary later thickness in the opposite direction: namely that hydration of the upper mantle should increase upwards (see below), which would progressively lower the solidus.

The sum of the effects indicates an uncertainty of $\sim$10% in our calculated boundary layer thickness of 69 km.

7.3.1.3 A leaky barrier

The solidified lithosphere impedes outgassing of the mantle (Fig. 7.14). However, the ridges and isolated volcanoes such as Hawaii provide an easy escape route for gas molecules (e.g., Taylor, 1986a).

7.3.2 Consequences of melting and dehydration in the upper mantle

Buffering along the peridotite solidus constrains upper mantle temperatures (Fig. 7.14), while also providing small amounts of partial melt, low velocities, and a slow variation of T with depth below the base of the solid lithosphere. Buffering disconnects dT/dZ from the flux arriving from below. Because the peridotite solidus is affected by the amount of water present, and slab surfaces are hydrated, slab behavior during subduction needs to be considered.

At 320 km, the eastward dipping slabs terminate. Fig. 1.9 suggests that their tips break off and sink. This depth is associated with the deepest origins of inclusions in diamonds, when a single inclusion with multiple phases is studied (see Chapter 1). Rock samples, borne upwards by magmas, come from shallower depths (above $\sim$220 km: Fig. 1.3). Shear is clearly present below 320 km, and probably extends to the seismic boundary at 410 km, but the frictional heating in this region does not produce buoyant magmas that carry deep samples upward. Thus, the material below 320 km is solid, and more refractory than peridotite above 320 km.

An obvious explanation of the dearth of magmas from below 320 km is mantle dehydration. As the temperature climbs towards the deeper region of shear and heat production, water thermal expansivity greatly increases, whereas α of silicates is nearly constant. Increasing pressure has a similar disparate effect on α. Thus, the buoyancy of water (or water vapor) relative to silicates or their melts is enhanced with depth, and so any water or vapor that is deeply generated, rises. The deeper regions must be dryer.

Progressive dehydration of the upper mantle with depth is supported by many observations.

1. The role of water released from slabs in producing andesites is well-known (e.g., Taylor, 1986b).
2. Back-arc basins release water at shallower depths (e.g., Taylor, 1986b).
3. The uppermost $\sim$3 km of oceanic crust (sediments, pillow lavas, and sheeted dikes) have more water and are cold, $\sim$273 K at their surface. As the slabs descend, these cold surface layers are in contact with much hotter sub-lithospheric mantle, $\sim$1475 K. Conduction rapidly warms this previously melted veneer to ambient temperature, releasing most of the water in the slabs, while dryer, central regions of slabs warm slowly (e.g., Hauck et al., 1999).
4. Independent of thermal models, the time for thermal equilibration of slabs can be estimated from flux data (Fig. 7.1). By $\sim$20 Ma, slabs moving away from the ridges have cooled from their formation at the solidus to steady-state conditions. The reverse procedure (warming) occurs during subduction. The corresponding distance along the plate from the erupting ridges is $\sim$1200 km, which is the length of the longest and shallowest east-dipping slabs (Fig. 1.9). The steepest of the east-dipping slabs are $\sim$500 km long: including the broken tips suggests a total length of $\sim$1000 km. The steeper, west-dipping slabs are $\sim$400 km long where they cross into the shear zone, and are generally shorter than 900 km. By the time all slabs reach 320 km, their central regions could still be cold, but their upper surface layers would have equilibrated, if not melted.

5. Seismic detection of slabs requires that these differ substantially from the surroundings. The upper surface is chemically distinct from peridotite, but the lower surface is considered to be similar to upper mantle material. Thus, detection of the lower boundary requires substantial temperature differences. Assimilation, which follows thermal equilibration, removes any vestigial boundary. As noted above, the tips of the longest slabs in Fig. 1.9 are only ~20 Ma old. If slabs oddly stayed cold, then ages >100 Ma should be seen, as on Earth's surface.

6. Rapid recycling of slabs in the upper mantle and transition zone is indicated independently from earthquake foci (Section 7.2.4). Bulbous slabs below 660 km, deduced from tomographic images, are not supported. For additional discussion of the limitations of seismic tomography, see Foulger et al. (2013).

7.3.2.1 Buffering of temperatures below the thermal boundary layer

Given the speed of water ascent, the base of the lithosphere is not water saturated. The "moist" solidus of Takahashi (1986) reasonably matches the temperature at the continental base, and therefore is used down to 320 km. Below 320 km, the temperature should be higher since water was removed from this region and shear heating is present.

It is possible that material below 320 km is more refractory, since calcium silicate inclusions in diamond samples formed at this depth (see Chapter 1). Ca silicates melt at high T and P (Huang et al., 1980). However, diamonds were brought to the surface in CO_2 gas-backed explosions, and do not necessarily represent typical mantle. Since calcium silicates are not found in xenoliths, we infer that peridotite represents the upper mantle. Hence, the dry peridotite solidus of Zhang and Herzberg (1994) sets the upper limit from 320 km to the 410 km seismic discontinuity.

Mantle composition below 410 km is poorly constrained. This seismic discontinuity is attributed to a structural transition in the major phase olivine, in which case the dry peridotite solidus sets the upper limit on temperatures down to 660 km. However, a chemical change could contribute to the 410 km discontinuity. This possibility is suggested by the transition zone receiving only the steepest of westward dipping slabs and by the unusual chemistry of the deepest material (Ca silicates). Thus, subduction could induce mixing of upper mantle peridotite with a more refractory underlying composition, as follows:

7.3.2.2 Recycling and heterogeneity

The crustal mass, formed from melt that migrated to the ridge, composes only 2% of the Earth's mass above ~320 km. This fraction is consistent with partial melting of the LVZ under steady-state conditions, since experiments produce basalt upon 2–4% partially melting of peridotite (e.g., Falloon et al., 2008).

Because very little oceanic crust exists that is older than ~125 Ma, this duration approximates the time needed to pave the ocean floor. Consequently, ~10 paving cycles that each involve ~69 km thick plates are needed to "fill" the upper mantle by subduction, and ~7 more paving cycles are needed to "fill" the transition zone (Table 7.1 lists layer masses). Recycling all mass above 660 km thus takes a minimum of 2 Ga, which is the age of the oldest ophiolites. Hence, recycling is very slow, and so homogeneity of the upper mantle is unexpected. With partial melting and incomplete recycling, a mixture of material above 660 km is expected. Importantly, the material which is most different, the 7 km veneer of basalt, diabase and gabbro, is relatively wet, low melting, and is exposed to high T immediately upon subduction. Hence, the upper mantle should not be grossly heterogeneous.

Recycling not completely homogenizing the upper mantle is consistent with magma compositions and petrology experiments. From Na and Fe contents of MOR glasses, Presnall and Gudfinnsson (2008) have argued that the magmas were buoyed upwards from a consistent depth of 40−50 km having temperatures of 1520−1550 K, but that mantle down to ∼140 km participated in producing melt. Niu and O'Hara (2008) agree with the shallow holding depths, but argue that chemical variations arise from differences in mantle sources and melting processes. Both seem correct: the process of partially melting in the LVZ produces fairly uniform compositions, but with a degree of heterogeneity. Greater heterogeneity is indicated by ocean island basalts, which have signatures of melted pyroxenite, not just peridotite, and have a deeper origin than ridge basalts, depths of 100 km or so (e.g., Herzberg, 2006; Lambart et al., 2013).

7.3.3 Thermal conduction in the shear zone

The conductive gradient in the shear zone is controlled by the flux entering the base of the continents. From Chapter 6, the flux $\Im$ under thick continents is low, ∼ 12 mW m^{-2}. Chapter 6 showed that the flux entering the continents is larger for thinner continents, which is consistent with our model of shear in the LVZ (Fig. 7.11).

At high temperatures indicated near 320 km (Fig. 7.14), the HT term in Eq. (7.19) must be considered. This increase in K with T is based experimental measurements of olivine (Fig. 7.13), which is fairly transparent in the overtone region (Hofmeister, 2019). Data on olivine were collected nearly to the melting point, permitting coefficient H to be constrained (Table 7.3). Data on the pyroxenes did not extend to such high T, so their H is unknown. The Cartesian form of Fourier's equation, Eq. (7.1), is used due to the uncertainty in H and $\Im$, and Earth's modest curvature at this depth.

Given the ambiguities, we assume that the flux is generated near the center of the shear zone (365 km) and that roughly equal amounts flow in both directions. Near 2000 K and ambient pressure, peridotite thermal conductivity is ∼3 W m^{-1}K^{-1} (Fig. 7.13). Pressure increases K at a rate of 4% GPa^{-1} (e.g., Hofmeister, 2019), which provides K ∼ 4 W m^{-1}K^{-1} and dT/dZ ∼ 3 K km^{-1} in the shear zone. As depths climb from 320 km, temperature should increase by ∼1 K per km.

If Takahashi's (1986) solidus describes T at 320 km, then the increase in T for all shear occurring at 365 km remains well below the dry solidus (Fig. 7.14). Distribution of shear throughout the zone lowers T at 365 km, while maintaining the 3 K km^{-1} gradient at 320 km. Due to the uncertainties, a range of temperatures is sketched in Fig. 7.14. We reevaluate temperature in the transition zone in Chapter 8, considering lower mantle and core conditions.

7.4 Summary

Despite 50 years of investigations, elementary mathematical errors in plate models have gone unrecognized. Consequently, these flawed models have been incorporated in many studies, leading to misunderstanding the thermal state of the Earth as well as its internal workings. Plate models do not describe behavior of the oceanic lithosphere, because this layer is not subsiding as a response to cooling. Rather, the depth of the ocean floor is controlled by the tilt angles of subducting oceanic lithosphere which competes wtih its nearly horizontal merging with sub-continental lithosphere (e.g., the Atlantic). Likewise, flux across the ocean floor does not record changes in the plates with time. Instead, the flux represents the present-day Earth, which is nearly steady-state, but is not spherically symmetric.

The main problem with plate models is conceptual: plates do not have a set thickness. The plate bottoms are not chemically distinct from the underlying mantle, but are defined by the melting point of peridotite. Hence, thickness is determined by the flux, which is nearly constant except for the $\sim 12\%$ of the ocean floor area adjacent to the ridges, and also by complex thermal conductivity of the near-surface layers. Our thermal models for surface conditions yield a depth for the lithosphere-asthenosphere boundary that is compatible with seismic measurements. This finding, combined with the solidus of peridotite that is measured at different water contents, constrains temperatures down to ~ 365 km, while showing that the upper mantle has low amounts of water at shallow depths, but is anhydrous by 320 km, which is the limit of the low velocity zone.

Measured flux mostly consists of a constant basal value of $50\,\text{mW m}^{-2}$ provided by shear in the LVZ. But, due to the tilt of the plates, the buoyant melt at their base flows up to the ridges, thereby augmenting ridge heat loss via both advection and release of latent heat upon solidification. Our deductions are supported by estimation of various energies and powers associated with the motions. The behavior is consistent with our mechanism of differential spindown that is supplemented by imbalances in forces on the lithosphere during Earth's complex orbit about the Sun (Chapter 5). The scant data near the poles, which provides lower flux than the equatorial zones, supports our inferences.

Seismic data on separations of the double Earthquake zones are fully compatible with the plates thinning as they warm during subduction. The warming rate is connected with speed whereas the time to warm is proportional to thickness of the plate just before it subducts. These behaviors are logical consequences of higher temperatures at depth and conduction. The process involves progressing warming of the bottom until it re-equilibrates with the surroundings and then is assimilated The relatively wet and chemically distinct top crust first melts, and then follows the same sequence as the bottom. Using seismic data, we show that the plates are recycled mostly above 320 km. Only the few plates that are thick when they subduct, steep, and fast can reach 660 km. Our calculated recycling times resemble the ages of the oldest ophiolites, which provides corroboration. Hence, the upper mantle is distinct mechanically from the lower mantle, and the thermal states are likewise unconnected. The large seismic discontinuity at 660 km appears to chemical in nature, as there is no means of mixing the layers.

This chapter did not model transport of heat and material during magmatism in detail. These processes are intimately tied with Earth'a chemical evolution, and are covered in Chapter 9, after we decipher the current thermal structure of the deep layers in Chapter 8.

References

Anderson, D.L., 2007. New Theory of the Earth, 2nd ed. Cambridge University Press, Cambridge.

Anderson, D.L., Sammis, C., 1970. Partial melting in the upper mantle. Phys. Earth Planet. Inter. 3, 41–50.

Bell, D.R., Rossman, G.R., 1992. Water in the Earth's mantle: role of nominally anhydrous minerals. Science 255, 1392–1396.

Branlund, J.M., Hofmeister, A.M., 2012. Heat transfer in plagioclase feldspars. Am. Miner. 97, 1145–1154.

Brudzinski, M.R., Thurber, C.H., Hacker, B.R., Engdahl, E.R., 2007. Global prevalence of double Benioff zones. Science 316, 1472–1474.

Carlson, R.L., Johnson, H.P., 1994. On modeling the thermal evolution of the oceanic upper mantle: an assessment of the cooling plate model. J. Geophys. Res 99, 3201–3214.

Carlson, R.L., Raskin, G.S., 1984. Density of the ocean crust. Nature 311, 555–558.

Carslaw, H.S., Jaeger, J.C., 1959. Conduction of Heat in Solids, second ed Oxford University Press, New York.

Chaplin, M., 2018. Water structure and science. http://www1.lsbu.ac.uk/water/water_structure_science.html (accessed 01.09.18.).

Christensen, N.I., Mooney, W.D., 1995. Seismic velocity structure and composition of the continental crust: A global view. J. Geophys. Res. 100, 9761–9788.

Comte, D., Dorbath, L., Pardo, M., Monfret, T., Haessler, H., Rivera, L., et al., 1999. A double-layered seismic zone in Arica, northern Chile. Geophys. Res. Lett. 26, 1965–1968. Available from: https://doi.org/10.1029/1999GL900447.

Courtillot, V., Davaille, A., Besse, J., Stock, J., 2003. Three distinct types of hotspots in the Earth's mantle. Earth Planet. Sci. Lett 205, 295–308.

Criss, R.E., Hofmeister, A.M., 2016. Conductive cooling of spherical bodies with emphasis on the Earth. Terra Nova 28, 101–109.

Criss, E.M., Hofmeister, A.M., 2017. Isolating lattice from electronic contributions in thermal transport measurements of metals and alloys and a new model. Int. J. Mod. Phys. B 31, No. 175020. Available from: https://doi.org/10.1142/S0217979217502058.

Davis, E.E., Lister, C.R.B., 1974. Fundamentals of ridge crest topography. Earth Planet. Sci. Lett. 21, 405–413.

Doglioni, C., Panza, G., 2015. Polarized plate tectonics. Adv. Geophys. 56, 1–167.

Doin, M., Fleitout, L., 1996. Thermal evolution of the oceanic lithosphere: an alternative view. Earth Planet. Sci. Lett. 142, 121–136.

Dyson, F., 2004. A meeting with Enrico Fermi. Nature 427, 297, 28–29.

Falloon, T.J., Green, D.H., Danyushevsky, L.V., McNeill, A.W., 2008. The composition of near-solidus partial melts of fertile peridotite at 1 and 15gpa: implications for the petrogenesis of MORB. J. Petrol. 49, 591–613.

Foulger, G.R., 2010. Plates vs Plumes: A Geological Controversy. Wiley-Blackwell, Chichester, UK.

Foulger, G.R., Panza, G.F., Artemieva, I.M., Bastow, I.D., Cammarano, F., Evans, J.R., et al., 2013. Caveats on tomographic images. Terra Nova 25, 259–281.

Fujita, K., Kanamori, H., 1981. Double seismic zones and stresses of intermediate depth earthquakes. Geophys. J. R. Astron. Soc. 66, 131–156.

Galenas, M.G., 2008. Transport Properties of Mid-Ocean Ridge Basalt From the East Pacific Rise (BS Honors thesis), University of Missouri-Columbia, 18 pp.

Gibson, G.K., Giauque, W.P., 1923. The third law of thermodynamics. Evidence from the specific heats of glycerol that the entropy of a glass exceeds that of a crystal at the absolute zero. J. Am. Chem. Soc. 45 (1), 93–104. Available from: https://doi.org/10.1021/ja01654a014.

Granot, R., 2016. Palaeozoic oceanic crust preserved beneath the eastern Mediterranean. Nat. Geosci. 9, 701–705. Available from: https://doi.org/10.1038/ngeo2784.

Gray, D.R., Gregory, R.T., 2003. Ophiolite obduction and the Samail Ophiolite: the behaviour of the underlying margin, Geol. Soc., 218. Special Publications, London, pp. 449–465. Available from: https://doi.org/10.1144/GSL.SP.2003.218.01.23.

Gregory, R.T., Criss, R.E., 1986. Isotopic exchange in open and closed systems. Rev. Miner. 16, 91–128.

Gregory, R.T., Taylor Jr., H.P., 1981. A n oxygen isotope profile in a section of Cretaceous oceanic crust, Samail Ophiolite, Oman: Evidence for $\delta^{18}O$ buffering of the oceans by deep (>5 km) seawater-hydrothermal circulation at mid-ocean ridges. J. Geophys Res. 86, 2737–2755.

Hasterok, D., 2013. A heat flow based cooling model for tectonic plates. Earth Planet. Sci. Lett. 361, 34–43.

Hauck II, S.A., Phillips, R.J., Hofmeister, A.M., 1999. Variable conductivity: Effects on the thermal structure of subducting slabs. Geophys. Res. Lett. 26, 3257–3260.

Herzberg, C., 2006. Petrology and thermal structure of the Hawaiian plume from Mauna Kea volcano. Nature 444, 605–609.

Hofmeister, A.M., 2019. Measurements, Mechanisms, and Models of Heat Transport. Elsevier, Amsterdam, pp. 201–250 (see Chapter 4 on methods; Chapter 7 on mineral and rock data, and Chapter 11 on a radiative diffusion description of heat conduction).

Hofmeister, A.M., Branlund, J.M., 2015. Thermal conductivity of the Earth. In: 2nd Edition Schubert In Chief, G. (Ed.), Treatise in Geophysics, V. 2. Elsevier, The Netherlands, pp. 584–608. , Mineral Physics (G.D. Price, ed.).

Hofmeister, A.M., Criss, R.E., 2005a. Earth's heat flux revisited and linked to chemistry. Tectonophysics 395, 159–177.

Hofmeister, A.M., Criss, R.E., 2005b. Mantle convection and heat flow in the triaxial Earth. In: Foulger, G.R., Natland, J.H., Presnall, D.C., Anderson, D.L. (Eds.), Melting anomalies: Their Nature and Origin. Geological Society of America, Boulder, CO, pp. 289–302.

Hofmeister, A.M., Criss, R.E., 2006. Comment on "Estimates of heat flow from Cenozoic seafloor using global depth and age data" by M. Wei and D. Sandwell. Tectonophysics 428, 95–100.

Hofmeister, A.M., Dong, J.J., Branlund, J.M., 2014. Thermal diffusivity of electrical insulators at high temperatures: evidence for diffusion of phonon-polaritons at infrared frequencies augmenting phonon heat conduction. J. Appl. Phys. 115, 163517. Available from: https://doi.org/10.1063/1.4873295.

Hofmeister, A.M., Sehlke, A., Avard, G., Bollasina, A.J., Robert, G., Whittington, A.G., 2016. Transport properties of glassy and molten lavas as a function of temperature and composition. J. Volcan. Geotherm. Res. 327, 380–388.

Hohertz, W.L., Carlson, R.L., 1998. An independent test of thermal subsidence and asthenosphere flow beneath the Argentine Basin. Earth Planet. Sci. Lett. 161, 73–83.

Honda, S., Yuen, D.A., 2004. Interplay of variable thermal conductivity and expansivity on the thermal structure of the oceanic lithosphere II. Earth Planets Space 56, e1–e4.

Huang, W.-L., Wylie, P.J., Nehru, C.E., 1980. Subsolidus and liquidus phase relationships in the system CaO-SiO_2-CO_2 to 30 kbar with geological applications. Am. Miner. 65, 285–301.

Iidaka, T., Furukawa, Y., 1994. Double seismic zone for deep earthquakes in the Izu-Bonin subduction zone. Science 263, 1116–1118.

Incopera, F.P., DeWitt, D.P., 2002. Fundamentals of Heat and Mass Transfer, fifth ed. John Wiley and Sons, New York.

Ionov, D.A., Ashchepkov, I.V., Stosch, H.-G., Witt-Eickschen, G., Seck, H.A., 1993. Garnet peridotite xenoliths from the Vitim volcanic field, Baikal Region: The nature of the garnet-spinel peridotite transition zone in the continental mantle. J. Petrol. 34, 1141–1175.

Kao, H., Rau, R.J., 1999. Detailed structures of the subducted Philippine Sea plate beneath northeast Taiwan' A new type of double seismic zone. J. Geophys. Res. 104, 1015–1033.

Kawamoto, T., Holloway, J.R., 1997. Melting temperature and partial melt chemistry of H_2O-saturated mantle peridotite to 11 Gigapascals. Science 276, 240–243.

Kious, J., Tilling, R.I., Kiger, M., Russel, J., 1995. This Dynamic Earth: The Story of Plate Tectonics. (online ed.), United States Geological Survey, Reston, Virgina, USA. ISBN 0-16-048220-8. http://pubs.usgs.gov/gip/dynamic/historical.html. (accessed 08.03.19.).

Lambart, S., Laporte, D., Schiano, P., 2013. Markers of the pyroxenite contribution in the major-element compositions of oceanic basalts: Review of the experimental constraints. Lithos 160–161, 14–36.

Mazzullo, S.J., 1971. Length of the year during the Silurian and Devonian periods: new values. Geol. Soc. Am. Bull. 82, 1085–1086. Available from: https://doi.org/10.1130/0016-7606(1971)82[1085:LOTYDT]2.0.CO;2.

McKenzie, D.P., 1967. Some remarks on heat flow and gravity anomalies. J. Geophys. Res. 72, 6261–6273.

McKenzie, D., 2018. A geologist reflects on a long career. Ann. Rev. Earth Planet. Sci. 46, 1–20.

McKenzie, D., Jackson, J., Priestley, K., 2005. Thermal structure of oceanic and continental lithosphere. Earth Planet. Sci. Lett. 233, 337–349.

Miao, S.Q., Li, H.P., Chen, G., 2014. Temperature dependence of thermal diffusivity, specific heat capacity, and thermal conductivity for several types of rocks. J. Therm. Anal. Calorim. 115, 1057–1063.

Mooney, W.D., Laske, G., Masters, T.G., 1998. CRUST 5.1: a global crustal model as $5° \times 5°$. J. Geophys. Res. 103, 727–747.

Müller, R.D., Sdrolias, M., Gaina, C., Roest, W.R., 2008. Age, spreading rates, and spreading asymmetry of the world's ocean crust. Geochem. Geophys. Geosyst. 9, paper Q04006. Available from: https://doi.org/10.1029/2007GC001743.

Naif, S., Key, K., Constable, S., Evans, R.L., 2013. Melt-rich channel observed at the lithosphere–sthenosphere boundary. Nature 495, 356–359.

Niu, Y., O'Hara, M.J., 2008. Global correlations of ocean ridge basalt chemistry with axial depth: a new perspective. J. Petrol. 49, 633–664.

Parker, R.L., Oldenburg, D.W., 1973. Thermal model of ocean ridges. Nat. Phys. Sci. 42, 137–139.

Parsons, B., Sclater, J.G., 1977. An analysis of the variation of ocean floor bathymetry and heat flow with age. J. Geophys. Res. 82, 803–827.

Pearson, D.G., Canil, D., Shirey, S.B., 2014. Mantle samples included in volcanic rocks: xenoliths and diamonds. In: Turekian, K., Holland, H. (Eds.), Treatise on Geochemistry, vol. 2. Elsevier, Amsterdam, pp. 169–253.

Perfit, M., Cann, J., Fornari, D., Engels, J., Smith, D., Ridley, W., et al., 2003. Interaction of sea water and lava during submarine eruptions at min-ocean ridges. Nature 425, 62–65.

Pollack, H.N., Hurter, S.J., Johnson, J.R., 1993. Heat flow from the Earth's interior: analysis of the global data set. Rev. Geophys. 31, 267–280.

Presnall, D.C., Gudfinnsson, G.H., 2008. Origin of the oceanic lithosphere. J. Petrol. 49, 615–632.

Riguzzi, F., Panza, G., Varga, P., Doglioni, C., 2010. Can Earth's rotation and tidal despinning drive plate tectonics? Tectonophysics 484, 60–73.

Robertson, E.C., Hemingway, B.S., 1995. Estimating heat capacity and heat content of rocks. Open-file Report 95–622. U.S. Geological Survey, Reston VA.

Robie, R.A., Hemingway, B.S., 1995. Thermodynamic Properties of Minerals and Related Substances at 298.15 K and 1 Bar (10^5 Pascals) Pressure and at Higher Temperatures. U.S. Geological Survey Bulletin 2131. U.S. Government Printing Office, Washington D.C.

Schmerr, N., 2012. The Gutenberg discontinuity: melt at the lithosphere-asthenosphere boundary. Science 335, 1480–1484.

Selby, S.M., 1973. Standard Mathematical Tables. The Chemical Rubber Co, Cleveland OH.

Smith, S.W., Knapp, J.S., McPherson, R.C., 1993. Seismicity of the Gorda Plate, structure of the Continental Margin, and an Eastward Jump of the Mendocino Triple Junction. J. Geophys. Res. 98, 8153–8171.

Stein, C.A., Stein, S.A., 1992. A model for the global variation in oceanic depth and heat flow with lithospheric age. Nature 359, 123–128.

Stern, R.J., 2005. Evidence from ophiolites, blueschists, and ultrahigh-pressure metamorphic terranes that the modern episode of subduction tectonics began in Neoproterozoic time. Geology 33, 557–560.

Takahashi, E., 1986. Melting of a dry peridotite KLB-1 up to 14 GPa: Implications on the origin of peridotitic upper mantle. J. Geophys. Res. 91, 9367–9382.

Taylor, B.E., 1986a. Magmatic volatiles: isotopic variation of C, H, and S. Rev. Miner. 16, 185–226.

Taylor Jr., H.P., 1986b. Igneous rocks II: isotopic case studies of circumpacific magmatism. Rev. Miner. 16, 273–318.

Taylor, B., Martinez, F., 2003. Back-arc basin basalt systematics. Earth Planet. Sci. Lett. 210, 481–497.

Transtrum, M.K., Machta, B.B., Brown, K.S., Daniels, B.C., Myers, C.R., Sethna, J.P., 2015. Perspective: sloppiness and emergent theories in physics, biology, and beyond. J. Chem. Phys. 143, No. 010901.

Tsukahara, H., 1980. Physical conditions for double seismic planes of the deep seismic zone. J. Phys. Earth 28, 1–15.

Vérard, C., 2017. Statistics of the Earth's topography. Open Access Library J. 4, No. E3398. Available from: https://doi.org/10.4236/oalib.1103398.

Vieira, F., Hamza, V., 2018. Global heat flow: new estimates using digital maps and GIS techniques. Int. J. Terrestrial Heat Flow Appl. Geotherm. 1, 6–13. Available from: https://doi.org/10.31214/ijthfa.v1i1.6.

Weatherall, P., Marks, K.M., Jakobsson, M., Schmitt, T., Tani, S., Arndt, J.E., et al., 2015. A new digital bathymetric model of the world's oceans. Earth Space Sci. 2, 331–345. Available from: https://doi.org/10.1002/2015EA000107.

II. The thermal state and evolution of Earth

Wei, M., Sandwell, D., 2006. Estimates of heat flow from Cenozoic seafloor using global depth and age data. Tectonophysics 417, 325–335.

Wiens, D.A., McGuire, J.J., Shore, P.J., 1993. Evidence for transformational faulting from a deep double seismic zone in Tonga. Nature 364, 790793.

Zhang, J., Herzberg, C., 1994. Melting experiments on anhydrous peridotite KLB-1 from 5.0 to 22.5 GPa. J. Geophys. Res. 99, 17729–17742.

Zhao, D., Matsuzawa, T., Hasegawa, A., 1994. Morphology of the subducting slab boundary in the northeastern Japan arc. Phys. Earth Planet. Inter. 102, 89–104.

Websites (accessed 15.03.18.)

Visible Earth: a catalog of NASA images and animations of our home planet.
https://visibleearth.nasa.gov/.
General Bathymetric Chart of the Oceans (GEBCO).
https://www.gebco.net/.
Thermal diffusivity data.
http://epsc.wustl.edu/~hofmeist/thermal_data/.

Appendix

The moving rectangular rod

Plate model formulae resemble a solution to a different type of cooling problem. Carslaw and Jaeger (1959, p. 148) consider a semi-infinite rod moving along x with velocity u (Fig. 7.3D). Only the ratio of the perimeter over the area enters into their formula, and so we can adapt their solution to a rectangular cross-section. The perimeter to area ratio of a plate with width $y >> L$ is:

$$\frac{2(L+y)}{Ly} = \frac{2}{L} + \frac{2}{y} = \frac{2}{L} \tag{7.22}$$

The rod is assumed to have an initial temperature T_{mtl}, constant D, and constant K while losing its heat into a surrounding medium at $T = 0\,°C$ by ballistic radiation. The loss is described by a surface conductance κ, the above physical properties, and the geometric ratio. The loss parameter is:

$$Q = \frac{2}{L}\frac{\kappa}{\rho C} = \frac{2\,\kappa D}{L\,K} \tag{7.23}$$

For a long, wide rectangular rod only losing heat from the upper surface, Q will differ somewhat from Eq. (7.23), but will depend similarly on the physical properties. Temperature of a rod which cools radiatively as it moves is governed by:

$$\frac{\partial^2 T}{\partial x^2} - \frac{u}{D}\frac{\partial T}{\partial x} - \frac{Q}{D}T = \frac{1}{D}\frac{\partial T}{\partial t} \tag{7.24}$$

Carslaw and Jaeger's (1959) steady-state solution is:

$$T(X) = T_{mtl} \exp\left\{\frac{x}{L}\left[\frac{uL}{2D} - \left(\frac{u^2L^2}{4D^2} + \frac{QL^2}{D}\right)^{1/2}\right]\right\} \tag{7.25}$$

The moving rod must lose heat ($Q \neq 0$) or it does not cool. Thus, advection cannot be dismissed, as assumed in plate models.

Comparing Eqs. (7.5) and (7.8) to (7.25) reveals that the heat loss factor of QL^2/D in Carslaw and Jaeger's moving rod solution occupies the same position as $j^2\pi^2$ in the plate models. Although Q may need modification to better describe cooling of the plate from the upper surface only, the integer multiple of π in plate models is definitely not a heat loss factor. Clearly, McKenzie's (1967) derivation, where solutions to two different problems are combined, is invalid. Changing a sign or using odd terms cannot fix the problem.

A slightly different type of plate model

For completeness, the recent model of Hasterok (2013) is examined here, because constant basal flux is assumed, after the Chablis model of Doin and Fleitout (1996), rather than assuming constant basal temperature, after McKenzie (1967).

Hasterok claims that his equation A1:

$$\frac{\partial \theta}{\partial \eta} = \frac{\partial^2 \theta}{\partial \eta^2} \tag{7.26}$$

where $\eta = Z/(4Dt)^{1/2}$ and $\theta \propto T + $ constant, is a non-dimensional version of Fourier's heat equation. Non-equivalence to Fourier's heat Eq. (7.1) is demonstrated either by integrating Eq. (7.26), which gives $\theta = \partial\theta/\partial\eta$, or by manipulating Eq. (7.1). The correct form for a heat equation that uses variables η and θ is:

$$\left[\eta - 2 + \frac{1}{\eta}\right]\frac{\partial \theta}{\partial \eta} = \frac{\partial^2 \theta}{\partial \eta^2} \tag{7.27}$$

Solutions to Hasterok's wrong equation (A1 = 7.26) are $\theta = $ constant, or $\theta \sim \exp(\eta)$, or some combination thereof. Hence, Hasterok's equation A2 [$\theta = erf(\eta)/erf(\lambda)$ where $\lambda = Z_1/(4\kappa t)^{1/2}$ for small z_1] does not solve his equation A1. To test if his A2 is valid, we substitute his result into Eq. (7.27) and apply the chain rule. Canceling like terms gives:

$$0 = \frac{erf(z/\sqrt{4Dt})}{erf^2(z_1/\sqrt{4Dt})}e^{\left(\frac{z_1^2}{4Dt}\right)}\frac{2z_1}{\sqrt{4Dt}}\left[\frac{\partial z_1}{\partial t} - \frac{z_1}{2t}\right] \tag{7.28}$$

The only term which can be zero for all Z or Z_1 and time is that enclosed in the square brackets. Thus for Hasterock's "solution" A2 to solve the heat equation requires: $Z_1 \propto t^{1/2}$. Consequently, λ is some constant and thus $erf(\lambda)$ and $\exp(\lambda^2)$ are also constants. The solution to Hasterok's variant of the Chablis model thus reduces to:

$$T = A + B erf\left(\frac{z}{2\sqrt{Dt}}\right) \tag{7.29}$$

where A and B are constants. Eq. (7.29) is identical to the half-space cooling model, Eq. (7.2).

The remainder of Hasterok's appendix, his equation 3, and beyond in his main text rest on the errors described above. His appendix contains additional algebraic errors. First, his equation A3 has a factor of πDt which instead should be $(\pi Dt)^{1/2}$. Second, his equations A6, A7 and 3 have the factor $2(Dt)^{1/2}$, which is wrong and should be replaced by $(\pi Dt)^{1/2}$. Third, he wrongly evaluated the quotient $\lambda/\mathrm{erf}\,\lambda$. As λ approaches zero, the quotient is not unity as he asserts, but instead is:

$$\lim_{\lambda \to 0} \left(\frac{\lambda}{\mathrm{erf}(\lambda)} \right) \to \frac{1}{2}\pi^{1/2} \tag{7.30}$$

No model with such glaring defects can elucidate processes in the Earth.

8

Thermal structure of the lower mantle and core

"As true as steel, as plantage to the moon, As sun to day, as turtle to her mate, As Iron to Adamant, as Earth to Center." *William Shakespeare*, **1609**.

The temperature profile inside the Earth is essential to connect seismic velocities with laboratory data, and to understand interior behavior. However, constraining Earth's present thermal state is challenging for several reasons. Foremost, samples from below ~ 300 km are lacking (Chapter 1). Heat flux is only measured on the surface and only

Heat Transport and Energetics of the Earth and Rocky Planets
DOI: https://doi.org/10.1016/B978-0-12-818430-1.00008-2

records emissions from the outer layers at the present time due to Earth's insulating nature (Chapter 2). A fortunate consequence is that the large cooling length and time scales permit evaluation of temperatures deep inside the Earth independently of the cooling of the outermost layers to space. Temperatures of the continental and oceanic crusts, and the lithosphere and upper mantle, are covered in Chapters 6 and 7.

Convection of the lower mantle is not possible, for reasons discussed by Hofmeister and Criss (2018) and summarized in Chapter 3. Briefly, the lower mantle is strong and under hydrostatic compression, so the yield strength criterion for convection in an extremely viscous medium is not met. Neither is the Rayleigh number criterion met, when this formula is modified to account for the three transport properties in solids taking on greatly disparate numerical values. Rayleigh had assumed gas-like behavior, where kinematic viscosity equals both thermal diffusivity and mass diffusivity, which also holds for small molecule liquids such as water, but not for large molecules (see chapters 5 and 6 in Hofmeister, 2019). Unlike gases that convect under any thermal gradient, and liquids that convect when exposed to a high thermal gradient, solids will melt. Force, not heat, is needed to move solids (Chapter 5). Regarding the core, this is heated from above and has a strong gradient in gravity downwards. Each of these conditions is stabilizing against convection. Therefore, interior temperatures are governed by conduction, but are greatly influenced by the location of melt and relevant phase boundaries. Again, we note that isentropes (constant S) cannot be substituted for adiabats (constant q) for a self-gravitating body (chapter 1 in Hofmeister, 2019).

Importantly, heat flows away from the location of its generation. The lower mantle is locally hot because radionuclides should be negligibly present in the metallic core, and because these were removed from the upper mantle by formation of the continental crust. The surface is cold, which defines one direction of heat flow from the lower mantle, and melting in the core consumes and stores heat, and so heat also flows down into the top of the core. A local temperature maximum thus exists in the lower mantle.

Second, co-existence of melt and solid iron alloy at the inner core boundary (ICB) sets the temperature through buffering (Criss and Hofmeister, 2016; Section 1.3.2). This situation is not steady-state because internal heat from the mantle is crossing the core mantle boundary (CMB) and progressively melting more material. However, melting is slow and so in discussing the present temperatures with scant information on Earth's interior, steady-state is a useful approximation here. Evolutionary behavior is explored in Chapter 9.

The third constraint on the thermal state of the deep interior is that temperatures in the upper mantle cannot exceed the dry peridotite solidus. Other constraints exist, but are equivocal, and rest on comparison with chondritic meteorites, which are part of the outer Solar System, not the inner, and not the Earth. The ~ 200 km region immediately above the CMB, which appears to be transitional and partially melted (e.g., Simmons and Grand, 2002) is useful, but its composition is unknown.

A geotherm is constructed in this chapter using the limited constraints on the interior. In addition, the procedures in Chapter 6 are adopted: Section 8.1 provides analytical formula for a spherical geometry to permit exploring tradeoffs between thermal conductivity, temperature, and internal heating. Section 8.2 estimates plausible amounts of internal heating in the lower mantle. Section 8.3 discusses the constraints on the temperature connected with chemical composition of the core and lower mantle. Section 8.4 describes thermal conductivity at high pressure and temperature, focusing on dense insulators, with some discussion of metals. Section 8.5 presents two possible geotherms for the lower mantle and core. Section 8.6 provides discussion.

8.1 Theory

Due to the low ellipticity of the Earth, spherical symmetry reasonably describes the interior. The core can be treated as a single entity due to buffering in a two-phase system (Section 8.3), which permits using a two layer model. Absence of melt in the transition zone suggests a temperature at a depth of 660 km (Chapter 7). Otherwise the outer layers need not be considered in modeling deep the interior, due to the immensity of Earth (Chapter 2; Criss and Hofmeister, 2016).

8.1.1 Steady-state approximation

From data on iron meteorites (e.g., Mittlefehldt et al., 1998), the core is essentially devoid of heat-producing elements. This is consistent with their lithophilic character. Hofmeister and Criss (2013) solved the case for steady-state where all the heat generation resides in the lower mantle (subscript mtl) above a barren core and constant thermal conductivity (K):

$$T - T_{top} = \frac{\Psi_{mtl}}{6K_{mtl}} \left[s_{top}^2 - s^2 + 2s_{core}^3 \left(\frac{1}{s_{top}} - \frac{1}{s} \right) \right] \quad \text{for } s_{core} \leqq s \leqq s_{top} \tag{8.1}$$

where Ψ is internal heat generation, and s is radius. "Top" denotes the seismic discontinuity at radius $s = 5700$ km (660 km depth), which defines the lower mantle. The CMB is at $s = 3480$ km (2891 km depth).

Once steady-state is achieved, the mantle temperature depends in no way on thermal transport properties of the core. This finding stems from the nature of conduction in the sphere (Chapter 2).

The core temperature from Eq. (8.1) is everywhere constant, and takes on the value evaluated at $s = s_{core}$ of:

$$T_{core} = T_{top} + \frac{\Psi_{mtl}}{6K_{mtl}} \left[s_{top}^2 - 3s_{core}^2 + 2\frac{s_{core}^3}{s_{top}} \right] = T_{top} + 1.82 \times 10^{12} \frac{\Psi_{mtl}}{K_{mtl}} \tag{8.2}$$

The numerical value on the RHS pertains to physical properties being based on meters. Eq. (8.2) shows that tradeoffs exist among the four different properties that govern lower mantle temperatures.

Flux, obtained by differentiation, is governed by:

$$\Im(s) = -\frac{\Psi_{mtl}}{3} \left[\frac{s_{core}^3}{s^2} - s \right] \quad \text{for } s_{core} \leqq s \leqq s_{top} \tag{8.3}$$

Steady-state conditions constrain the flux at the two boundaries:

$$\Im(s_{core}) = 0; \quad \Im(s_{top}) = -1.46\Psi_{mtl} \qquad (8.4)$$

where the numerical factor pertains to Ψ in $\mu W\ m^{-3}$, for $\Im$ in $W\ m^{-2}$. Thus, heat generated in the lower mantle leaks into the transition zone at a constant, but non-negligible rate.

Some useful constraints can be set by comparing results from Eq. (8.4) to the flux carried by conduction:

$$\Im(s) = -K\frac{\partial T}{\partial s} = -K\frac{\Delta T}{\Delta s} \qquad (8.5)$$

The RHS assumes constant K over the distances being considered.

8.1.2 More realistic conditions

Prior to reaching steady-state, heat leaks in both directions and so the lower mantle has a temperature maximum. This is local due to buffering occurring in overlying and underlying layers. We now discuss consequences of our assumptions.

Because refractory phases have widely varying thermal conductivity, a series approximation is relevant for a rock:

$$\frac{d}{K} = \sum X_i \frac{d_i}{K_i} \rightarrow \frac{1}{K} = \sum \frac{X_i}{K_i} \qquad (8.6)$$

where d is grain-size and X is the fraction of each type of the i minerals composing the lower mantle. Validity of this formula was confirmed by Merriman (2019). The harmonic mean on the RHS is based on sub-equal grain-size and shows that the insulating phase generally dominates, unless the mixture is predominately the conductive phase. Hence, various monomineralic compositions for the lower mantle are considered.

The opposite case of a barren mantle and radioactive core is described by Eq. (2.14). For this unlikely circumstance, core conduction is relevant, and the temperature profile is more complicated. But because the ratio of the two internal heat contents affects departures in temperature from Eq. (8.1), and buffering controls temperature in the 2-phase core, approximating the core as barren core suffices.

Thermal conductivity depending on T and s requires a much more complex model. From the above, it is only variations of the mantle that pertain, which are discussed in Section 8.4.

Without knowing the composition of the lower mantle, assuming constant K and Ψ_{mtl} is a reasonable approach. However, in investigating the CMB, we considered the effect of pressure on K.

The next few sections describe available constraints and the approximations needed.

8.2 Constraints on internal heat generation

We start by assuming that internal heat in the lower mantle is generated by radioactive decay. One goal is to evaluate whether radioactive emissions are sufficient to sustain the current core temperatures, or if another source of heat is needed, such as differential spin of the core-mantle boundary (discussed below).

From Chapter 6, the present-day, volumetric heat generation (in $\mu W\ m^{-3}$) is given by:

$$\Psi_{mtl} = \left[0.095 \times (\%K) + 0.266 \times (ppm\ U) + 0.071 \times (ppm\ Th)\right] \frac{\rho}{2700} \qquad (8.7)$$

where density (ρ) is in $kg\ m^{-3}$, and all concentrations are for the three bulk elements as reported by conventional analysis. Using a lower mantle density of $4000\ kg\ m^{-3}$, and elemental concentrations from Lodders and Fegley (1998) suggests $0.02\ \mu W\ m^{-3}$ if the lower mantle has a chemical composition like that of either an average chondritic meteorite or an ultramafic rock. Obtaining a similar numerical value for the sub-continental lithosphere from thermal modeling (Chapter 6) supports using $\Psi_{mtl} = 0.02\ \mu W\ m^{-3}$ as our reference value.

A range must be considered, since the mantle composition is unknown. One possibility is Taylor's (1982) composition for the Moon, which has little K, but considerable U and Th. For this case, Ψ_{mtl} is $0.045\ \mu W\ m^{-3}$, which is not significantly different. Considering carbonaceous chondrites gives a fairly low value of $0.013\ \mu W\ m^{-3}$, but this estimate is also on the same order-of-magnitude. The upper continental crust has $1\ \mu W\ m^{-3}$, but such a high concentration of radioactive elements is unexpected for the lower mantle. Importantly, carbon should be present in substantial amounts (Hofmeister and Criss, 2013), which would provide less heat in the lower mantle than the reference value. The extended range of $\Psi_{mtl} = 0.001$ to $0.3\ \mu W\ m^{-3}$ considered below covers the possibilities.

8.3 Constraints on chemical composition and melting temperatures

Without samples, meteorite compositions are used to suggest possibilities. See Chapters 1 and 11 for additional discussion of these materials.

8.3.1 The core

Iron meteorites consist of several distinct Fe-Ni metal phases, which range between 4 and 60 wt% Ni, such that most have between 5 and 20 wt% Ni (reviewed by Mittlefehldt et al., 1998; Grady and Wright, 2003). Alloys consist of domains of various phases, each of which can also have domains. Meteoritic alloys also have ubiquitous inclusions of semi-conducting and insulating phases: troilite (FeS), schribersenite [(Fe,Ni)$_3$P], and graphite (C). Inclusions of carbides such as (Fe,Ni,Co)C$_3$, as well as FeCr$_2$S$_4$, are common. Occasionally, CuFeS$_2$, CrN, or FeCr$_2$O$_4$ are present. Silica and phosphates exist but are rare.

Bulk chemical analyses on iron meteorites are difficult to perform by electron microprobe due to the ubiquitous presence of carbon. However, chemical compositions of the alloy phases, tabulated by Mittlefehldt et al. (1998) are compatible with the elements in the non-metallic included phases. Specifically, Co ~ 0.1 wt% $>$ Cu $>$ Cr $>$ As $\sim$ Mo $\sim$ Pd, where each greater sign indicates an order of magnitude difference. Regarding the light elements, S is near 1 wt%, P ranges from 0.1 to a few wt%, whereas C is below 0.2 wt%. Nitrogen is variable, but reaches 1 wt% in the Ni-rich component of the iron alloy. The various iron meteorites are not much different chemically, so trends in the minor elements are used to distinguish types. Types IAB and IIE are considered primitive, while all others have characteristics indicative of melting. The gradation of iron meteorites into stony irons, along with the scarcity of silicate, oxide, and phosphate phases suggest that these three types of inclusions result from secondary processing, after accretion.

The above data on iron meteorites suggest a core composition of Fe alloy that contains large amounts of Ni, minor additional transition metal elements, and substantial amounts of the light elements S, P, C, and N in that order. The latter are mostly derived from the secondary phases as the core melts with time.

Many high pressure melting experiments have been performed, but not on a complex composition like that of iron meteorites. Melting at core pressures have been reached, starting with pure Fe^0 (Li and Fei, 2014). Extremely high temperatures once considered for the core stem from such experiments on pure Fe^0, and belief in conversion of all gravitational potential to heat during accretion, which is not possible (Chapter 4).

Phase equilibria for Fe-Ni alloys have not been much explored much: Ni contents considered are at the lower limit of those observed, and additional transition metals such as Co have not been included. The experiments on 5 wt% Ni by Morard et al. (2011) show that the eutectic melting occurs at 2750 K in the Fe-Ni-S system. This temperature is within experimental uncertainty of eutectic melting in the Fe-S system, and suggests the transition metal cation chemistry is probably not important to melting in the core. The light elements, in contrast, have a measureable effect: eutectic melting is at 2750 K for $Fe-Fe_3S$ (Chudinovskikh and Boehler, 2007; Mori et al., 2017), at 3000 K for $Fe-Fe_3C$, and at 3200 K for Fe-FeO (Morard et al., 2017). Oxygen in the core is inconsistent with meteorite compositions and chondritic abundances, and is not considered further.

Phosphorous has not been explored at high pressure. At ambient pressure, concentrations above 0.45 wt% in cast iron (Fe alloys with significant carbon and minor sulfur) have a eutectic at 1100 K (Abbasi et al., 2007). Without phosphorous, cast iron has a eutectic at 1400 K. Thus, presence of the light element P in the core should greatly lower its solidus. Based on the parallel eutectic curves for the systems explored (Morard et al., 2017; Mori et al., 2017), a eutectic for the Fe-P system at pressure should be parallel to that of the Fe-S system, but could be lower by 300 K.

Dissolved N_2 does not provide a eutectic at 1 atm, but instead lowers the melting point of Fe in proportion to its impurity content (e.g., You et al., 2018). How the light elements work together under high pressure is unknown, but further reduction in the eutectic is plausible. Thus, the Fe-S system sets an upper limit on core temperatures. For a reasonable lower limit, we subtract 300 K from the higher pressure measurements on Fe-S of Mori et al. (2017) which extends nearly to the ICB. Extrapolating indicates a solidus of 3800 K at the ICB radius of 1220 km.

It is not known whether the outer core is 100% melted. The density contrast at the ICB is 6%, which is not large. Tradeoffs make it difficult to ascertain both the chemistry of the outer core and proportions of solid and liquid phases. Melting should be substantial, based on the absence of shear waves in the outer core, but the presence of even a small amount of solids buffers its temperature to the solidus.

If the Earth were not compressed, the outer core would be isothermal. The same process applies for depth-dependent melting. The solidus temperature increases with depth:

however, heat is not flowing from a colder to hotter regions. Instead, heat from the slightly hotter mantle just above the CMB is transformed into latent heat and stored as the core is incrementally melted at the CMB. Due to buffering, the incrementally supplied heat from the mantle progressively moves the ICB inwards.

8.3.2 The mantle at 660 km and below

Seismological comparisons do not provide a definitive answer for a multi-component system, particularly as temperature varies as well as composition (Chapter 1). It is unlikely that the lower mantle has the identical chemical composition as the upper mantle since the plates are recycled mostly above 300 km (Chapter 7).

Diamond could be an important lower mantle constituent, considering solar abundances and its refractory nature (Hofmeister and Criss, 2013). Also, chondrites contain refractory inclusions (CAIs) that have been linked to the pre-Solar nebulae. From Brearly and Jones (1998), the common phases are hibonite ($CaAl_{12}O_{19}$, with up to 8 wt% Ti and other cations), perovskite ($CaTiO_3$), spinel (nearly $MgAl_2O_4$), melilites (from $Ca_2Al_2SiO_7$ to $Ca_2MgSi_2O_7$), anorthite ($CaAl_2Si_2O_8$), and low Fe pyroxene (fassaite, $Ca(Mg,Fe,Al)(Si,Al)_2O_6$ with Ti in the reduced 3 + state). A lower mantle that is much richer in C, Ca, Al, and Ti, but poorer in Si, than the upper is plausible.

If interior temperatures are high, the lower mantle likely is a mixture of oxides, diamond (or metastable graphite), with some silicates. Silicate minerals have the lowest melting temperatures and thermal conductivity. On this basis, the geotherms will be constructed for two cases: a refractory oxide lower mantle vs a more reactive silicate composition.

Solidi of various peridotite samples measured by different research groups are compatible (Fig. 7.14; also see Morard et al., 2017). These materials are found in the upper mantle and restrict temperatures at the top of the lower mantle (pressure = 22.5 GPa; depth = 660 km) to 2400 K for dry peridotite (e.g., Zhang and Herzberg, 1994). Melting of oxides are much higher (see e.g., Boehler, 2001) and far above the melting of Fe alloy. This information is not used in our construction, due to oxides being better thermal conductors and therefore setting the minimum temperatures in the lower mantle (see below).

8.3.3 The region D" above the core

The ~200 km thick region above the core, which is part of the lower mantle, has drawn much interest. Seismic velocities in D" are low (Simmons and Grand, 2002) and have been interpreted as indicating partial melting. Due to limitations on resolution, tradeoffs exist between the thickness of the layer and its velocity contrast with that of the lower mantle. For example, a 40 km thick melt layer could have shear velocities 8% lower than the mantle. Due to this tradeoff, the composition of D", or of the possible, thin ultralow velocity zone, are ambiguous.

The previous view that core temperatures are very high (>5000 K), based on melting of pure Fe^0 and hot accretion, led to the view that the core and mantle reacted to form D". This is not consistent the much higher melting temperatures of silicates and oxides compared to iron alloys, nor with cold accretion of the interior, which is needed to produce the high spin rate of the Earth (Chapter 4). More likely, the ultralow velocity zone is a variant of slag, the buoyant constituent produced by melting iron ore in a blast furnace.

8.4 Thermal conductivity in the mantle and core

Without samples of the lower mantle, its thermal conductivity must be estimated. We utilize the measurements and models from Hofmeister (2019), which are summarized in Chapter 3, and the discussion of chemical composition above. Core materials are covered here for use in Chapter 9.

8.4.1 Effect of temperature

8.4.1.1 Electrical insulators

Thermal conductivity of silicates decreases from room temperature (Fig. 7.13). For dark and/or grainy samples, K asymptotes to a constant by ~ 1500 K. At higher T, K increases for single-crystals, such as olivine. This behavior is due to vibrations in the near-infrared diffusing heat at high T (Chapter 3). Fits are given in Table 7.4 to Eq. (7.33). Due to the approximations made to describe the effect of pressure (below), the upturn with T is neglected here.

The thermal diffusivity of oxides depends simply on temperature: a power law is followed, which stems from the small number and low frequency of their vibrations (Hofmeister, 2019; Chapter 3). Thermal conductivity for representative phases (Fig. 8.1) was computed from D, C_P, and density as a function of T by Hofmeister (2014). The data are well-represented by:

$$K(T) = AT^{-b} + HT \tag{8.8}$$

Even with the strong, power law decrease, oxides conduct heat more efficiently at high T than do silicates.

Insulating or semi-conducting elements have very high K that follows a power law, but coefficient H in Eq. (8.8) is negligibly small. This behavior is connected with the absence of fundamental IR modes, and so only overtone/combinations and impurity modes in the near-IR are available to diffuse heat. Graphite is shown in Fig. 8.2 because diamond is off-scale with $K = 820$ W m^{-1}K^{-1} (from laser flash data on natural diamond of Hofmeister, 2019). Much higher values are reported for diamond in the literature due to spurious radiative transfer affecting this extremely transparent material. From Eq. (8.6), very large proportions of diamond are needed for this phase to have a substantial effect on K of a mixture.

Solid-solutions and/or impurity ions lower D and K near 298 K substantially and non-linearly. The difference in D or K decreases as T increases, as is shown graphically for many silicates by Hofmeister et al. (2014). Similarly, K varies little among oxide phases above ~ 1200 K (Fig. 8.1). Table 8.1 lists the high temperature approximations for K, based on data in Hofmeister (2019).

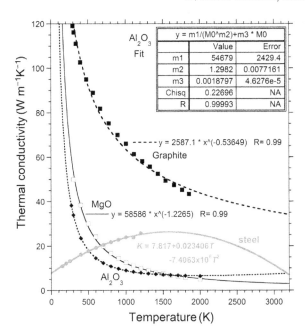

FIGURE 8.1 Dependence of thermal conductivity on temperature for refractory materials and an iron alloy. Various fits are listed. Measurements of these and related materials are summarized by Hofmeister (2014; 2019).

TABLE 8.1 Thermal conductivity approximations for the lower mantle and core.

Material	K at high T	Bulk modulus	Average K	τ^{a}
	W m^{-1}K^{-1}	GPa	W m^{-1}K^{-1}	Ga
Garnet/peridotite	2.8	170	4.2	270
Olivine	3	130	4.9	270
HT term[b]	0.5	n.a.	~0.5	~270[b]
Periclase	5	160	7.5	155
Corundum	6.5	250	8.6	155
Graphite	45	160	68	18
Diamond	290	440	342	1
Steel	27	160	40	45

[a]Cooling scale time estimated from L^2/D.
[b]Radiative diffusion in the near-IR for olivine (Table 7.4). The pressure dependence is weak and so is neglected. Because fundamentals and overtone modes exchange energy, the process is diffusive, with slow speeds. Ballistic processes, in contrast, are near light speed.

8.4.1.2 Alloys

Thermal conductivity of pure elements decreases strongly with T, whether metallic (Fig. 3.6B) or semiconducting (Fig. 8.1). In contrast, alloys have thermal conductivity that increases with T (Fig. 8.1) because D is flat or increases with T (Criss and Hofmeister, 2017), whereas heat capacity always strongly increases. Invar, steel, iron meteorites, CoMo, brass, and bronze all behave in this manner. Heat being diffused by IR overtones, combinations and impurity modes provides D and K increasing with T (the HT term) regardless of the bonding type. Thus, K of steel reasonably represents the core.

8.4.2 Effect of pressure

Pressure enhances heat conduction, through densification. From our model for diffusion of radiation (Hofmeister, 2019; Chapter 3), insulators follow:

$$\frac{\partial(\ln K)}{\partial P} = 4.6\frac{\gamma_{th}}{B_T} \cong 4\frac{\partial(\ln V)}{\partial P} \tag{8.9}$$

where B_T is the isothermal bulk modulus, V is volume, and γ_{th} is the thermal Grüneisen parameter $= \alpha B_T/(C_V\rho)$, which uses volumetricthermal expansivity, α. For metals, a numerical factor of 2.65 replaces the factor of 4.6 for the middle term. The factor of 4 on the RHS is approximate and is based on γ_{th} for metals being close to 2 whereas for insulators, γ_{th} is near unity, and that γ_{th} is roughly independent of T and P.

The generic form for $K(P)$ in Eq. (8.9) is appropriate given the approximations. The numerical factor of 4 is a little less than the average, but this approximation makes integrating Eq. (8.7) simple. The reason is that $B_T(P) = B_T(0) + 4P$ for many dense materials (e.g., Knittle, 1995). With these simplifications:

$$K(T,P) = K(T)\left[1 + \frac{P}{B_T(0)}\right] \tag{8.10}$$

We have based our model on the isothermal bulk modulus because B_S is the isentropic bulk modulus, not the adiabatic bulk modulus, and because heat is moving. Fig. 8.2 shows the trends for typical materials over the entire Earth, using the constant, high T values for K discussed above. The inferred values for K of the core are high, but not inordinately so, since these are similar to K of diamond at ambient conditions. For comparison, Table 8.1 also includes K for diamond at 1500 K from Sukhadolau et al. (2005).

8.4.3 Constant K approximation

From Figs. 8.1 and 8.2 combined, constant thermal conductivity of 4.5 W m^{-1}K^{-1} for a silicate composition, 8 W m^{-1}K^{-1} for oxides, and 68 W m^{-1}K^{-1} for graphite reasonably represent the combination of P and T effects in the lower mantle. Graphite is considered, even though diamond is the stable phase, in view of the harmonic mean of Eq. (8.6) and also because a carbide phases could occupy the CMB region.

Regarding the core, K of steel increases with both temperature and pressure, so a constant value may not be appropriate for the core.

8.5 Geotherms for the lower mantle and core

Earth cooling only to depths of ~660 km over 4.5 Ga (Criss and Hofmeister, 2016) has decoupled its exterior and interior temperatures to a good extent, which permits

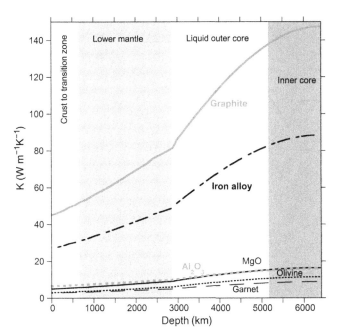

FIGURE 8.2 Variation of thermal conductivity with pressure from Eq. (8.10), constructed from the high temperature values in Table 8.1. Very different types of material are shown, to represent possible behavior in the lower mantle. Each material is shown for all three zones for comparison.

deciphering temperatures from the surface down and the center up. A reasonable merge supports the validity of the estimates. Because the continents are warm rafts that shed their heat to space above ~ 220 km (Chapter 6), only the geotherm for the oceanic lithosphere (Chapter 7) need be considered in our merge. This is well-defined to depths of ~ 320 km, due to buffering by melting in the low velocity zone (LVZ) and restrictions on upper mantle mineralogy from xenoliths.

Buffering in the partially molten outer core sets its temperature. Because the process is ongoing, melting is not complete and so the solidus equals the outer core geotherm. The Fe-S eutectic is clearly too high due to the presence of the element P in meteorite analogues, which lowers cast iron eutectics considerably (Abbasi et al., 2007). Hence, we estimate the outer core geotherm (Fig. 8.3)) as 300 K lower than that of Mori et al. (2017). Even lower core temperatures are possible. Inner core temperatures should be constant. Except at the CMB, flux in the core is negligibly small.

The lower mantle is not melted, except possibly in the zone close to the core (layer D"). This 40 km thick zone should be buffered to its melting temperature, which is unknown.

The top of the lower mantle, if substantially silicate, constrains its maximum temperature to the dry peridotite solidus. This assumption is robust at the base of the upper mantle, providing a maximum of 2400 K at 660 km for our "hot" geotherm for a silicate rich lower mantle (Fig. 8.3. Joining this with the oceanic geotherm at 320 km requires a steep slope below depths of 320 km. That shown in Fig. 8.3 is consistent with $K = 3.3 \text{ W m}^{-1}\text{K}^{-1}$ for a silicate composition and the maximum $\Im = 12 \text{ mW m}^{-2}$ at ~ 320 km (Figs. 7.13 and 7.14).

Flux produced by shear near the surface should decrease steadily from 320 down to 410 km, and should be negligible by 660 km, where the Earth is a very strong solid. A slight contribution from radionuclides in the transition zone is possible, because the

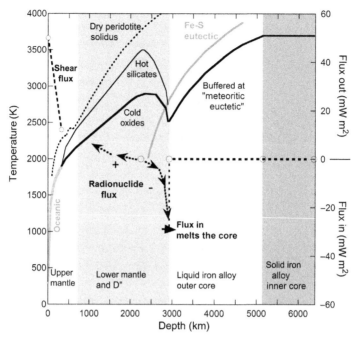

FIGURE 8.3 Conductive geotherm constructed for the Earth, considering buffering in regions of co-existing melt and solid. See text for details. Gray rectangles separate major regions. Flux (right axis) where constrained is shown as circles; light dotted lines show estimates. Temperature (left axis) for oceanic lithosphere from Ch. 7. Dotted line = peridotite solidus from Fiquet et al. (2010) linked to dry solidus of Zhang and Herzberg (1994). Dark gray curve = Fe-S system solidus from Mori et al. (2017). Heavy black line = geotherm constrained the estimated solidus of a meteoritic core, joined smoothly to the oceanic geotherm. Because the core is undergoing melting, temperatures must decrease in the lower mantle towards the core. The temperature peak exists because flux changes direction, but the depth where $\Im = 0$ is uncertain. Light black line = estimated temperature for the lower mantle, based on near melting at 660 km. Slopes at 660 and 2890 km are based on estimated K, which is particularly uncertain in the CMB region.

descending slabs recycle continental debris to ~660 km, as indicated by earthquakes (Fig. 1.9). More likely, internally generated heat provided by the lower mantle augments flux from the seismic discontinuity at 660 km up to ~410 km. The gradient from depths of 320−660 km depends on the sum of shear and radionuclide fluxes in this region. Temperatures for a "cold" lower mantle are obtained by extending the sub-oceanic geotherm to greater depths. Low $\partial T/\partial z$ for the "cold" geotherm is consistent with an oxide composition since such materials have higher K (Figs. 8.1 and 8.2).

Due to the lack of data, we estimate temperatures below 660 km by extrapolating the curves in the transition zone downward. Somewhere inside the lower mantle the flux must be zero, which is accompanied by a maximum in temperature, since the lower mantle is shedding its heat to both the barren core below and to the cooling and radiating surface above. Where the maximum temperature occurs depends on K of the mantle and the flux in each direction. The hot silicate mantle with lower K should have a very steep decline near the CMB, as shown. The flux into the core is equivocal, but we begin by estimating the present flux from the amount of core that was melted over geologic time, using known its latent heat (450–500 J g^{-1}: Aitta, 2006). Heat stored in the core due to is heat capacity is not relevant because the core was warmed up before it melted. Our estimate from melting of 6.2 TW ($\Im = 0.04$ W m^{-2} at the CMB) probably overestimates the current flux, because radioactivity decreases with time.

Alternatively, we assume that the power going out of the lower mantle equals that going into the core. The flux out of <12 mW m^{-2} (power < 0.39 TW) suggests that $\Im$ into the core is <0.32 mW m^{-2}, where we have accounted for changes in shell area with radius. Lower flux is possible. Fig. 8.3 shows this value, and provides gradients using the lower estimate for $\Im$ at the CMB. Even so, a lop-sided thermal profile is expected for the lower mantle.

Table 8.2 lists the two possible geotherms, and includes that of the oceanic lithosphere from Chapter 7 for completeness. Because very little is truly known about the lower mantle, T in this region is more uncertain than in the core. Temperatures can lie in between the hot and cold geotherms for a mixture and the shapes can vary for a graded composition. Slightly higher and lower geotherms for the lower mantle are possible. A very asymmetric peak could exist, but temperatures cannot exceed 4000 K for a silicate composition. Regarding the "cold" oxide lower mantle, its minimum cannot be lower than ~ 2500 K, which is estimated by combining data on phase equilibria on Fe-P and Fe-S at the CMB.

8.5.1 Core temperatures upon reaching steady-state

Fig. 8.4 shows core-mantle boundary temperatures calculated from Eq. (8.2) for various materials, as a function of the internal heat generation. Thermal conductivity values vary substantially, and are inversely correlated with estimated times to achieve steady-state (Table 8.1).

A diamondiferous lower mantle would be currently steady-state, and a graphitic composition would be nearly so. Although these endmember compositions are very unlikely, the trends in Fig. 8.4 are useful in understanding the tradeoffs involved in our characterization of Earth's deep interior. Note that during steady-state, progressive melting of the core is no longer ongoing, and so the core should have segregated. In this case, T at the CMB should equal the solidus at the ICB. For a diamond composition, a chondritic value for Ψ_{mtl} is needed to reach the solidus, whereas 10 $\times$ that amount is required for a segregated core. For a graphitic lower mantle, lower concentrations are needed for a 2-phase core, but these are in the range of chondritic. Clearly, K and Ψ_{mtl} are inversely correlated for any given composition.

Based on the above, a reasonable temperature profile for a silicate composition requires much less internal heating than an oxide composition would. This finding is consistent with Fig. 8.3, which showed that at the same (chondritic) Ψ_{mtl}, a silicate lower mantle would be hotter than an oxide mantle. Trade-offs create the need to consider two cases for Earth's geotherm.

8.5.2 Contribution from super-rotation of the core

Circulation of the molten outer core is considered to provide Earth's magnetic field, but a long standing debate is the source and magnitude of the associated energy. As occurs

TABLE 8.2　Proposed geotherms.

Depth	Oceanic lithosphere
km	°K
0	273.15
2	335.65
3	355.65
4	375.65
5	395.65
6	415.65
7	435.65
8	446.74
10	469.5
20	595.4
30	740.16
40	899.7
50	1069.1
60	1244
68*	1410

Depth	Oxide lower mantle; meteoritic core	Silicate lower mantle; Fe-S core
km	°K	°K
69	1411	1411
77.2	1430	1430
91	1490	1490
122	1600	1600
159	1690	1690
199	1760	1760
239	1800	1800
283	1860	1860
320*	1880	1900
325	1900	
420	2000	2160
660*	2150	2400
2000	2800	

Depth	Oxide lower mantle; meteoritic core	Silicate lower mantle; Fe-S core
km	°K	°K
2150	2860	
2171		3450
2271		3500
2321		3500
2350	2900	
2421		3450
2550		3350
2600	2890	
2670		3200
2841	2700	2950
2891*	2520	2750
2919.4	2520	2820
3024.5	2620	2920
3153.4	2740	3040
3281.2	2850	3150
3416.1	2950	3250
3621.3	3080	3380
3773.7	3180	3480
4015.6	3300	3600
4235.8	3410	3710
4441.5	3500	3800
4656.6	3580	3880
5150*	3700	4000
6371	3700	4000

Notes: Both cases begin with the oceanic lithosphere. Depths corresponding to the bottom of a layer are marked with "*."

for mantle convection, radionuclide emissions are an inadequate power source for the dynamo. Based on chemical compositions of meteorites, the power is negligible. To address the missing energy, potassium has been proposed to partition into the core. Experimental data show that this lithophilic element requires special conditions such as an O-rich liquid for such incorporation (Gessmann and Wood, 2002). In a seeming contradiction, others have considered volatilization of potassium during accretion, in order to address differences between meteorite compositions and Earth's rocks. Volatilization is

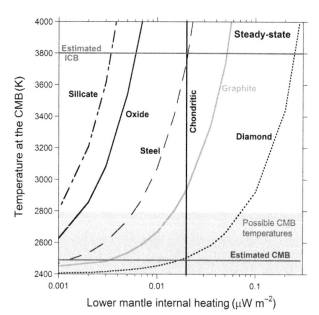

FIGURE 8.4 Steady-state temperatures at the CMB as a function of lower mantle internal heat generation. From left to right, thermal conductivities and cooling times (Table 8.1) increase and decrease, respectively. The vertical line marks our reference composition. The gray box shows temperatures between the Fe-S system eutectic at the CMB and the peridotite solidus at 660 km. Dark horizontal lines are our best estimate for the current CMB and ICB temperatures.

inconsistent with the uniformity of potassium isotopes in the Earth, Moon, and meteorites (Humayun and Clayton, 1995). Both hypotheses are attempts to reconcile ambiguous compositional models of the Earth with equivocal models of planetary processes.

A possible source for energy in the core does exist, and that is drag produced during differential spin, which generates heat and alters motions. Super-rotation of the inner core has been probed in seismologic studies. The inferred angular velocity is small: probably below 0.1 degrees yr^{-1} (Dehand et al., 2003). However, the corresponding tangential velocity of 2 km yr^{-1} is *enormous!* Differential spin of the core being as much as $10^5 \times$ westward drift of the plates is excessive. Seismological studies cannot resolve a tiny differential spin of the inner core that is on the order of measured westward drift velocities. Study of the shape of the Earth and its core point to very little super-rotation occurring today (Criss, 2019).

Although spin down of the core may exist, it is currently very small. Chapter 9 addresses whether this process may have been important in the past.

8.6 Discussion and further work

Our constructed geotherm for the Earth's interior rests on the melting temperature of the 2-phase core. Refining the model requires additional constraints on physical properties such as density and eutectic melting of Fe-Ni-S-P-C-N compositions. The effects of high Ni concentrations may be important, as are those of multiple light elements that are present in meteorites and are constituents of commercial steel and iron alloys. Comparing seismic sound velocities to those calculated for realistic, complex alloys at their eutectic melting temperatures could provide further constraints, and should reduce the uncertainties in constructing a core geotherm.

Improved understanding of the core is essential to bracket the thermal state of the lower mantle. It is clear that the lower mantle serves as the core's insulating blanket, but it is a blanket with an internal heater. Neither its insulating capability nor heating power are known. The trade-offs in the geotherm pertain to interpreting seismic velocities. Presently, the mineralogic models of the lower mantle assume a steady increase in temperature, which cannot be true since the core has much lower heat production than the lower mantle. Improved mineralogic models for seismic velocities, which focus on the proportions of oxides vs silicates and include a local maximum in temperature are needed.

Super-rotation of the core can be constrained from the aspect ratio of the inner and outer core, based on the analytical shape model of Criss (2019). This rate is needed to infer how much heat that spin-down could provide deep in the interior, to better understand forces and motions deep in the Earth, and to better constrain conditions of formation .

References

Abbasi, H.R., Bazdar, M., Halvaee, A., 2007. Effect of phosphorus as an alloying element on microstructure and mechanical properties of pearlitic gray cast iron. Mater. Sci. Eng. A444, 314−317. Available from: https://doi.org/10.1016/j.msea.2006.08.108.

Aitta, A., 2006. Iron melting curve with a tricritical point. J. Stat. Mech. Theory Exp. 12, . Available from: https://doi.org/10.1088/1742-5468/2006/12/P12015Online at: stacks.iop.org/JSTST/2006/P12015.

Boehler, R., 2001. High pressure experiments and the phase diagram of lower mantle and core materials. Rev. Geophys. 38, 221−245.

Brearly, A.J., Jones, R.H., 1998. Chondritic meteorites. Rev. Miner. 36, Ch. 3 (398pp).

Chudinovskikh, L., Boehler, R., 2007. Eutectic melting in the system Fe−S to 44 GPa. Earth Planet. Sci. Lett. 257, 97−103.

Criss, R.E., 2019. Analytics of planetary rotation: improved physics with implications for the shape and super-rotation of Earth's Core. Earth Sci. Rev. 192, 471−479.

Criss, E.M., Hofmeister, A.M., 2017. Isolating lattice from electronic contributions in thermal transport measurements of metals and alloys and a new model. Int. J. Mod. Phys. B 31 (175020). Available from: https://doi.org/10.1142/S0217979217502058.

Criss, R.E., Hofmeister, A.M., 2016. Conductive cooling of spherical bodies with emphasis on the Earth. Terra Nova 28, 101−109.

Dehand, V., Creager, K.C., Karato, S.I., Zatman, S., 2003. Earth's Core: Dynamics, Structure, Rotation. American Geophysical Union, Washington, DC, pp. 5−82.

Fiquet, G., Auzenda, A.L., Siebert, J., Corgne, A., Bureau, H., Ozawa, H., et al., 2010. Melting of peridotite to 140 Gigapascals. Science 329, 1516−1518.

Gessmann, C.K., Wood, B.J., 2002. Potassium in the Earth's core? Earth Planet. Sci. Lett. 200, 63−78.

Grady, M.M., Wright, I.P., 2003. Elemental and isotopic abundance of carbon and nitrogen in meteorites. Space Sci. Rev. 106, 231−248.

Hofmeister, A.M., 2014. Thermal diffusivity and thermal conductivity of single-crystal MgO and Al_2O_3 as a function of temperature. Phys. Chem. Miner. 41, 361−371.

Hofmeister, A.M., 2019. Measurements, Mechanisms, and Models of Heat Transport. Elsevier, Amsterdam, pp. 201−250 (in particular, see Chapter 7 on mineral and rock data, and Chapter 11 on a radiative diffusion description of heat conduction).

Hofmeister, A.M., Criss, E.M., 2018. How properties that distinguish solids from fluids and constraints of spherical geometry suppress lower mantle convection. J. Earth Sci. 29, 1−20. Available from: https://doi.org/10.1007/s12583-017-0819-4.

Hofmeister, A.M., Criss, R.E., 2013. How irreversible heat transport processes drive Earth's interdependent thermal, structural, and chemical evolution. Gwondona Res. 24, 490–500.

Hofmeister, A.M., Dong, J.J., Branlund, J.M., 2014. Thermal diffusivity of electrical insulators at high temperatures: evidence for diffusion of phonon-polaritons at infrared frequencies augmenting phonon heat conduction. J. Appl. Phys. 115, 163517. Available from: https://doi.org/10.1063/1.4873295.

Humayun, M., Clayton, R.N., 1995. Potassium isotope cosmo-chemistry: genetic implications of volatile element depletion. Geochim. Cosmochim. Acta 59, 2131–2148.

Knittle, E., 1995. Static compression measurements of equations of state. In: Ahrens T.J. (ed.) *Mineral Physics and Crystallography: A Handbook of Physical Constants.* American Geophysical Union, Washington D.C., p. 98–142.

Li, J., Fei, Y., 2014. Experimental constraints on core composition. In: Turekian, K., Holland (Eds.), Treatise On Geochemistry, vol. 2. Elsevier, The Netherlands, pp. 169–251. , 527-557.

Lodders, K., Fegley, B.J., 1998. The Planetary Scientist's Companion. Oxford University Press, Oxford.

Merriman, J.D., 2019, On the Ability of Rocks to Conduct Heat in the Lithosphere [PhD Dissertation]: Columbia, University of Missouri, 256 p.

Mittlefehldt, D.W., McCoy, T.J., Goodrich, C.A., Kracher, A., 1998. Non-chondritic meteorites from asteroidal bodies. Rev. Miner. 36, 1529–6466.

Morard, G., Andrault, D., Antonangeli, D., Nakajima, Y., Auzende, A.L., Bouland, E., et al., 2017. Fe–FeO and Fe–Fe$_3$C melting relations at Earth's core–mantle boundary conditions: implications for a volatile-rich or oxygen-rich core. Earth Planet. Sci. Lett. 473, 94–103.

Morard, G., Andrault, D., Guignot, N., Siebert, J., Garbarino, G., 2011. Melting of Fe–Ni–Si and Fe–Ni–S alloys at megabar pressures: implications for the core–mantle boundary temperature. Phys. Chem. Miner. 38, 767–776.

Mori, Y., Ozawa, H., Hirose, K., Sinmyo, R., Tateno, S., Morard, G., et al., 2017. Melting experiments on Fe–Fe$_3$S system to 254 GPa. Earth Planet. Sci. Lett. 464, 135–141.

Shakespeare, W. 1609. Troilus and Cressida (Reprint of the player's text: The Shakespeare Society of New York, New York).

Simmons, N.A., Grand, S.P., 2002. Partial melting in the deepest mantle. Geophys. Res. Lett. 29, . Available from: https://doi.org/10.1029/2001GL013716paper 1552.

Sukhadolau, A.V., Ivakin, E.V., Ralchenk, V.G., Khomich, A.V., Vlasov, A.V., Popovich, A.F., 2005. Thermal conductivity of CVD diamond at elevated temperatures. Diamond Relat. Mater. 14, 589–593.

Taylor, S.R., 1982. Lunar and terrestrial crusts: a contrast in origin and evolution. Phys. Earth Planet. Inter. 29, 233–241.

You, Z., Paek, M.-K., Jung, I.-H., 2018. Critical evaluation and optimization of the Fe-N, Mn-N and Fe-Mn-N systems. J. Phase Equilib. Diffus. 39, 650–677.

Zhang, J., Herzberg, C., 1994. Melting experiments on anhydrous peridotite KLB-1 from 5.0 to 22.5 GPa. J. Geophys. Res. 99, 17729–17742.

PART III

Thermal evolution of Earth and other rocky bodies

Thermo-chemical evolution of the Earth

"For the treatment of the whole universe in all its regions and during all of time we have, nevertheless, no adequate model, and to obtain ideas as to its complete nature can only rely on the roughest methods of scientific induction." *(R.C. Tolman, 1934).*

Although the Solar System contains many large rocky bodies, only Earth has an active surface with horizontal tectonic motions, although many rocky bodies have exhibited vertical transfer of mass and heat (volcanism) at some point in their history. Volcanism is a subset of tectonics, but the converse is not true. Therefore, Earth's history is unique. In addition, only the outermost ~ 300 km of Earth has been sampled, which limits the accuracy of evolutionary models. Furthermore, magmatism and outgassing couple Earth's chemical and thermal evolution.

Modern computers facilitate the construction of complex, multi-parameter models. But without validation, this approach can be misleading. Therefore, this chapter takes a minimalistic approach, where the models presented are only as sophisticated as available information warrants, per Tolman (1934).

Importantly, previous inferences of interior concentrations of the heat producing elements (K, U, Th) have assumed steady-state, but this condition does not exist throughout large rocky planets (Chapters 2 and 6). Part of Earth's interior heating being gravitational (Chapters 5 and 7) and part being radiogenic overrides any connection of surface flux with mantle heat production. Therefore, previous estimates for K, U, or Th contents for the bulk Earth or upper mantle are flawed and are not discussed further. We instead utilize measurements of rocks, minerals, and meteorites, but this chapter cannot do justice to the copious geochemical and isotopic studies on Earth materials. Fortunately and importantly, current independence of the temperatures deep in the interior from that of the outer layers (Chapter 8), suggests that this situation has existed over most, if not all, of Earth's history. Hence, the present chapter evaluates the thermo-chemical history of the lower mantle and core independently of that of the oceanic and continental crusts, the lithosphere, and the upper mantle, which strongly interact and are strongly influenced by gravitational forces.

Section 9.1 covers the connection of initial conditions with energy sources and processes. Section 9.2 evaluates the thermal energy currently stored in the Earth which leads to a chemical model for the main layers of the Earth. An origin for changes in K, Th, and U concentrations with time is proposed. Section 9.3 describes the evolution of the slowly warming lower mantle and core and the complicated thermo-chemical evolution of the layers above ~ 660 km. Section 9.4 summarizes.

9.1 Key drivers of planetary evolution

Regardless of how the planets form, and how they evolve, the following energy sources are potentially large. Any given planet may possess a subset of these sources:

- Gradual dissipation of spin provides energy and relative motions of affected layers
- Forces leading to spin-orbit coupling also create heat, early-on
- Imbalance in the external gravitational forces experienced by a planet orbiting the Sun create torque and heat, particularly on the near-surface layers
- Intercepted projectiles can apply force and impact heating to upper zones
- Internal radioactivity generates interior heat locally, which both builds and dissipates over time

Several large yet commonly overlooked planetary energy sources are gravitational. Gravity being a strong force on large scales can provide not only vertical, but horizontal motions, as well as shear. Dissipation of Earth's spin is currently documented in the westward drift at 6 cm y^{-1} of the whole lithosphere and so this gravitational source clearly applies to Earth's upper layers, and is part of plate tectonics (Chapters 5 and 7). Another possible locus for spin dissipation is the outer core, which, being melted, is weaker than the surrounding solid layers (Chapter 8 and below).

Although advection is not a source of energy, it is important to the redistribution of heat and matter on planetary scales. We distinguish magmatism from outgassing, because the former creates internal layers, whereas the latter can escape planetary interiors.

More information is available on the Earth than on any other planet: constraints on the main layers were covered in Chapters 1, 6, 7 and 8. These are boundary conditions. Initial conditions are extremely important, and are discussed next.

9.1.1 Why accretion was cold

The currently popular fractal models for formation of the planets have many serious and recognized shortcomings. Key inconsistencies are that the fractal models do not conserve angular momentum, treat impacts as constructive despite these being inherently destructive, and lack a mechanism for formation of grains. Popular models provide no explanation for planetary spin, which is a key and ubiquitous attribute of planets.

Fractal models are based on flawed ideas. One is the key concept of a thin disk envisioned by Laplace, which conflicts with the observed gravitational shape of planetary objects being the oblate spheroid of Newton and Maclaurin. Second, contraction of the nebula is assumed to directly convert gravitational potential energy into heat, via impacts starting with those of gas molecules. Viscous dissipation cannot occur in rarified nebula gas because generation of heat by friction requires both dense matter and differential motions. Collisions of molecules or atoms are inelastic, but these do not generate retained heat; rather they generate blackbody radiation, which describes the temperature of the medium and leaves as it is generated (Hofmeister, 2019). Temperatures near 8 K in star-forming regions are maintained by young stars embedded therein.

"Hot" accretion is part of these fractal models. It is popularly believed that hot gas shot out of the Sun's equatorial region, after Laplace, and then condensed into solid phases. Much effort has been put into determining the condensation sequence of minerals with temperature and vapor pressure in this scenario, which relates to distance from the Sun. Comets being a mixture of ice and refractory phases (Zolensky et al., 2006), and other observations, refutes the expected thermal gradient. The alternative, that mineral dust predates accretion, is consistent with the existence of pre-solar grains in all types of chondritic meteorites (Bernatowicz and Zinner, 1997), and also with spectroscopic analyses of the dust in outflows of aging stars, which contain the same phases found in pre-solar grains and chondritic meteorites (reviewed by Molster and Waters, 2003). Condensation is relevant to dying stars, which are losing mass, but not to young stars, which gain mass as they form, and are governed by different processes.

We have developed a new model for the physics of "cold" accretion that lacks the problems and inconsistencies of the currently popular "hot" fractal model. Chapter 4 covers accretion of nearly formed planets. Chapter 11 proposes an aeolian sedimentary origin for chondrules and chondrites. Chapter 5 covers processes in the protoEarth. The present chapter explores thermal evolution of the Earth that follows a cold start, relying on the thermal models of the main layers (Chapters 6, 7, and 8).

9.1.2 Limitations on gravitational heat sources set by cold accretion

Energy and angular momentum conservation require that the gravitational potential associated with any given pocket of nebula that collects to form a planet must be converted to spin. However, calculating the initial spin of a planet is under-constrained because planets form within a torus of nebula dust and gas that was orbiting the still forming Sun. Of importance to planetary evolution, is that after a spinning body is formed, it can lose its spin through friction. Viscous dissipation produces heat over time, which may be substantial. As discussed in Chapter 7, westward drift of the lithosphere shows that spin-down currently operates at shallow levels.

Criss and Hofmeister (2016) considered some heating from accretion to explain the extremely high core temperatures from high-pressure experiments at that time. Subsequent experiments have yielded much lower melting points (Sect. 8.5), so this primordial reservoir is unnecessary (Section 9.2).

With cold accretion, immediate segregation of the core is not possible, because motions within an otherwise gravitationally stable, central compressed region require melting. In addition, gravitational acceleration is zero at Earth's center, and so if homogeneous accretion occurred, the center of the core could not be involved and would retain a large proportion of silicates, as in meteorites. To address these issues, and to explain how grains initially aggregate, Chapter 4 proposed that an iron proto-core formed first. Magnetic attraction augments gravitational draw while welding processes and the ductility of iron mitigates the inherently destructive nature of particle collisions. Once a proto-core is formed it will serve as a local center of gravitational attraction, and will hold onto the fragments of non-metallic dust grains produced during impact. Sorting outside the proto-core is possible, due to finite gravitational acceleration and admixtures with ice (Fig. 5.2; Section 9.3).

In view of heterogeneous accretion, partial sorting, entropy considerations, and conservation laws, core formation did not generate substantial heat (Hofmeister and Criss, 2015). Core formation is not instantaneous, and so cannot be reasonably described by comparing initial and final energies. Timing of core formation is not known, since terrestrial samples are from the upper ~ 300 km only, with no communication with the core. Isotopic data on meteorites (e.g., Kleine et al., 2009) are interpreted to indicate that Earth's core formed early and rapidly, within the first ~ 30 to ~ 125 Ma of its 4.5 Ga history. However, iron meteorites have been linked to bodies that are significantly smaller than the Earth, such as asteroids. Because size largely controls thermal evolution (Chapter 2), these data must be cautiously applied to the immense Earth. An alternate interpretation is provided in Section 9.3.

Some heat would be generated during the late stages of planetary accretion when objects drawn towards the immense Sun were intercepted by the planets, analogous to bugs striking a windshield of a moving car. Heat generated in this manner is shallow, and so late stage impacts provided energy and mass to Earth's outermost layers only, if at all (Section 9.3).

9.1.3 Heat produced by radioactive decay

The discovery of radiogenic heat was key to overturning Kelvin's incorrect hypothesis for the 90 Ma age of the Earth. Advancement of the geologic sciences at the turn of the century being predicated on this key discovery of the Curies has led to the view that radioactive decay composes *all* of Earth's interior energetics. Many have utilized the equivalence of meteorite heat production with Earth's surface emissions in geochemical models. However, Criss and Hofmeister (2016) showed that heat from below 660 km cannot reach the surface over geologic time. Also, surface emissions of Earth include a large and possibly dominant component of shear heating (Chapters 5 and 10; Table 7.3).

Nonetheless, over geologic time even low concentrations of K, U, and Th will heat the regions of Earth where they occur. Thus, we endeavor to constrain internal heating production in the different regions in Section 9.2.

Short-lived isotopes are a possible very early heat source. This heat source is poorly constrained and mostly predates planet assembly. Short-lived isotopes are not considered further.

9.2 Inference of chemical composition through comparison to energy storage

9.2.1 Thermal energy stored in the Earth

Thermal energy currently stored in the various layers in Earth depends on their heat storage and temperature. Values in Table 9.1 were calculated using masses from Anderson (2007) and the geotherms of Table 8.2. Notably, Earth's total thermal energy is $\sim 60 \times$ larger than the rotational energy of its current spin (0.214×10^{30} J). However, a car on a freeway has more kinetic energy than the lithospheric plates (Table 7.2). Dissipation is important: powers (Π) listed in Table 9.1 represent averages over the age of the Earth, as the time evolution is not constrained.

9.2.2 Meteoritic reservoirs

Several different reservoirs are essential, because some solids have radioactive elements and others do not. Chondritic meteorites are considered, since several achondrites are documented ejecta from fully formed planets. The existence of volatiles also needs to be taken into consideration (Fig. 9.1).

Two reservoirs bearing radioactive elements are considered, due to their distinct attributes. Primitive chondritic meteorites are rich in silicates and potassium, but are relatively poor in uranium. Calcium aluminate inclusions (CAIs) occur in most chondritic meteorites, including the enstatite chondrites (Fagan et al., 2000). CAIs are distinguished by being rich in oxides, U, and Th. The formation age of CAIs at 4.567 Ga is close to that of 4.564

TABLE 9.1   Estimates of thermal energy stored in the present day Earth and average power needed to produce it over geologic time.

Layer	Mass[a] 10^{27} g	C_P J g^{-1} K^{-1}	Avg. T K	Heat stored[b] 10^{30} J	Average Π[b,c] TW
Crustal	0.030	0.8	800	0.019	<0.13
UM + TZ	1.063	0.8	2000	1.7	11.9
Lower mantle	2.938	0.8	2700–3200	6.34–7.52	44–52.5
Both cores	1.941	0.6	3500	4.08	28.5
Core latent heat[d]	1.839	475 J g^{-1d}	n.a.	0.87[d]	6.2[d]
Below 660 km	5.331	n.a.	n.a.	11.3–12.5	79–87
Whole Earth	5.972	n.a.	n.a.	13.0–13.4	91–94

[a]From Anderson (2007).
[b]Assumed a ~0 K initial temperature; for ~300 K initial T, power and heat are ~10% less.
[c]This is an average value = heat stored divided by Earth's age of 1.433×10^{17} s.
[d]Complete melting of the outer core is assumed, which sets the upper limit on energy and power. Instead of C_P, this calculation uses latent heat per gram at core conditions from Aitta (2006).

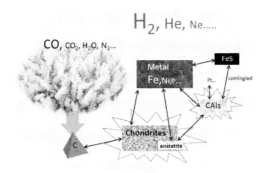

FIGURE 9.1   Schematic of the major reservoirs. A gas of H_2 with ~10% He and additional noble gases dominates the pre-solar nebula. Other gases: e.g., CO, CO_2, H_2O freeze on dust particulates. Black arrows indicate prominent co-existing reservoirs in meteorites. Box sizes reflect the importance (size) of the reservoirs, but are not in proportion. Starbursts indicate the presence of heat-producing elements in certain reservoirs. Carbon is a separate reservoir due to its production (gray arrow) from the abundant ices. The metal reservoir is mostly Fe-Ni alloy, but contains additional elements. FeS is distinguished due to ubiquitous occurrence of troilite, alternatively this could be considered as a reservoir of S. Refractory metals and the Pt group constitute a tiny reservoir. Ordinary chondrites and enstatite chondrites are mainly silicate phases and are richer in K, whereas the calcium aluminate inclusions are mainly oxides and are richer in U and Th.

Ga for certain chondrules (Amelin et al., 2010), whereas whole rock data on diverse chondritic meteorites give a substantially younger age of 4.555 Ga (Brearly and Jones, 1998).

Metallic iron is ubiquitous in meteorites, as is commensurate with Earth's large core. Because this phase has negligible K, U, and Th, its presence is immaterial to the proportions of the chondritic and CAI reservoirs, other than bulk compositions for meteorites including whatever metal is present. The variability of metal to rock proportions in all chondrite types (Brearly and Jones, 1998) shows that metal must be considered as an independent reservoir. Note that the CAIs have another metal reservoir, the Pt group elements with some highly refractory metals. No constraint exists on its proportion either.

Volatiles also lack heat producing elements. Temperatures and pressures are such that CO, CO_2, and H_2O in molecular clouds exist as ice along with vapor. Porosity of carbonaceous chondrites is high, averaging $\sim 20\%$, more than double the porosity in other chondrites (e.g., Consolmagno et al., 2008). This suggests that the carbonaceous chondrites once included substantial ices which sublimated and/or reacted, producing graphite and hydrated minerals, while leaving void space.

The elements S, P, and N, could each be considered as individual reservoirs. Troilite is ubiquitous in meteorites, and occurs in variably amounts as nearly pure and stoichiometric FeS, whereas P and N occur as minor elements in the metal (e.g., Mittlefehldt et al., 1998). On this basis, we consider an additional FeS reservoir, and lump P and N with the main metal reservoir.

9.2.2.1 Why ordinary chondrites are key

Homogeneity of the Solar System is unexpected for either hot or cold accretion. Instead, heterogeneity is documented in the diversity of chondritic meteorites and their constituents. Thus, representing the Earth requires an appropriate average composition.

Ordinary chondrites are the most sensible choice for the Earth's bulk composition because these are most abundant meteorites by far. Over 13,625 ordinary chondrites have been collected, representing 98% of all meteorite samples; moreover, chondrites represent 80% of observed falls (Scott and Krot, 2007). Even if this number were affected by a 4-fold duplication, ordinary chondrites would still constitute > 95% of all meteorites.

The bulk chemistry of ordinary chondrites is very similar to that of the rare enstatite chondrites, which are highly reduced, as is consistent with hydrogen gas dominating space and nebulae (Section 9.2.6). The relatively oxidized state of ordinary chondrites is attributed to processing (Section 9.3). Therefore, ordinary chondrites are the key to Earth's bulk composition whereas enstatite chondrites are key to the precursor mineralogy.

Carbonaceous chondrites have previously been used to represent the bulk Earth in many studies. However, these are rare, altered, and have very little iron. The presence of carbon is an insufficient reason because this reservoir, being tied to extremely abundant CO in space, is independent of any reservoir of solids. Similarity of carbonaceous chondrite to Solar abundances is also an insufficient reason, as follows: (1) The Sun has a much different major element chemical composition than any rocky body. Why should the minor elements be the same? (2) Solar spectra reflect only the composition of the photosphere, which could differ substantially from the composition of deeper zones. (3) Ample evidence exists for composition gradients across the solar system (Hofmeister and Criss, 2012).

9.2.2.2 Why presolar grains are not treated as a distinct reservoir

Pre-solar grains are far too tiny and scarce to affect a chemical average. Diamond is the main phase (e.g., Bernatowicz and Zinner, 1997), which is covered by the carbon reservoir (Fig. 9.1). Other common pre-solar phases (SiC, silicon nitride, iron metal, carbides) have negligible amounts of K, U, and Th and can be considered as part of other reservoirs. Corundum, which is relatively rare, and silicate grains, which are tiny (Nittler and Ciesla, 2016), are phases in the CAI and chondritic reservoirs, respectively. The relative abundances and sizes of pre-solar crystal types correlate with hardness, which is consistent with their surviving long journeys across space.

9.2.2.3 Amorphous vs crystalline dust in space?

Solids in chondritic meteorites, presolar grains, and comets are predominantly crystalline, although glass exists as tiny components (e.g., Brearly and Jones, 1998). This and other observations are counter to prevailing views regarding hot accretion and the dust constituents of the presolar nebulae:

Although the origin of chondrules (which are roundish with ~ 0.5 mm diameter and extremely diverse in mineralogy and textures, e.g., Scott and Wasson, 1975) is unknown and debated, chondrules are generally considered to have been melted. Problems exist with reconciling cooling rates with such formation, and with observed textures and mineralogies (some are refractory aluminates): plus, the heat source is a puzzle (e.g., Nittler and Ciesla, 2016). Chapter 11 documents a sedimentary origin.

Whereas the consensus view of the astronomy community is that silicate dust in interstellar space is amorphous (e.g., Mumma and Charnley, 2011), definitive spectral evidence exists for crystalline Mg_2SiO_4 in the dust formed from ions ejected from stars (e.g., Molster and Waters, 2003). Far-IR peaks probed in ejecta spectra are sharp and weak, requiring grain-sizes of several μm from comparison with laboratory measurements, e.g., of Hofmeister (1997). Near-IR spectra of dust around young stellar objects has overtone peaks consistent with crystalline silicates, requiring grain sizes $>50\,\mu$m for resolution (Bowey and Hofmeister, 2005). These objects have broad peaks in the mid-IR, like the interstellar medium, where the fundamental and strong Si-O stretch absorbs. Sub-micron grain sizes are required for a wide variety of minerals to produce Si-O absorption peaks in laboratory studies which are "pointy" (Hofmeister, 1997; 2019). It is long known that obtaining well-resolved peaks in the mid-IR spectra from dispersions of silicates in KBr requires grinding to sub-micron sizes and dilute samples.

Logically, the dust from dying stars drifts into interstellar space, and then eventually into star forming regions. Collisions in the stellar winds ejecting the particles may reduce their sizes. If this material were soft, as glassy material tends to be, it would not survive as indicated by the constitution of isotopically confirmed pre-solar grains, see above.

9.2.2.4 The oxidation state from data on gas and ice in astronomical objects

Domination of the enormous volatile reservoir by H_2 gas shows that highly reducing conditions existed during accretion of the planets. This is consistent with the mineralogy of enstatite chondrites. Constituents of ice in space set additional constraints on the composition of the presolar nebula, whereas data on comets (e.g., Mumma and Charnley, 2011) provide information on chemistry during or after the accretion process.

Fig. 9.2 shows that CO ice dominates space between the stars and regions forming stars. The ratio of CO/H for our Sun is likewise small, $\sim 10^{-4}$ (e.g., Lodders and Fegley, 1998, p. 76). Amounts of CO_2 or H_2O ice vary in protostar environments, such that an average protostar has $CO_2/CO \sim 0.2$ and water ice is $\sim 15\%$ of the total ices (Mumma and Charnley,

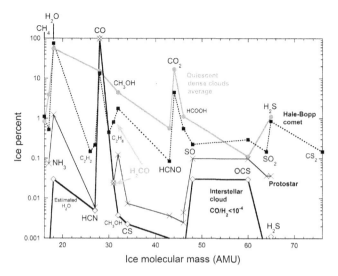

FIGURE 9.2 Types of ices in molecular clouds and comets. The y-axis is logarithmic. The quiescent cloud data is the average from table 3 in Mumma and Charnley (2011): all other objects are individual studies listed in their table 4. The ratios of other ices to water ice were recalculated to 100% total ices. Halley's comet has similar proportions. Species with high concentrations are labeled.

2011, their Table 3). Species rich in S occur as small amounts for both environments. Thus, highly reduced conditions describe star forming regions, and remnants of gas/ice remaining after formation.

In contrast, quiescent clouds (Mumma and Charnley, 2011) have substantial amounts of CO_2, H_2O and methanol, as do comets, and their patterns of abundance for the other ices are similar (Fig. 9.2). Under dense conditions, minor species can be detected, whose compositions indicate that reactions have occurred among the abundant phases. These dense objects still are reduced, but not as reduced as the more dilute environments, where interactions occur infrequently. This comparison points to dust in the presolar nebula being highly reduced prior to accretion, as in the enstatite chondrites, and suggests that the ordinary chondrites have been oxidized via reactions occurring during densification and accretion.

Proportions of the ices in Fig. 9.2 point to the C and S reservoirs being large, whereas N is minor. The presence of much larger amounts of N in the Sun point to its photosphere not well representing a rocky planet.

9.2.3 Radioactive heating in the lower mantle

Flow of heat towards Earth's surface from the lower mantle is impeded by (1) great distance, (2) low thermal diffusivity, and (3) the upper mantle generation of frictional heat, since heat can only flow from hotter to colder regions. To estimate heat leakage from the lower into the upper mantle, the present-day geotherm (Fig. 8.3) indicates that roughly ~ 4 mW m^{-2} crosses the 660 km boundary, which is equivalent to ~ 1 TW. Heat generated in the lower mantle remains below 660 km to a good approximation.

Over geologic time, the average power required to warm the lower mantle and core and to melt all of the outer core ranges from 79 to 87 TW (Table 9.1). To deduce the amounts of heat producers, results for the two geotherms of Chapter 8 are next compared to values in Table 1.3 for the average power for the whole Earth possibly generated by various meteorite compositions. Multiplying the numbers in Table 1.3 by 0.9, to account for the mass of the non-participating upper mantle, gives the power arising below 660 km. Thus, a lower mantle composed of ordinary chondrites (i.e., 800 ppm K; 38 ppb Th ; 11 ppb U from Lodders and Fegley, 1998, their Table 16.11) would produce an average power of 79 TW over 4.5 Ga, which is consistent with our cold geotherm and partial melting of the outer core (buffering).

To discuss a wider range of meteorite types, we parameterized the average power (in TW) over history for the whole Earth:

$$\Pi_{ave,Earth} = 0.0949 \times [ppm\ K] + 1.44 \times [ppb\ U] + 0.138 \times [ppb\ Th] \qquad (9.1)$$

A visualization of the dependence of average power on the heat producing elements (Fig. 9.3) was obtained by approximating ppb Th as $3.5 \times$ ppb U. Contours are parallel lines because Π_{ave} linearly depends on concentration. Thermal storage is shown as minima and maxima, which allow for the two model geotherms and that the lower mantle and core provide $\sim 90\%$ of the whole Earth.

Because Earth's core size is larger than that suggested by the metal content of most meteorites, thermal storage values must be slightly lower than those of meteoritic reservoirs, excepting enstatite chondrites, whose high Fe proportions (Table 9.2) permits direct comparison. In addition, the estimated thermal storage assumes negligible heat crosses into the transition zone, so the power shown is a minimum. Hence, thermal storage in the Earth is met by the ordinary chondrites and possibly by enstatite chondrites, but not by a composition solely of carbonaceous chondrites or solely of CAIs. Meteorite types could be mixed to yield sufficient power, but from Fig. 9.2, all such mixtures must include a large proportion of ordinary chondrites.

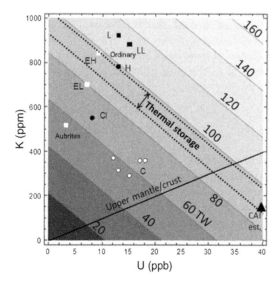

FIGURE 9.3 Contour plot of the average power generated over history for the observe range of K and U in meteorites. Eq. (9.1) was used, assuming that Th concentrations are 3.5 times those of U. Contours are every 20 TW. Dotted lines show the range of power stored in the Earth, that is needed to raise its temperature to the lower mantle geotherms shown in Fig. 8.3. These are recalculated for a lower mantle with the mass of the whole Earth. Symbols = meteorite compositions as labeled, mostly from Lodders and Fegley (1998). CAI compositions are from Jacobsen et al. (2008). Other studies suggest K and U contents may vary by a factor of two. Fine line = historic correlation of upper mantle and crustal rocks obtained by Wasserburg et al. (1964).

TABLE 9.2 Meteorite reservoirs and proposed compositions of the lower mantle and core.

Element wt%[a]	Ordinary chondrite[b]	Core sorted[c]	Lower mantle settled[d]	Lower mantle reacted[e]	Lower mantle Enst. chon[f]	CAIs[g]	Protocore metal[h]
C	See notes	≡3.4	≡3.4	≡3.4	≡3.4	≡3.4	<0.2
O	36.5		42.0	48.8	45.6	42.5	
Na	0.645		0.74	0.86	1.0		
Mg	14.4		16.6	19.2	18.8	6.0	
Al	1.10		1.27	1.47	1.4	17.4	
Si	17.7		20.4	23.7	27.4	12.6	
P[c]	0.109	0.36	0.06				0.47
S[c]	2.08	7.20					i
Ca	1.26		1.52	1.69	1.5	16.4	
Ti	0.0645		0.08	0.09	0.1	0.9	
Cr	0.356		0.21	0.47	0.5		<0.01
Mn	0.244		0.14	0.32	0.3		<0.001
Fe[c]	24.0	83.77	11.86			0.8	90.06
Co[c]	0.0687	0.25	0.04				0.50
Ni[c]	1.44	5.02	0.84				8.75
K	0.084	~0	0.10	0.11	0.12	~0.030	~0
Th[a]	40	~0	46	53	53	~360	~0
U[a]	13.9	~0	16	18.5	12.5	~50	~0
Ni/Fe	0.06	0.06	0.07	-	-	-	0.10
metal	11.2	~85	0	0	22[f]	low	100
FeS	5.7	~12	0	0	12[f]	~0	i

[a]All values are in wt%, except for Th and U, which are in ppb.

[b]Computed from the table on H, L and LL types in Lodders and Fegley (1998), except that C is assumed.

[c]Elements associated with sulfide and metal phases were assumed to enter the core in the same proportions as that of the bulk ordinary chondrites. In enstatite chondrites, Cr and Mn are in the rocky minerals, consistent with iron meteorite compositions.

[d]Assumes the iron and sulfides in the chondrites settled into the core without chemical reactions.

[e]Assumes reactions put all the iron and associated elements into the core.

[f]Metal and sulfide in the meteorites are not included in the suggested mantle composition.

[g]Average of data (7–15 measurements on CAIs in the C-rich Allende meteorite (Jacobsen et al., 2008). Fe-Ni and Pt-group alloys are observed (e.g., MacPherson, 2003). Sulfides can be FeS, but CaS and other lithophilic cations are observed (Fagan et al., 2000). Thorium was provided from U and Pb analyses. Other studies provide averages of 25 and 65 ppb U, and similar K and Th values, although K is usually near the limit of detection.

[h]Calculated from the proportions of iron types given by Scott and Wasson (1975): Ni contents from ~500 samples. Co and P contents from Mittlefehldt et al. (1998) on a similarly large database. Ni, Co, and P contents for 27 irons by Jaresowich (1990) give nearly the same averages. Mn and Cr contents are low (Sugiura and Hoshino, 2003).

[i]Troilite is ubiquitous, but a proportion of FeS to metal does not seem to exist in the literature. Assuming proportions in the bulk chondrites are representative, the core should contain 10 wt% S.

Notes: Totals are set to 100 ± 0.1 wt%. Elements more abundant than 0.05 wt% are listed, along with heat-producers. Carbonaceous chondrites have on average 3.4 wt% C, which we use. Solar abundances are 8.6 wt% C. Troilite is nearly pure and stoichiometric FeS.

For K, U, and Th concentrations of an average ordinary chondrite (Table 9.2), radioactive decay generates 0.0196 μW m^{-3} according to Eqs. (2.25) and (8.7), which is repeated here:

$$\Psi_{mtl} = \left(0.095 \times [wt\% \ K] + 0.266 \times [ppm \ U] + 0.071 \times [ppm \ Th]\right)\frac{\rho}{2700} \qquad (9.2)$$

This formula provides results similar to those of van Schmus (1995), but differs from Hasterok et al. (2018), who do not correctly account for beta decay of potassium, and used a coefficient for potassium that is 82% of our value.

Similar values are obtained for ordinary chondrites, type EH, and the CAIs whereas the rare carbonaceous chondrites have about $\frac{1}{2}$ the energy density. Calculated Ψ_{mtl} matches our reference value in Chapter 8 for the lower mantle of 0.02 μW m^{-3}, which was assumed to equal Ψ for the sub-continental lithosphere (Chapter 6). Thus, our meteoritic estimate for Ψ_{mtl} is independent of our reference value, and their agreement shows consistency.

9.2.4 Inferred chemical composition of the lower mantle and core

Meteoritic models must be viewed as inexact guides to a planetary interior. We construct a few models, focusing on the abundant elements and heat-producers, and recognizing that reduced material such as enstatite chondrites formed Earth, but such are insufficiently abundant to provide an accurate average chemistry, which is better represented by the chemically similar ordinary chondrites. These however, were oxidized during assembly (Section 9.3), which we attempt to address.

9.2.4.1 Possible lower mantle compositions

The key feature of enstatite chondrites is being reduced, whereby Fe is in the metals and sulfides only. An enstatite chondrite composition for the lower mantle (Table 9.2) is supported by spectroscopic identification of dust shed by dying stars as forsterite and end-member pyroxenes (Molster and Waters, 2003) and presolar olivine being close to the Mg-endmember.

Ordinary chondrites statistically represent the bulk Earth. These are divided into three types: high iron (H, 49% by number), low iron (L, 44%), and very low iron (LL, 7%). The range of iron alloy and sulfide among the three types is large (8.3−21.5 wt% metal + sulfide). In addition, the rocky minerals are also iron rich, and so the bulk samples have 21−29 wt% Fe + Ni, and 2 wt% S (e.g., Scott and Krot, 2007).

An average ordinary chondrite has 27.6 wt% Fe + Ni + S whereby 16.9% of the phases in ordinary chondrites are metal plus sulfides. If only the existing metal and sulfide phases presently in ordinary chondrites gravitationally settled, then ordinary chondrites could provide 0.60 × 10^{24} kg, or 31% of the current core's mass. If instead complete segregation occurred (i.e., the precursors were actually enstatite chondrites) then ordinary chondrites could provide a core mass of 1.12 × 10^{24} kg, or 58% of the current core's mass.

Because a harmonic mean is needed to describe the conductivity of a rock with multiple phases (Hofmeister, 2019), it is the low thermal conductivity (K) phase that dominates a mixture for most distributions, per Eq. (8.6). Chapter 8 results thus do not constrain the amounts of graphite and/or diamond, and so a modest C content was assumed, based on the solid Earth retaining fewer volatiles than the gassy Sun. Importantly, segregation of Fe ions in ordinary chondrites into the core would make the lower mantle high in Mg, which elevates K over a composition with Fe in silicates, even at a few % level (e.g., Pertermann and Hofmeister, 2006). An enstatite chondrite model would likewise provide a high K lower mantle. Cold accretion and reduced conditions point to the "cold" geotherm representing the lower mantle.

Possibly, CAIs represent a minor refractory reservoir. Their $\sim 1\%$ occurrence in enstatite chondrites (Fagan et al., 2000) slightly exceeds that of $\sim 0.1\%$ for most chondrites, whereas carbonaceous chondrites typically have about 10 vol% CAIs (Scott and Krot, 2007). The focus has been on variability, not average compositions, so we can only roughly estimate their effect.

Adding a small proportion of CAI material to enstatite chondrites would give a composition for the lower mantle that is very similar to that obtained from the "segregated" ordinary chondrites. Admixing CAIs would slightly elevate U, Al, Ti, Ca, and probably Th concentrations while slightly decreasing the K, Na, and Si, and probably Mg contents. The latter depends on the proportion of spinel to corundum. These inclusions have Ti^{3+} and are thus are highly reduced, as is the case for circumstellar ejecta. Considering CAI chemistry in CV types, chondrite models are short of the refractory element Ti. Again, reactions need not have occurred if the material accreting was reduced, as are the enstatite chondrites. Also, formation of the core first would have depleted iron from the sub-nebula, so the lower mantle could have accreted from only rocky dust, similar to that in enstatite chondrites.

9.2.4.2 Estimated chemical compositions for the core

Iron alloy being an independent reservoir makes estimating protocore *size* equivocal, but it is clear that heat producing elements are negligible. The amount of Fe alloy plus FeS in ordinary chondrites, indicate that the Earth began with a protocore of $\sim 1/2$ the mass of the current core, and $\sim 1/2$ of the existing core was added by gravitational settling. Enstatite chondrites have enough alloy and sulfide to form the whole core.

The IAB types are now considered to be magmatic (Benedix et al., 2000), so only the IIIAB type may be primitive. The IIIABs constitutes 24% of the collection. Except for high Ni contents (~ 8 wt%), these differ little from the others. Therefore, to estimate the composition of protocore metal, Table 9.2 averages the data on the most abundant elements (Ni, Co, and P), using proportions observed for the different types (Lodders and Fegley, 1998) and obtain the iron content by difference (Table 9.2). The resulting composition agrees with Buchwald's (1977) assessment, but is higher in Ni than previous models, which assume the minimum Ni content observed in meteorites.

The protocore must have had some FeS, since troilite is ubiquitous in iron meteorites, and ices with S are common in clouds (Section 9.2.2). In chondrites, the proportion of FeS to alloy is close to $1/3$ by weight. However, iron meteorites have less. Buchwald (1977) suggest ~ 1 wt% S, whereas Mittlefehldt et al. (1998, their table 3) suggests 3 wt%. Either is possible, as are higher concentrations of C or N.

If the protocore was indeed $\sim 1/2$ the mass of the modern core, then this composition should describe the current inner core. The outer core could have the same chemical composition, like iron meteorites, or could resemble the iron and sulfide components of the ordinary chondrites (Table 9.2). All things considered, S is the main light element in the outer core and probably in the inner core as well.

9.2.5 Radioactive heat generation in the upper mantle

Oceanic surface flux mostly represents shear heating (Chapter 5). Continental flux is controlled not by heat from the deep interior, but by the high concentrations in the upper-most ~35 km (Chapter 6). Meteoritic compositions must be used with trepidation, due to Earth's chemical evolution. Therefore, we devise alternative methods to constrain K, U, and Th in the upper mantle, for present and early chemical compositions. First, calculations based on the homogenous oceanic lithosphere are presented, then background information is provided to utilize data on the continental crust.

9.2.5.1 Constraints on the present upper mantle from MOR basalts via partitioning

Lithospheric plates (total thickness of ~65–70 km) have a thin, top coat of basalt, dike and gabbros (~7.5 km total) which were produced by a low degree of partially melting (<4%). These three layers are chemically similar and are not distinguished in this analysis, and likewise for a small amount of residua from melting.

The cold thin top of the plates melts shortly after descent since their high water content reduces the solidus. The majority of the plate (~90%) differs from the underlying mantle only by temperature, and so as plates descend into warmer regions, these simply warm up, becoming indistinct from the upper mantle. Thus, the plates thin as they descend (Chapter 7). The combination of thermal-chemical structure with slow descent results in thermal equilibration of most slabs between ~200 and 300 km. In contrast, fast and steep descent delays re-equilibration to 660 km, but this describes only a few, quite thick slabs in the Pacific (Chapter 7). All slabs are resorbed and recycled within the upper mantle, with a little material crossing into the transition zone.

Since mid-ocean ridge magmas are produced at shallow levels, above 200 km (Section 7.3.2), the K, U, and Th contents in the present-day upper mantle can be determined from the known chemical composition of the MOR basalts (Gale et al., 2013) and partitioning coefficients.

Coefficients measured by Salters et al. (2002) are used because they explored conditions linked to MOR magma production (e.g., Presnall and Gudfinnsson, 2008). Salters et al. (2002) values are similar to those of several studies, and lead to ~ppb concentrations of U and Th in the upper mantle (Table 9.3, top section). The very low values for U and Th calculated for the present day upper mantle are consistent with earlier extraction of the highly enriched continental crust (Section 9.2.5.5).

Our estimate is consistent with bulk chemistry of abyssal peridotites. These have K_2O below 500 ppm, based on its absence from the compilation of major elements (e.g., Warren, 2016) and a typical limit-of-detection for potassium in peridotite minerals (P. Carpenter, personal communication, 2019). An average Th content of 4.9 ppb was determined from 60 abyssal peridotites shown in Figure 11 of the review by Bodinier and Godard (2004). Data on 75 ophiolites shown in this figure provided 3.8 ppm Th, which is

TABLE 9.3 Survey of heat producing elements in Earth's components.

Region[a]	K, ppm	U, ppb	Th, ppb	Source or method
Lower mantle, excluding the core	1100	18	53	Ordinary chondrites (Section 9.2.6)
MOR basalts, average	1330	119	404	Gale et al. (2013)
Komatiites, $N = 8$	450	27	65	Arndt et al. (2008)
Present upper mantle	540	~2	~3	From MORB via partition coefficients[b]
Abyssal peridotites, average	<400		4.9	Bodiner and Godard (2014)[c]
Oceanic nodules, average	54	1.9	5.5	Wakita et al. (1967)
Ophiolite peridotites, average			3.8	Bodiner and Godard (2014)[c]
Orogenic peridotites, $N = 32$	49		15.4	ibid.; Hamilton and Mountjoy (1965)
Trachyte, $N = 22$	59,095	5900	18,000	Bowerman et al. (2007)
Lamprophyres, $N = 23$	55,695	8800	38,000	Peterson et al. (1994)
Shale composite	32,960	2660	12,300	Gromet et al. (1984)
Upper continental crust (0–35 km)	23,000	2700	10,500	Rudnick and Gao (2014)
Ultramafic lamprophyres, $N = 11$	19,920	5150	28,500	Tappe et al. (2004)
Acasta gneiss (3.9 Ga), $N = 23$	13,970	670	4200	Mojzsis et al. (2014)
Precambrian Canadian shield, $N = 88$	13,100	1100	7700	Shaw et al. (1986)
Kimberlites, average	10,400	3100	16,000	Lodders and Fegley (1998)
Carbonates, $N = 2038$	5150	1100	1990	Gao et al. (1998)
Continental lithosphere (0–130 km)	5000	500	2000	From thermal models of Ch. 6
Greenland anorthosites (2.9 Ga), $N = 18$	1985	<230	235	Dymek and Owens (2001)
Initial upper mantle (0–410 km)	1080	61	235	Dispersion and remixing[d]
Xenoliths, average	500–660	43	194	Pearson et al. (2014)[e]
Ca(Si,Ti)O$_3$ in diamond, $N = 7$ (~300 km)	~200	~38,000	~54000	Kaminsky et al. (2001); Walter et al. (2011)
CAIs, $N = 7$	~250	~50	~360	Jacobsen et al. (2008)

[a]Numbers in parentheses are layer depths. The number of samples (N) is specified if the averages were computed from tabulated data in the cited reference.

[b]Tabulated data on the bulk sample composition (Table A2 of Salters et al., 2002) were used for potassium, whereas averages of minerals (Table 1 of Salters et al., 2002) were used for U and Th. Uncertainties exist because olivine does not uptake U or Th and the mineral proportions were not provided.

[c]Average Th was obtained by digitizing figure 11 of Bodiner and Godard (2003). Potassium contents were not reported among the major elements in several compilations, which indicates this is below detection in microprobe analyses. Given the mineralogy of peridotites, K$_2$O contents are likely below 500 ppm and may be below 200 ppm (P. Carpenter, personal communication, 2019).

[d]Dispersing the upper continental crust or continental lithosphere results give similar values for the amount removed from upper mantle: respectively K = 570 or 470 ppm; U = 67 or 46 ppm; Th = 260 or 183 ppb. The average of these estimates was summed with a remix of the oceanic crust with the current upper mantle (which gave similar K, but much lower U and Th) to estimate the old upper mantle, before differentiation.

[e]From mafic minerals in Table 9 of Pearson et al. (2014), who gives two approximate K values.

in good agreement. Within uncertainty, averages obtained from rock data and our calculation from partition coefficients agree.

Applying Eq. (9.1) to our upper mantle estimate provides a flux of $\sim 3 \, mW \, m^{-2}$ across the oceanic lithosphere. This steady state estimate is $\sim 6\%$ of the measured flux distance from the ridges, which equals $50 \, mW \, m^{-2}$. Power estimated from shear (Table 7.3) accounts for almost all oceanic heat production, within the uncertainties of the estimates.

9.2.5.2 Relative buoyancy of heat producing elements

Uranium and thorium are commonly considered to be large ions, but potassium is larger. Importantly, the density of UO_2 and ThO_2 is much greater than that of K_2O and other X_2O compounds (Fig. 9.4). Much of the periodic table was perused in constructing Fig. 9.4, which further shows that K_2O is as buoyant as Na_2O. The alkali metal trend differing from that of the heavy metals is attributed to ionic vs. covalent bonding.

All heat producing elements are strongly lithophilic and thus are enriched in the continental crust. Equivalent phases with concentrations of K similar to that of U (or Th) should be more buoyant, by roughly a factor of $\sim 5 \times$. Many other considerations are relevant, such as cation sites in minerals, partitioning coefficients, and greater abundance of K (ppm vs ppb). In particular, occurrence of U and Th in higher valance states than K increases their relative density.

Upward concentration of K relative to other alkalis, recognized by Heier and Adams (1964), is supported by Fig. 9.4 and by the increasing abundance of true granites relative to granodiorites with time (R.F. Dymek, personal communication, 2019). A time dependence of K content compared to U and Th exists but is affected by remobilization, as deduced by Lambert and Heier (1963). The recently compiled database on 500 granitic rocks (Artemieva et al., 2017) corroborates that the time dependent content of K, U, and Th in continental crust is complicated. Yet, trends exists: plotting the data in Table 9.3 as a

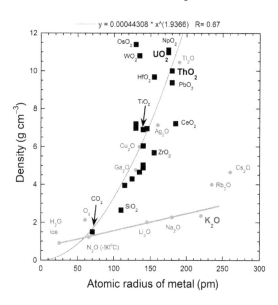

FIGURE 9.4 Dependence of density on atomic radius for X_2O and XO_2 compounds. Empirical radii from Slater (1964). Density from compilations and websites. Other determinations of atomic radii are similar. Oxides of the similar ions Au, Fr, Nd, and La occur with XO or X_2O_3 stoichiometry, as do most other elements. Low abundance actinides have little data, but should behave similarly to ThO_2 or UO_2. Oxides with small cations follow linear trends, due to their substitution stretching the oxygen sub-lattice.

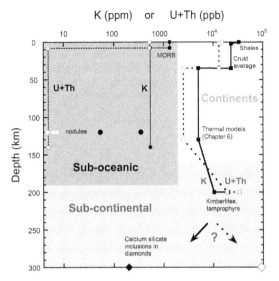

FIGURE 9.5 Dependence of the concentrations of heat producing elements on depths, based on the summary of Table 9.3. The focus is on averages of a large number of measurements and/or regions. Dark gray box represents MOR basaltic crust and its region of extraction. Depths vary from ~140–200 km, where the latter has been linked to ocean islands (see Chapter 7). Nodules could be residua. White box shows the continents, whose roots can reach almost 250 km. Shales (surface), rock, and thermal averages are shown. Lavas bringing up xenoliths are shown, where depths are estimated. Diamond inclusions are not rocks, but suggest the changes of K compared to U + Th with depth.

function of depth (Fig. 9.5), shows the sub-oceanic crust has very little K, U, and Th, whereas the continents have much more, and suggests that mantle below where MORBs are derived shows a reversal in K vs U + Th proportions. Lateral heterogeneities are expected, since continental (lithosphere) drift places the continents over different parts of the mantle.

9.2.5.3 Abundances vs. ratios of heat producing elements in the rock sample

Regarding heat and its flow, concentrations of K, U, and Th are essential per Eqs. (9.1) and (9.2). Ratios are convenient, but information is lost by using this approach, as follows:

It is known that ratios of K/U ~10000 or Th/U ~3.6 change with time. It is also known that rocks of different silica content have different ratios, e.g., Heier and Adams (1964). Fig. 9.6 shows that silicate rocks cluster around K/U = 9700, such that a weak dependence of silica content is observed. Sub-oceanic mantle rocks tend to have high ratios, whereas continental xenoliths have low ratios, even though these commonly rich in potassium (Table 9.3). Continental rocks have high K/U particularly if old (e.g., Acasta gneiss). In contrast to silicates rich in Mg or Fe, rocks and phases rich in Ca have ratios of K/U far below 10^4 from Wasserburg et al. (1964). Carbonate-rich rocks have very low ratios. Ordinary chondrites and CAIs diverge from the line in the opposite direction.

Fig. 9.7A shows individual measurements of K/U as a function Th for continental rocks with deep origins. Papers with many measurements on different regions were utilized. The individual samples vary quite a bit within each region such that K/U decreases as Th increases. Although this trend is connected with U being proportional to Th, the different slopes suggest that the Th/U ratio varies among the datasets. Interestingly, trends for all three areas all point to the position of the ordinary chondrites, implicating the importance of this meteorite reservoir. In this depiction, the CAI reservoir differs, but seems related to

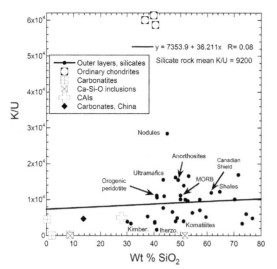

FIGURE 9.6 Dependence of K/U ratio on silica content from various compilations. The goal was to represent different rock types, different localities, and different ages. Mantle derived rocks are likely over-represented in this survey. The fit shown is to the silicate rocks only. Sources are listed in Tables 9.2 and 9.3. Additional sources are Adetunji et al. (2018), Andreeva et al. (2007), Verhulst et al. (2000), Zhao et al. (2018), and references therein.

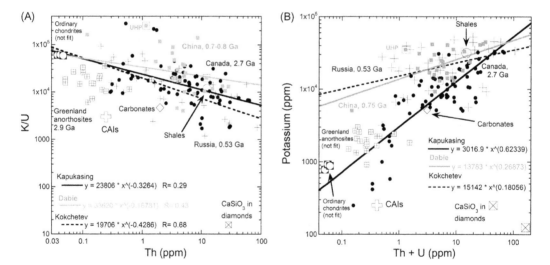

FIGURE 9.7 Dependence of heat producers on each other for continental crustal rocks. Data sources are in Tables 9.2 and 9.3 for the averages. Meteorites ages are 4.5 Ga. These inclusions from ∼300 km were not dated and a wide range of ages exists in the literature for other types of inclusions (Nestola et al., 2017). Individual rock data are from Ashwal et al. (1987); Shatzsky et al. (1999); Bryant et al. (2004); Dymek and Owens (2001). The deepest rocks from Dabei are shown with open squares. (A) Dependence of K/U on Th content. CaTiO₃ inclusions are not shown, but have Th ∼92 and K/U ∼1.8. (B) Dependence of K content on the sum of Th and U. These fits have R of 0.43−0.8. CaTiO₃ inclusions are rich in the heavy elements.

old anorthosites. The deepest mineral inclusions known seem to suggest that Th may reside in the transition zone.

One might concluded from Fig. 9.7A that another meteorite reservoir exists with very high Th and low K/U. A different circumstance is indicated in plots of abundances, i.e., of ppm K against the sum of Th and U (Fig. 9.7B), or against each of Th or U, which are similar and thus not shown. Younger rocks have more potassium at any given Th or U content, despite the fits not being well constrained. Anorthosites, which have a restricted major element chemistry, lie on the trend for old Canadian rocks. Had the anorthosites been included, the Archean trend would point to the ordinary chondrites. Fig. 9.7B indicates evolution of the crust from a precursor that is largely ordinary chondrite material with a small amount of admixed CAIs, discussed further below.

9.2.5.4 Constraints on the upper mantle from the continental crust

In principle, the composition of the early mantle can be ascertained from mathematically rehomogenizing the continental crust and today's mantle, if the depth of extraction were known. Bounds can be set on both, but some difficulties exist. An obvious problem is heterogeneity.

Regarding depths, mantle xenoliths originate above 250 km (Chapter 1). Although subduction involves greater depths, down to 660 km, very little plate material reaches these depths presently, and most plates are resorbed by 300 km (Chapter 7). Occurrence of a seismic discontinuity at the intermediate depth of 410 km, along with cooling of the outer layers with time, suggests that the upper mantle is the zone of extraction. The possibility of deeper extraction to 660 km (the transition zone) is considered below.

The average composition of the upper continental crust (above 35 km) is established by rock data. Table 9.3 lists a refinement after decades of effort. The lower continental crust composition remains debated, due to sampling limitations. Therefore, we base this discussion on results for the sub-continental lithosphere from thermal models (Chapter 6). For a 130 km thick continental lithosphere, calculations in Chapter 6 set an upper limit on U and Th: our maximums for these elements are listed in Table 9.3, along with a K content that provides the average measured surface flux. Roughly, the heat producing elements in the deep continental crust and lithosphere are a factor of 5 lower than in the upper crust (Fig. 9.5). Due to volumetric considerations, the substantial contribution of the lithosphere is used to calculate the early mantle composition.

We first calculate a composition, assuming that no radioactive elements remained after extraction. This is not true, so the value provided in Table 9.3 is actually the sum of this result with a composition for the present upper mantle, determined from remixing the present oceanic crust with the present upper mantle. Our result is supported by low concentrations of K, U, and Th in sub-continental mantle xenoliths. These are only slightly lower than our estimate, and the ratios of the elements are similar, both of which are consistent with these being young (1−3 Ga) compared to the oldest continental rocks (3−4 Ga). However, while our estimate for K in old upper mantle matches that of ordinary chondrites, U and Th are $3\times$ and $5\times$ higher, respectively. Did a deeper extraction occur, or did we omit something from our calculation, such as heterogeneity?

An omission is suggested by Fig. 9.7B, where the old continental rocks show evidence of regional effects during extraction, and by Table 9.3, which shows that rocks conveying mantle xenoliths upwards are rich in all three elements. The latter data implicates heterogeneity, since the kimberlite eruptions are connected with CO_2 volatiles, yet carbonate rocks have low K relative to U and Th (Fig. 9.6). Heterogeneity is demonstrated by the large variations of K, U, and Th contents of the rocks in Table 9.3. Although this list is far from complete, it should represent the known ranges of the important heat producers, since it encompasses samples from different depths, different ages, different types, and different localities. Lastly, vertical heterogeneity (Section 9.2.5.5) is substantiated by diamond inclusions from ~ 300 km depths having huge concentrations of U and Th, but little K.

Heterogeneity prohibits improving our estimate. However, available data are consistent with extraction of the continental and oceanic crusts from above 410 km, where K, U, and Th contents of the original upper mantle were close to those in an average ordinary chondrite. Petrology points to shallower origins of MORB and ocean island basalts, that are more in line with re-equilibration depths of plates mostly above 300 km (Fig. 1.7; Chapter 7).

The data are also consistent with vertical zone refining with time, such that K is concentrated upwards, whereas U and Th are sent downwards, in the affected zone. Fig. 9.5 suggests a zonation that is compatible with compositional (Rudnick and Gao, 2014) and thermal models (Chapter 6). Kimberlites and lamprophyres represent material moving upwards.

Recycling of crustal material is important (Armstrong, 1991). But, vertical communication is limited, since the continents do not subduct, the ocean crust melts at low temperatures, the layer of sediments is very thin, and very little of the ocean plate mass crosses 410 km. In view of the present day geotherm (Figs. 7.14 and 8.3) and cooling of the upper mantle, differentiation is currently a very slow process.

9.2.5.5 Constraints on the upper mantle from Fe/Mg ratios

Olivine, which is considered to be the most abundant mineral in the upper mantle, has iron in the Fe^{2+} state, with an atomic ratio of Fe/Mg of $\sim 1/9$, which is equivalent to Fe/Mg $\sim 1/4$ by weight. Ordinary chondrites likewise contain olivine, rather than the forsterite endmember. Removing the metal and sulfide components (the "settled" composition of Table 9.2) represents heat producers in the upper mantle, but contains too much iron for mantle olivine. Two possibilities come to mind: (1) The upper mantle is between the "settled" and "reacted" compositions, where mixing should be constrained by the $1/4$ weight ratio. (2) The upper mantle is stratified in Fe/Mg, with Fe increasing downwards.

Importantly, all rock samples are derived above ~ 250 km and so Fe/Mg $= 1/4$ by weight describes this region only. Regarding diamond inclusions, depths of ferromagnesian phases have been estimated assuming phases in separate inclusions are in equilibrium, which is counter indicated (Nestola et al., 2017). Nonetheless, (Mg,Fe)O is commonly included in the diamonds hosting $CaSiO_3$ and thus should be likewise from the upper mantle. Compositions vary geographically, such the ratio of Fe/Mg equals or exceeds $1/4$ by weight (Fig. 9.8). The variation could be due to either age or depth, but in either case, some parts of the upper mantle are richer in Fe than the nominal value from olivine. Given this heterogeneity, which may be vertical, the composition of the upper mantle could be as rich in Fe as the "settled" ordinary chondrite composition. We favor this possibility because the meteorites that exist today should represent the late stage of planetary accretion. Essentially everything else has been deeply buried, as is evident in the rarity of reduced meteorites, whose oxidation state is consistent with that of the pre-solar nebulae.

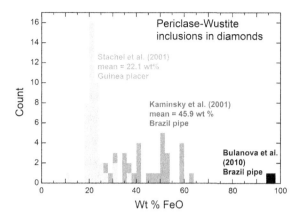

FIGURE 9.8 Histogram of the iron content of simple XO phases in diamonds. This material is nearly stoichiometric, suggesting reducing conditions, and has few cations other than Fe and Mg. The diamonds in the three studies shown (Bulanova et al., 2010; Kaminsky et al., 2001; Stachel et al., 2001) also had inclusions with calcium silicate. Inclusions from Africa and Canada are similar to the Guinea placer deposit.

9.2.6 The transition zone and a summary

The above depiction of the upper and lower mantles rests on these being accreted from the same materials. But, as sketched in Fig. 9.1, the various reservoirs are independent. Plus, processing occurred during accretion and an iron protocore formed first, to literally "get the ball rolling." The lower mantle is proposed to be reduced, and low in Fe^{2+}, like the enstatite chondrites, whereas the upper mantle has more iron and is more oxidized. Little iron metal being in the lower mantle is consistent with heterogeneous accretion: the core forming first would have depleted the local sub-nebula of iron, leaving the rock components. But after this close-in metal-depleted region collapsed to form the lower mantle, the remaining sub-nebula would have had substantial iron metal.

For the bulk rocky upper mantle to be like the average ordinary chondrite, more iron must reside above 660 km than is observed in generic Fo90 olivine. We tentatively suggest that a gradation in Fe^{2+} exists from the base of the continents to near $\sim 220-660$ km. This gradation need not be smooth or continuous. The 410 km seismic discontinuity could well be chemical, in which case the transition zone is rich in Fe^{2+}, which balances the known composition of olivine from above 250 km. The 660 km discontinuity is almost certainly chemical, given its abruptness and the bending of the pointy slab tips at this apparently impenetrable boundary (Fig. 1.9). A chemical gradation or layering stemming from density variations limits the depths of sample extraction. In our depiction, the transition zone is transitional in iron content.

9.3 Thermal evolution of the Earth

Cold accretion points to a different thermo-chemical history than that for hot accretion models. We sketch the time-line and temperatures, using the available evidence. Constraints are too few for quantitative modeling.

9.3.1 Assembly timeline and early processes

Isotopes, not ages, are measured, so chronology is affected by modeling. Age determinations from Pb-Pb measurements and models are utilized for direct and accurate comparison. Table 9.4 summarizes the oldest ages so far determined, and compares these to the Fe^{2+} content of the major silicate phases in the meteoritic samples, which is a measure of oxidation. Data on ordinary chondrites show a surprising spread in the high precision Pb-Pb ages. Importantly, 8 of the 58 measurements in this essentially Gaussian distribution (Fig. 9.9) have ages older than those of the CAIs, which have been considered the oldest components of meteorites. A very old age of 4565 was also determined for a silica-rich, porous achondrite (Srinivasan et al., 2018), but because Al-Mg isotopes were measured, we did not include this measurement in Table 9.4. Although uncertainties exist, CAIs are apparently not the oldest material in the Solar System.

The data reveal that not only do the reduced and oxidized chondrites overlap in age, but moreover that overlap in ages also involves meteorites which lack chondrules but instead have material that had been melted. To make sense of the data, we need to know what chondritic material likely represents.

TABLE 9.4 Early chronology of the Earth and terrestrial planets from Pb-Pb measurements and models.

material (object)	Age, Ma[a]	Reference[b]	Fe content[c]	Process/event/region[d]
Ordinary chondrites[e]	4591	See text	~Fo80	Beginning of upper mantle accretion
CAIs (CV)[f] $N = 4$	4567.3	Connelly et al. (2017)	Fo98	Ejecta from surface of lower mantle
Chondrules (Acfer)	4564	Amelin et al. (2010)	Fo99.5	Ejecta from surface of lower mantle
Enstatite (Stillwater)	4563	Gilmour et al. (2009)	En99	Ejecta from surface of lower mantle
Chondrules (CB)	4562.7	Krot et al. (2005)	Fo99	Ejecta from surface of lower mantle
Various (H) $N = 20$[e]	4559 ± 14	Bouvier et al. (2007)	Fo80−85	Ejecta from accreting upper mantle
Achondrites (angrite)	4556.2	Baker et al. (2005)	Fo55	Ejecta after Mercury was spalled deeply
Various (LL) $N = 17$[e]	4553 ± 13	Bouvier et al. (2007)	Fo68−76	Ejectafrom accreting upper mantle
Various (L) $N = 13$[e]	4540 ± 17	Bouvier et al. (2007)	Fo75−79	Ejecta from accreting upper mantle
Ordinary chondrites[e]	4523	See text	various	Accretion of rocky planets complete
Shergottites (Mars) $N = 4$	4504	Bellucci et al. (2018)	Fe rich	Rock made from old magma inside Mars
Achondrite (eucrite)	4483	Tera et al. (1997)	En50	Ejecta from the once large Mercury
Zircon (Moon)	4417	Nemchin et al. (2009)	-	Ejected from prior lunar crust
Zircon (Australia)	4400	Valley et al. (2014)	-	Ejected from prior continental crust

[a]The last digit is uncertain. Measurements with large uncertainties are not included because these blur the distinctions provided by high precision measurements.
[b]Most recent literature, which cites values from earlier work.
[c]On similar meteorites, from Scott and Krot (2007) and other review papers.
[d]The upper mantle and transition zone are not distinguished in this chronology; plus, the meteorites studied need not be those of Earth.
[e]Data on whole rock, chondrules, phosphates, residues, and leachates from 9 different ordinary chondrites. N indicates the number of measurements. One age (4609 Ma) was considered to be an outlier by the authors. However, including the outlier and data with large uncertainties in the averages made negligible difference. Bjurböle was included in the LL category based on Fo content. The average for 51 precise measurements is 4552 ± 15, and the accurate data range from 4523 to 4591 Ma.
[f]Names in parentheses provide details, such as either meteorite type, locality where collected, or the parent body.

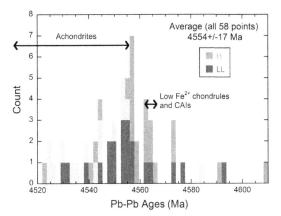

FIGURE 9.9 Histogram of precise Pb-Pb ages from 9 different ordinary chondrites prepared using the tables in Bouvier et al. (2007). A single study was used for the most accurate comparison possible. Ages for type L are overall lower than those for type H, but the ranges overlap and the averages for the three types are within uncertainty of each other. Double arrow shows the range for the reduced and refractory chondritic components, while the single arrow marked achondrites points to the Basaltic Achondrite Best Initial age. See Table 9.4.

9.3.1.1 *Chondrule and chondrite formation during cold accretion*

Chondritic meteorites are a highly porous assembly of diverse materials, consistent with incorporation of ice on the dust grains which collected to form the growing planets. Chondrule characteristics are inconsistent with an origin as melt droplets (Section 9.2.2.3). The variety of minerals in chondrules combined with their fairly round shapes and restricted grainsize distribution point to chondritic meteorites being analogous to aeolian sandstones: evidence is presented in Chapter 11. This proposal is based on winds swirling around the surface during the local collapse that formed the planets and spun them up. Thus, we propose that chondrules are assembled and shaped during collisions of dust grains in the winds of accretion, becoming consolidated along with matrix and metal into chondrites, which are rocks assembled as the chondrules, metal, ice, and dust were deposited the surface of the growing protoplanets. Note that very little material leaves the forming planet because the influx knocks much of the outflux back onto the surface. During and after this violent process, a tiny proportion of the aeolian deposit was ejected: Solar directed impacts are implicated, as follows:

Ejection of differentiated, compact, and low porosity achondrites from full sized Mars (radius = 3390 km; surface acceleration = 3.7 m s^{-2}) is well documented by the SNC meteorites; analogous evidence exists that brecciated, less durable material was ejected from our smaller Moon. The latter ejections occurred during infrequent impacts postdating accretion. Impacts have become increasingly rare with time, but were common in the Hadean, as evidenced by the lunar surface (e.g., Stöffler et al., 2006). Frequency of impacts during formation would have been at least as high. As is well-known from studies of ancient crater-saturated surfaces, the later impacts add material which buries the former additions (e.g., Squyres et al., 1997). Hence the earliest material is rare, and moreover, burial causes diagenesis (Section 9.3.1.3).

Because the Sun is far more massive, it continued to form long after small sub-nebulae collapsed to form the various protoplanets. As the immense dusty nebula was gathered by the Sun, impactors intercepted by the protoplanets added some material and heat, while ejecting rocks near the surface at the time of impact.

9.3.1.2 Interpretation of chondrite ages

The data in Table 9.4 and Fig. 9.8 are consistent with the above depiction. Importantly, meteorites represent various rocky bodies, such that ejections occurred at different times from their growing and changing surfaces. To make sense of the overlapping ages in the diverse meteorite collection, we first focus on chondrites.

The rarity of the enstatite chondrites is consistent with this reservoir being precursor material which formed the lower mantle but was then buried. Hence, enstatite chondrites are old, along with other reduced material, e.g., CAIs and forsterite chondrules. Although this reduced material is the building block of the planet, material that has been processed to some degree describes the majority of the meteorite collection. Achondrites have been extensively processed (Section 9.3.1.4).

The porosity of chondrites is attributed to (1) interstices accompanying close packing, (2) ices being incorporated along with dust on planetary surfaces, and (3) impacts ejecting near-surface material that had been partially compacted. This depiction accounts for the mixtures of phases constituting the primitive meteorites. The collection is $\sim 95\%$ ordinary chondrites, which are chemically similar to enstatite chondrites, but more oxidized. Reaction of H_2O with enstatite to form Fe-rich olivine is covered in Section 9.3.1.2. The Fe^{2+} content of olivine in chondrites increases as the age decreases. Thus, oxidation progressed with planetary growth.

The data of Bouvier et al. (2007) indicate that the peak of ordinary chondrite formation was at 4555 Ma. However, because of burial, early material is undersampled. Also, high influx during accretion knocks material back down. In contrast, the slower influx directed towards the Sun at the end of its accretion would eject material from the outer layers. Hence, the peak in ages of the ejected material reflects burial on the old age side, but an exponential decline in influx on the young age side. This applies to all terrestrial planets, but since the impactors were focused on the Sun, the cross-sections and fluxes received increased inwards. As is evidenced by relative sizes of planetary cores to mantles and Solar System trends (Fig. 4.4), many meteorites were ejected from Mercury (Section 9.3.1.4).

We propose that the ordinary chondrites represent the accumulation of Earth's upper mantle with time, which is 10% of Earth's mass. The little constrained transition zone, constituting an additional 7.5%, likely grew during this stage, as suggested by the global uniformity of the 660 km seismic discontinuity. The time span was 175 Ma from Table 9.4. The growth rate provides some perspective. For the upper mantle only, the rate is estimated as 4×10^{15} kg y^{-1}. For a comparison, Americans generate 2×10^{11} kg of municipal waste each year, so the rate of worldwide trash production is $\sim 0.05\%$ of the estimated accretion rate. With slow accumulation, heat from impacts can be dissipated.

9.3.1.3 Oxidation during assembly releases gas

The main ices in nebulae, H_2O, CO_2, and CO, freeze at 273, 195, and 68 K, respectively. Although the starting conditions are ~ 8 K, collisions of dust in the winds of accretion should create heat and dislodge ice. Because CO is abundant, this ice would nucleate on Fe^0, where collisions of this complex with copious H_2 gas permit a reaction:

$$\{Fe^0 + CO\} + H_2 \rightarrow FeC + H_2O \tag{9.3}$$

Both Ni and Fe metal are known catalysts (e.g., the Haber process). The reaction of Eq. (9.3) helps to explain H_2O being more abundant than CO_2 upon densification of nebula into clouds and comets (Fig. 9.2), and the ubiquitous presence of graphite and carbides in iron meteorites. Chapter 11 provides further discussion. Low temperatures limit reaction rates, so the earliest stages (i.e., production of the core and the lower mantle) should involve mostly dust collection and compaction.

As temperatures climb, perhaps due to Solar ignition or radiative transfer decreasing in the densifying nebula, CO is the first species to vaporize, whereas highly polar and reactive water is retained, as observed for comets (Fig. 9.2). Because carbonic acid is effective at rusting iron and rusting is exothermic (see websites) and therefore highly favored, another reaction may oxidize iron as accretion proceeds:

$$Fe^0 + H_2O \rightarrow FeO + H_2 (\text{promoted by dissolved } CO_2) \qquad (9.4)$$

$$FeO + MgSiO_3 \text{ (enstatite)} \rightarrow (Mg, Fe)SiO_4 \text{ (Fe-rich olivine)} \qquad (9.5)$$

$$Mg_2SiO_4 \text{ (forsterite)} + (Mg, Fe)SiO_4 \rightarrow 2(Mg_{1.5}Fe_{0.5}SiO_4) \text{ (meteoritic olivine)} \qquad (9.6)$$

These reactions are simplifications of a very complex electrolytic corrosion process that releases hydrogen gas (see websites; Graedel and Frankenthal, 1990). Hydrogen gas is light and diffuses upward, which is promoted by compaction as material is added. This loss occurs at shallow levels, as a gas-producing reaction is not promoted by pressure. Departing H_2 gas carries some heat of reaction away. Substantial heating will vaporize CO_2 ice, which will rise through the porous substrate. This process is consistent with H_2O dominating comets. The presence of acidifying H_2S, SO, and SO_2, as occurs in comets (Fig. 9.2; see also Fig. 10.6), promote electrolytic corrosion (Graedel and Frankenthal, 1990).

Our proposal is supported by compositional data. As shown in Table 9.4, the different types (H, L, LL) are associated with different ranges of Fe/Mg in the olivine. From Table 9.2, an average ordinary chondrite has about $^1/_2$ of the iron alloy of an enstatite chondrite, which on average are composed of 57 mol% $MgSiO_3$ + 33 mol% Fe + 10 mol% FeS. Reacting about $^1/_2$ of the metal in an enstatite chondrite via Eqs. (9.3) and (9.4) yields Fo77, which is close to the average of $\sim$Fo79 for the ordinary chondrites, by number proportions. Rusting much iron metal in an enstatite chondrite makes L and LL chondrites, whereas rusting less metal makes type H. These reactions are consistent with the observed proportions of rock to metal in ordinary chondrites (see Lodders and Fegley, 1998, their table 16.11, and Ringwood, 1961).

Also, the proportion of forsterite to enstatite in the precursor material affects the amount of Fe^{2+} in the olivine of the ordinary chondrites. Many reactions are possible, since rusting actually puts Fe^{2+} ions into solution, which can ion exchange with Mg in forsterite and other minerals. Analogous reactions involving troilite could also occur, and would similarly evolve gas, while oxidizing iron and contributing to the variability. Loss of H_2 gas and volatilizing ices, which depart as vapor, mitigates heating of the surface.

Rusting of the surface of planets occurred as these formed. However, the current upper mantle is not as iron rich as an average ordinary chondrite: the composition at least for the upper 250 km is Fo90, rather than Fo79. The correlation of younger ages with higher Fe contents suggests that an unstable density profile was formed, and furthermore that the more magnesian olivines likely formed early on, as the transition zone accumulated, but were buried, see Section 9.3.2.

9.3.1.4 Achondrites and the spalling of mercury

The overlap in ages of oxidized chondrites with achondrites may point to accretion and differentiation occurring simultaneously. Another possibility exists: Hofmeister and Criss (2012) suggested that the largest family of achondrites (HEDs) and many related irons, stony-irons, and angrites came from Mercury. This loss was caused by Mercury being close to the Sun so it was heavily bombarded at the end of accretion. The other terrestrial planets would have received a much lower influx, since the cross section would go as the orbital radius squared, and clearly retained much rocky material.

The ages of achondrites point to Mercury being spalled between 4483 and 4556 Ma (Table 9.4). These meteorites have low porosity and have been differentiated, in accord with the great depths of excavation for Mercury. The old age of the angrites is due to these being early additions to Mercury, which became susceptible to ejection by impact only after the upper layers were spalled away. On this basis, the structure of Mercury just after formation can be reconstituted using high precision dates of achondrites.

9.3.1.5 Flux melting

The presence of water substantially lowers the melting temperatures of silicate rocks, as is well-known. On this basis, we suggest the 4565 MA old, high silica meteorite found on the Moon (Srinivasan et al., 2018), was flux melted. Quench melt exists around the pores. Tridymite is stablilized by incorporation of Na and Al, which were detected in this sample, and by hydroxyl incorporation, which is consistent with the low microprobe total for this phase. The orthopyroxene has Fe/Mg near 1/10, which suggests little rusting. Scant iron metal exists. Given the unique chemistry, matrix material that is unlike the key phases, and the age, this sample was melted during an impact in part due to the presence of ices. The oxygen isotopes (Srinivasan et al., 2018) suggests proto-Mercury as the source (Fig. 1.5).

9.3.2 The lower mantle and core

From the above, heat generation below 660 km is limited to radioactivity in the mantle. The age of ~ 4590 Ma is linked to completion of lower mantle growth, but how long this early stage took in unknown. However, for cold accretion, this duration is not terribly important: ~ 4590 Ma can be taken as the cold start below 660 km with a constant and cold temperature in the lower mantle and core.

Considerable uncertainty in ascertaining thermal evolution of Earth's deep zones arises from ambiguity in values for thermal conductivity. If the material is iron free, then $K(T)$ can be ascertained from data on minerals, particularly enstatite. But if a small amount of Fe^{2+} exists, then heat is transported diffusively at high frequencies. For this case, conduction in the lower mantle would be very efficient (Chapter 3) but a constraint is needed on the Fe^{2+} content. Thermal conductivity in the core is important prior to melting. Values are affected by impurity content, and thus constraining the amount of Ni is crucial.

To probe the thermal history of the deepest regions of the Earth with any certainty requires reconstructing Mercury from the meteoritic data. Otherwise, one could assume different heating rates. However, a bit can be said, based on the findings in this book.

9.3.2.1 Lower mantle temperatures

The geotherm (Fig. 8.3) shares many features with the calculated thermal evolution (Fig. 2.4), which considered radionuclide decay only while assuming an initial temperature of 1000 K. However, the core temperatures should rise with time as well as the mantle temperature. The maximum mantle temperature calculated for today (i.e., at an elapsed time of 4.56 Ga) is almost 4000 K. Although this temperature is likely overestimated, it is in reasonably good agreement with the proposed geotherm (Table 8.2), and supports that the interior of the Earth is warming up, while the upper ~ 660 to 1000 km cools to space (Criss and Hofmeister, 2016).

The lower mantle has apparently not melted. The density gradient is smooth and stable, with no suggestion of differentiation. Temperatures (Fig. 8.3) are far from steady-state, and so the lower mantle will continue to warm. At the present time, the outer regions are cooler than the inner, so heat from the lower mantle is leaking towards the surface. This loss limits temperatures in the lower mantle and melting of the core.

9.3.2.2 Age of the liquid core

From Table 9.1, 8.7×10^{29} J is stored as latent heat in a 100% melted outer core. From Fig. 8.3, the power entering the core or generated in the core is ~ 3 TW, suggesting that a fully liquid core would have required ~ 920 Ga to form, but this age is far older than the Earth. Complete melting is impossible. Thus, the outer core is a sludge, which is consistent with buffering and progressive melting with time. Existence of both solid and liquid phases in the outer core means that its viscosity is much higher than that of a completely melted iron alloy. Furthermore, existence of a high viscosity sludge is consistent with negligible super-rotation (see Chapter 8).

Estimating the age of melting requires constraints on the proportion of crystallites in the outer core, and an accurate value for flux. However, the ballpark age above suggests that very little melt exists. The warming of the lower mantle and core could provide another constraint. This calculation requires solidi on a wider range of compositions, and particularly with varying Ni content.

9.3.3 The transition zone, upper mantle, and crust

A better chance exists for understanding the thermal history of Earth's outer layers, for which samples exist. The present state was covered in Chapters 5–7, where we argued that the gravitational force imbalance during Earth's non-planar orbit combined with differential spin and a moderately thick lithosphere cause plate tectonics. The thickness of the lithosphere is largely controlled by heat produced during shearing of differential rotation, but is also affected by additional factors which are changing with time. Hence, modeling the changes of the upper layers is daunting. Much information is lacking, and many questions exist, which pertain to fairly recent behavior, such as:

- When did organized plate tectonics with subduction begin?
- When did differential rotation begin?
- How long has the Wilson cycle existed?
- When did the continental crust first form?

- Are the plates thinning or thickening (on average) with time?
- Is subduction past the 410 km discontinuity a recent event?

Even more questions exist regarding the behavior of the early Earth. A few of these issues are discussed below, using the new findings in this book. The focus is the initial state, which sets the stage for the Earth we know. Because the Earth started cold, early heating would have occurred where friction was generated or radionuclides were concentrated. As shown in Chapter 10, these reservoirs are similar in magnitude and in their approximately exponential decline with time.

9.3.3.1 Magmatism and differentiation

Magmatism, which conveys radioactive isotopes plus latent heat rapidly upwards while advecting heat (Chapter 3), links Earth's thermal and chemical evolution. If a zone becomes hot enough to produce magma, the buoyant liquid will work its way upwards. As observed today, large magma chambers cannot support their roofs. Thus, substantial volumes of melt will ascend as the material above sinks into the liquid. Spin loss is potentially the largest source of heat, should be most important early on, and is a focus of Chapter 10. Tidal forces are an additional source of deformation, and would have been strong early on when the Moon was close in.

The ascent of magmas will concentrate materials with the lowest melting temperatures towards the surface, while leaving more refractory material such as MgO and aluminates at depth. Magmatism carries material that is enriched in the heat-producing elements upwards for several reasons: radioactive-enriched material would have a higher temperature than their surroundings, plus these cations preferentially partition into magmas. This behavior underlies the growth of the continents.

9.3.3.2 Outgassing

Outgassing is not only a much more buoyant process than magmatism, but gas can leave the solid Earth and even the atmosphere carrying heat with it, whereas liquid remains. The violent nature of this process is connected with the high thermal expansion coefficients of vapor. Liquid water is $1000 \times$ more expansive than silicate melts, and water vapor expansion is larger than liquid (see website list). Degassing should have been very important to the early Earth because water and CO_2 ices would have been incorporated with dust during accretion, as in primitive comets: above we proposed that a corrosive process converted precursor enstatite chondrites to ordinary chondrites.

Outgassing helped to stratify the Earth in density for several reasons. (1) Pressure creates a stable density configuration, even of heterogeneous solids. Pressure affects the transition zone strongly, but the upper mantle only weakly. (2) The strength of rocks prevents density sorting in a depths, but if fracturing occurs, their strength is reduced.

9.3.3.3 Hadean warming and differentiation

From Table 9.3, the continents have been derived from material that was initially positioned below 250 km, down to at least 410 km, and perhaps to 660 km. From Table 9.4, accretion was heterogeneous, where incorporation of Fe^{2+} in olivine increases upwards. This density gradient was not a stable configuration, and compression is unimportant at shallow levels. The solidus of olivine is reduced by Fe^{2+} incorporation to $\sim 1300\,°C$ (1600 K) near the middle of the binary. Melting is further promoted by H_2O.

But for anything to happen, Earth had to warm up. Although radionuclides may have once been distributed fairly uniformly, this is not the case for heat generated during spin-down. First, rocks are strong under compression, so Earth's deep layers spun essentially congruently then and now. Second, the outside layers are much weaker because these formed with ice and were not compacted fully. Third, the outside layers spin faster and are subject to more stress. Thus, friction was preferentially generated in the outer layers from slippage between the grains, so Earth was heated more to the outside. Heat was also added during interception of impactors.

Warming describes the Hadean, which was followed by differentiation. The lack of rocks points to vigorous sorting and upheaval, which explains the current mantle being Fo90, rather than Fo84 to Fo60 of ordinary chondrites. We suggest that the late arriving LL and L chondrites are now in the transition zone, whereas the earlier arriving H chondrites are towards the surface.

9.3.3.4 Was the Archean hot?

Eruption of komatiite lavas which are Mg-rich with a higher melting temperature (Chapter 1). has previously been interpreted as an indication of a hot early Earth. However, only radionuclide emission and accretionary heat, after Kelvin's contracting stars, were considered, and the view has been that the whole Earth cools to space, both of which are in error (Chapter 2).

The Archean was indeed warmer than the Proterozoic, but this due not to a hot initial Earth, but largely to the dissipation of spin in the Hadean to the outer layers, with peak radioactive decay being simultaneous. From magma temperatures (Fig. 9.10), peak temperatures occurred in the early Archean. Preservation of substantial volumes of gneisses by ~ 2.9 Ga, shortly after the bulk of komatiite production, cooroborates the timing of stabilization and radial cooling of the outer layers. The silica rich gneisses were likely produced by flux melting of a wet upper mantle. Importantly, the surface temperature of the Earth is controlled by the Sun to just above the freezing point of water (Chapter 10). Earth uniquely possessing continents appears connected with its particular orbital radius, and its size being sufficiently large to retain light molecules (Chapter 10).

Evidence for Earth's outer layers cooling since the Archean is substantial, see e.g., Chapters 1 and 2. The cooling rate for the upper mantle at $\sim 0.2\,K\,Ma^{-1}$ is not that much slower than the early warming rate at $\sim 1.3\,K\,Ma^{-1}$ (Fig. 9.9). The cooling rate indicates an average loss of heat from the upper mantle of ~ 3.8 TW, which is compatible with our estimate of today's radioactive power as 1.5 TW (Section 9.2.5). The bulk of surface cooling is spin dissipation.

9.4 Discussion and further work

We have pieced together a picture of assembly of the Earth and its chemical and thermal evolution based on production of kinetic energy (spin), not heat, during its gravitational assembly, and on diverse evidence from meteorites and rocks. With this approach,

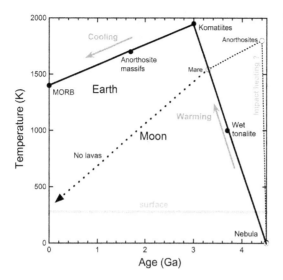

FIGURE **9.10** Temperatures of volumetrically important magmas that are associated with specific ages. For data, see e.g., Arndt et al. (2008), Ashwal (1993), Weinberg and Hasalová (2014).

many inconsistencies and puzzles arising in the currently popular depictions of Earth and its changes with time are moot. For example, there is no need to sequester potassium in the core or to volatilize this element preferentially. If Earth were so hot as to volatilize potassium, then Earth could not have retained noble gases, water, or CO_2, and all the elements more volatile than K would be depleted to greater degrees. Improved accuracy in K-contents of ultramafic rocks and mantle xenoliths would be helpful, as would an average CAI composition, and more focus on the important heat-producers, rather than the early heat-producers.

A cold icy beginning with heating of the outer layers in the Hadean, differentiation following, existence of peak temperatures for the layers, and subsequent cooling of the outside addresses the serious issue of radioactive heating being insufficient by itself to warm the whole mantle over geologic time. Warming the outer layers by spin down is consistent with the weakness of these little compressed regions and explains the well-known heterogeneity of the outer layers, both laterally and vertically. Impacts likely provide local heating, only.

Much remains to be examined in view of cold, heterogeneous accretion. Ages of meteorites need to be compared statistically, using the same isotope system. Experiments on ice-mineral reactions need to be run at low pressures. Thermal conductivity and thermal diffusivity data are needed below room temperature. To understand the differentiation of the Earth, the notion that the upper and lower mantles are chemically identical needs to be set aside, and the experimental focus needs to be on low pressure and modest temperature conditions, and on open systems. The latter is particularly important, as advection of water and other molecules not only affects heat and mass transfer, but H_2O drastically alters phase relations. The view that the mantle has much water, and particularly that its low viscosity basaltic magmas store water, is inconsistent with mobility, buoyancy, mass diffusivity, and reactivity of H_2O.

References

Adetunji, A., Olarewaju, V.O., Ocan, O.O., Macheva, L., Ganev, V.Y., 2018. Geochemistry and U-Pb zircon geochronology of Iwo quartz potassic syenite, southwestern Nigeria: constraints on petrogenesis, timing of deformation and terrane amalgamation. Precambrian Res. 307, 125–136.

Aitta, A., 2006. Iron melting curve with a tricritical point. J. Stat. Mech. Theory Exp. 12, Online at : stacks.iop.org/JSTST/2006/P12015. Available from: https://doi.org/10.1088/1742-5468/2006/12/P12015.

Amelin, Y., Kaltenbach, A., Iizuka, T., Stirling, C.H., Ireland, T.R., Petaev, M., et al., 2010. U−Pb chronology of the solar system's oldest solids with variable $^{238}U/^{235}U$. Earth Planet. Sci. Lett. 300, 343–350.

Anderson, D.L., 2007. New Theory of the Earth, second ed. Cambridge University Press, Cambridge.

Andreeva, I.A., Kovalenko, V.I., Nikiforov, A.V., Kononkova, N.N., 2007. Compositions of magmas, formation conditions, and genesis of carbonate-bearing Ijolites and carbonatites of the Belaya Zima alkaline carbonatite complex, eastern Sayan. Petrology 15, 551–574.

Armstrong, R.L., 1991. The persistent myth of continental growth. Aust. J. Earth Sci. 38, 613–630.

Arndt, N., Lesher, C.M., Barnes, S.J., 2008. Komatiite. Cambridge University Press, Cambridge.

Artemieva, I.M., Thybo, H., Jakobsen, K., Sørensen, N.K., Nielsen, L.S.K., 2017. Heat production in granitic rocks: global analysis based on a new data compilation GRANITE2017. Earth Sci. Rev. 172, 1–26.

Ashwal, L.D., 1993. Anorthosites. Springer Berlin Heidelberg, Heidelberg.

Ashwal, L.D., Morgan, P., Kelley, S.A., Percival, J.A., 1987. Heat production in an Archean crustal profile and implications for heat flow and mobilization of heat-producing elements. Earth Planet. Sci. Lett. 85, 439–450.

Baker, J., Bizzarro, M., Wittig, N., Connelly, J., Haack, H., 2005. Early planetesimal melting from an age of 4.5662 Gyr for differentiated meteorites. Nature 436, 1127–1131. Available from: https://doi.org/10.1038/nature03882.

Bellucci, J.J., Nemchin, A.A., hitehouse, M.J., Snape, J.F., Bland, P., Benedix, G.K., et al., 2018. Pb evolution in the Martian mantle. Earth Planet. Sci. Lett. 485, 79–87.

Benedix, G.K., McCoy, T.J., Keil, K., Love, S.G., 2000. A petrologic study of the IAB iron meteorites: constraints on the formation of the IAB-Winonaite parent body. Meteor. Pianet. Sci. 35, 1127–1141.

Bernatowicz, T.J., Zinner, E.K., 1997. Astrophysical Implications of the Laboratory Study of Presolar Materials. AIP, New York.

Bodinier, L., Godard, M., 2014. Orogenic, ophiolitic, and abyssal peridotites. In: Turekian, K., Holland, H. (Eds.), Treatise on Geochemistry, vol. 2. Elsevier, Amsterdam, pp. 103–167.

Bouvier, A., Blichert-Toft, J., Moynier, F., Vervoort, J.D., Albarede, F., 2007. Pb–Pb dating constraints on the accretion and cooling history of chondrites. Geochim. Cosmochim. Acta 71, 1583–1604.

Bowerman, M., Christianson, A., Creaser, R.A., Luth, R.W., 2007. A petrological and geochemical study of the volcanic rocks of the Crowsnest formation, southwestern Alberta, and of the Howell Creek suite, British Columbia. Can. J. Earth Sci. 43, 1621–1637.

Bowey, J.E., Hofmeister, A.M., 2005. Overtones and the 5-8 mm spectra of deeply embedded objects. Monthly Notices of the Royal Astronomical Soc 358, 1383–1393.

Brearly, A.J., Jones, R.H., 1998. Chondritic meteorites. Rev. Miner. 36, Ch. 3 (398pp).

Bryant, D.L., Ayers, J.C., Gao, S., Miller, C.F., Zhang, H., 2004. Geochemical, age, and isotopic constraints on the location of the Sino−Korean/Yangtze suture and evolution of the northern Dabie complex, east central China. GSA Bull. 116, 698–717. Available from: https://doi.org/10.1130/B25302.1.

Buchwald, V.F., 1977. The mineralogy of iron meteorites. Phil. Trans. R. Soc. Series A, Math. Phys. Sci. A286, 453–491. Available from: https://doi.org/10.1098/rsta.1977.0127.

Bulanova, G.P., Walter, M.J., Smith, C.P., Kohn, S.C., Armstrong, L.S., Blundy, J., et al., 2010. Mineral inclusions in sublithospheric diamonds from Collier 4 kimberlite pipe, Juina, Brazil: subducted protoliths, carbonated melts and primary kimberlite magmatism. Contrib. Miner. Petrol. 160, 489–510. Available from: https://doi.org/10.1007/s00410-010-0490-6.

Connelly, J.N., Bollard, J., Bizzarro, M., 2017. Pb−Pb chronometry and the early solar system. Geochim. Cosmochim. Acta 20, 345–363.

Consolmagno, G.J., Britt, D.T., Macke, R.J., 2008. The significance of meteorite density and porosity. Chemie der Erde: Geochemistry 68, 1–29.

Criss, R.E., Hofmeister, A.M., 2016. Conductive cooling of spherical bodies with emphasis on the Earth. Terra Nova 28, 101–109.

Dymek, R.F., Owens, B.E., 2001. Chemical assembly of Archean anorthosites from amphibolite- and granulite-facies terranes, SW Greenland. Contrib. Miner. Petrol. 141, 513–528.

Fagan, T.J., Krot, A., Keil, K., 2000. Calcium-aluminum-rich inclusions in enstatite chondrites (I): mineralogy and textures. Meteor. Planet. Sci 35, 771–781.

Gale, A., Dalton, C.A., Langmuir, C.H., Su, Y., Schilling, J., 2013. The mean composition of ocean ridge basalts. Geochem. Geophys. Geosyst. 14 (3), 489–518. Available from: https://doi.org/10.1029/2012GC004334.

Gao, S., Luo, T.-C., Zhang, B.-R., Zhang, H.-F., Han, Y.-W., Zhao, Z.-D., et al., 1998. Chemical composition of the continental crust as revealed by studies in East China. Geochim. Cosmochim. Acta 62, 1959–1975. Available from: https://doi.org/10.1016/S0016-7037(98)00121-5.

Gilmour, J.D., Crowther, S.A., Busfield, A., Holland, G., Whitby, J.A., 2009. An early I-Xe age for CB chondrite chondrule formation, and a re-evaluation of the closure age of Shallowater enstatite. Meteor. Planet. Sci. 44, 573–579.

Graedel, T.E., Frankenthal, R.P., 1990. Corrosion mechanisms for iron and low alloy steels exposed to the atmosphere. J. Electrochem. Soc. 137, 2385–2394.

Gromet, L.P., Dymek, R.F., Haskin, L.A., Korotev, R.L., 1984. The "North American shale composite": its compilation, major and trace element characteristics. Geochim. Cosmochim. Acta 48, 2469–2482.

Hamilton, W.B., Mountjoy, W., 1965. Alkali content of alpine ultramafic rocks. Geochim. Cosmochim. Acta 29, 661–671.

Hasterok, D., Gard, M., Webb, J., 2018. On the radiogenic heat production of metamorphic, igneous, and sedimentary rocks. Geosci. Frontiers 9, 1777–1794.

Heier, K.S., Adams, J.A.S., 1964. The geochemistry of the alkali metals. Phys. Chem. Earth 5, 253–379.

Hofmeister, A.M., 1997. Infrared reflectance spectra of fayalite, and absorption data from assorted olivines, including pressure and isotope effects. Phys. Chem. Miner. 24, 535–546.

Hofmeister, A.M., 2019. Measurements, Mechanisms, and Models of Heat Transport. Elsevier, Amsterdam (see Chapter 7 on mineral and rock data, and Chapter 8 on a blackbody radiation).

Hofmeister, A.M., Criss, R.E., 2012. Origin of HED meteorites from the spalling of Mercury: implications for the formation and composition of the inner planets. In: Hwee-San, L. (Ed.), New Achievements in Geoscience. InTech, Croatia, pp. 153–178. Available from: http://www.intechopen.com/articles/show/title/the-case-for-hed-meteorites-originating-in-deep-spalling-of-mercury-implications-for-composition-and.

Hofmeister, A.M., Criss, R.E., 2015. Evaluation of the heat, entropy, and rotational changes produced by gravitational segregation during core formation. J. Earth Sci. 26, 124–133.

Hofmeister, A.M., Criss, R.E., Criss, E.M., 2018a. Verified solutions for the gravitational attraction to an oblate spheroid: implications for planet mass and satellite orbits. Planet. Space Sci. 152, 68–81. Available from: https://doi.org/10.1016/j.pss.2018.01.005.

Jacobsen, B., Yin, Qing-zhu., Moynier, Frederic., Amelin, Yuri., Krot, Alexander N., Nagashima, Kazuhide., et al., 2008. ^{26}Al–^{26}Mg and ^{207}Pb–^{206}Pb systematics of Allende CAIs: canonical solar initial 26Al/27Al ratio reinstated. Earth Planet. Sci. Lett. 272, 353–364.

Jaresowich, E., 1990. Chemical analyses of meteorites: a compilation of stony and iron meteorite analyses. Meteoritics 25, 323–337.

Kaminsky, F.V., Zakharchenko, O.D., Davies, R., Griffin, W.L., Khachatryan-Blinova, G.K., Shiryaev, A.A., 2001. Superdeep diamonds from the Juina area, Mato Grosso State, Brazil. Contrib. Miner. Petrol. 140, 734–753.

Kleine, T., Touboul, M., Bourdon, B., Nimmo, F., Mezger, K., Palme, H., et al., 2009. Hf–W chronology of the accretion and early evolution of asteroids and terrestrial planets. Geochim. Cosmochim. Acta 73, 5150–5188.

Krot, A.N., Amelin, Y., Cassen, P., Meibom, A., 2005. Young chondrules in CB chondrites from a giant impact in the early solar system. Nature 436, 989–992. Available from: https://doi.org/10.1038/nature03830.

Lambert, I.B., Heier, K.S., 1963. The vertical distribution of uranium, thorium, and potassium in the continental crust. Geochim. Cosmochim. Acta 31, 377–390.

Lodders, K., Fegley, B.J., 1998. The Planetary Scientist's Companion. Oxford University Press, Oxford.

Mittlefehldt, D.W., McCoy, T.J., Goodrich, C.A., Kracher, A., 1998. Non-chondritic meteorites from asteroidal bodies. Rev. Miner. 36, 1529–6466.

Mojzsis, S.J., Cates, Nicole L., Caro, Guillaume., Trail, Dustin., Abramov, Oleg., Guitreau, Martin., et al., 2014. Component geochronology in the polyphase ca. 3920 Ma Acasta Gneiss. Geochim. Cosmochim. Acta 133, 68–96.

Molster, F.J., Waters, L.B.F.M., 2003. The mineralogy of interstellar and circumstellar dust. In: Hennings, Th. (Ed.), Astromineralogy, ed. Springer-Verlag, Berlin, pp. 121−171.

Mumma, M.J., Charnley, S.B., 2011. The chemical composition of comets—Emerging taxonomies and natal heritage. Annu. Rev. Astron. Astrophys. 49, 471−524.

Nemchin, A., Timms, N., Pidgeon, R., Geisler, T., Reddy, S., Meyer, C., 2009. Timing of crystallization of the lunar magma ocean constrained by the oldest zircon. Nat. Geosci. 2, 133−136.

Nestola, F., Jung, H., Taylor, L.A., 2017. Mineral inclusions in diamonds may be synchronous but not syngenetic. Nat. Commun. 14168. Available from: https://doi.org/10.1038/ncomms14168.

Nittler, L.R., Ciesla, F., 2016. Astrophysics with extraterrestrial Materials. Annu. Rev. Astron. Astrophys. 54, 53−93.

Pearson, D.G., Canil, D., Shirey, S.B., 2014. Mantle samples included in volcanic rocks: xenoliths and diamonds. In: Turekian, K., Holland, H. (Eds.), Treatise on Geochemistry, vol. 2. Elsevier, Amsterdam, pp. 169−253.

Pertermann, M., Hofmeister, A.M., 2006. Thermal diffusivity of olivine-group minerals. Am. Miner. 91, 1747−1760.

Peterson, T.D., Esperanca, S., LeCheminant, A.N., 1994. Geochemistry and origin of the Proterozoic ultrapotassic rocks of the Churchill Province, Canada. Miner. Petrol. 51, 251−276.

Presnall, D.C., Gudfinnsson, G.H., 2008. Origin of the oceanic lithosphere. J. Petrol. 49, 615−632.

Ringwood, A.E., 1961. Chemical and genetic relationships among meteorites. Geochim. Cosmochim. Acta 19, 159−197.

Rudnick, R.L., Gao, S., 2014. Chemical assembly of Archean anorthosites from amphibolite- and granulite-facies terranes, SW Greenland. In: second ed. Turekian, K., Holland, H. (Eds.), Treatise on Geochemistry, vol. 4. Elsevier, Amsterdam, pp. 1−51.

Salters, V.J.M., Longhi, J.E., Bizimis, M., 2002. Near mantle solidus trace element partitioning at pressures up to 3.4 GPa. Geochem. Geophys. Geosyst. 3, paper 6. Available from: https://doi.org/10.1029/2001GC000148.

Scott E.R.D., Krot A.N., 2007. Chondrites and their components. In Meteorites, Comets and Planets (ed. A. M. Davis) Chapter 1.07, Treatise on Geochemistry Update 1, Elsevier, online at http://www.sciencedirect.com/science/referenceworks/9780080437514

Scott, E.R.D., Wasson, J.T., 1975. Classification and properties of iron meteorites. Rev. Geophys. 13, 527−546.

Shaw, D.M., Cramer, J.J., Higgins, M.D., Truscott, M.G., 1986. Composition of the Canadian Precambrian shield and the continental crust of the Earth. In: Dawson, J.D., Carswell, D.A., Hall, J., Wedepohl, K.H. (Eds.), The Nature of the Lower Continental Crust: Geological Society Special Publications, 24. Geological Society of London, pp. 275−282.

Shatsky, V.S., Jagoutz, E., Sobolev, N.V., Kozmenko, O.A., Parkhomenko, V.S., Troesch, M., 1999. Geochemistry and age of ultrahigh pressure metamorphic rocks from the Kokchetav massif (Northern Kazakhstan). Contrib. Miner. Petrol. 137, 185−205.

Slater, J.C., 1964. Atomic radii in crystals. J. Chem. Phys. 41, 3199−3205. Available from: https://doi.org/10.1063/1.1725697.

Squyres, W., Howell, Colin., Liub, Michael C., Lissauer, Jack J., 1997. Investigation of crater "saturation" using spatial statistics. Icarus 125, 67−82.

Srinivasan, P., Dunlap, D.R., Agee, C.B., Wadhwa, M., Coleff, D., Ziegler, K., et al., 2018. Silica-rich volcanism in the early solar system dated at 4.565 Ga. Nat. Commun. 9, No. 3036. Available from: https://doi.org/10.1038/s41467-018-05501-0. Online at www.nature.com/naturecommunications.

Stöffler, D., Ryder, G., Ivanov, B.A., Artemieva, N.A., Cintala, M.J., Grieve, R.A.F., 2006. Cratering history and lunar chronology. Rev. Min. Geochem. 60, 519−596.

Stachel, T., Harris, J.W., Brey, G.P., Joswig, W., 2000. Kankan diamonds (Guinea): II. Lower mantle inclusion parageneses. Contrib. Miner. Petrol 140, 16−27.

Sugiura, N., Hoshino, H., 2003. Mn-Cr chronology of five IIIAB iron meteorites. Meteor. Planet. Sci. 38, 117−143.

Tappe, S., Jenner, G.A., Foley, S.F., Heaman, L., Besserer, D., Kjarsgaarde, B.A., et al., 2004. Torngat ultramafic lamprophyres and their relation to the North Atlantic alkaline province. Lithos 76, 491−518.

Tera, F., Carlson, R.W., Boctor, N.Z., 1997. Radiometric ages of basaltic achondrites and their relation to the early history of the solar system. Geochim. Cosmochim. Acta 61, 1713−1731.

Tolman, R.X., 1934. Relativity, Thermodynamics and Cosmology. Oxford University Press, Oxford.

Valley, J.W., Cavosie, A.J., Ushikubo, T., Reinhard, D.A., Lawrence, D.F., Larson, D.J., et al., 2014. Hadean age for a post-magma-ocean zircon confirmed by atom-probe tomography. Nat. Geosci. 7, 219−223.

Verhulst, A., Balaganskaya, E., Kirnarsky, Y., Demaiffe, D., 2000. Petrological and geochemical trace elements and Sr−Nd isotopes/characteristics of the Paleozoic Kovdor ultramafic, alkaline and carbonatite intrusion, Kola Peninsula, NW Russia. Lithos 51, 1−25.

Wakita, H., Nagasawa, H., Uyeda, S., Kuno, H., 1967. Uranium, thorium and potassium contents of possible mantle materials. Geochem. J. 1, 183−198.

Walter, M.J., Kohn, S.C., Araujo, D., Bulanova, G.P., Smith, C.B., Gaillou, E., et al., 2011. Deep mantle cycling of oceanic crust: evidence from diamonds and their mineral inclusions. Science 334, 54−57.

Warren, J.M., 2016. Global variations in abyssal peridotite compositions. Lithos. 248−251, 193−219.

Wasserburg, G.J., MacDonald, G.J.F., Hoyle, F., Fowler, W.F., 1964. Relative contributions of uranium, thorium, and potassium to heat production in the Earth. Science 143, 465−467.

Weinberg, R.F., Hasalová, P., 2014. Water-fluxed melting of the continental crust: a review. Lithos 212 − 215, 158−188.

Zhao, X., Hao, L., Wei, Q., Liu, Q., Zhou, J., Wang, S., et al., 2018. Origin of late triassic mafic−ultramafic intrusions in the Hongqiling Ni−Cu sulfide deposit, Northeast China: evidence from trace element and Sr−Nd isotope geochemistry. Can. J. Earth Sci. 55, 1312−1323. Available from: https://dx.doi.org/10.1139/cjes-2018-0041.

Zolensky, M.E., et al., 2006. Mineralogy and petrology of comet 81P/wild 2 nucleus samples. Science 314, 1735−1739.

Websites

Summary of databases on rocks.
Geochemical Earth Reference Model (accessed 07.06.19.).
https://earthref.org/GERMRD/datamodel/.
Data on atomic radii (accessed 08.06.19.).
https://en.wikipedia.org/wiki/Atomic_radius.
https://en.wikipedia.org/wiki/Atomic_radii_of_the_elements_(data_page).
Rusting/corrosion (accessed 18.06.19.).
https://www.thoughtco.com/how-rust-works-608461.
https://en.wikipedia.org/wiki/Rust.
Water structure and science by M. Chaplin (accessed 01.09.18).
http://www1.lsbu.ac.uk/water/water_structure_science.html.

CHAPTER

10

Thermal history of the terrestrial planets

Robert E. Criss and Anne M. Hofmeister

OUTLINE

"I ain't never see 'im myse'f, but I done seed dem what say dey hear tell er dem what is see 'im." *(Joel Chandler Harris, 1955), in "The Story of the Doodang".*

This chapter deduces the thermal history of the rocky bodies of the inner Solar System from indirect information, specifically by estimating their initial inventories of energy, and by evaluating their subsequent production and losses of heat. In order to compare the results inferred for various bodies to Earth, a less detailed approach is taken here than in Chapter 9, which utilized Earth's substantially larger database.

The rocky bodies probably accreted from a cold, contracting nebula, which is congruent with the observed conditions of astronomical nebulae and galaxies, and explains system-wide trends in the Solar System such as axial spin. Inventories per unit mass of heat generated over geologic time by radionuclides vary tenfold for these bodies, whereas their

Heat Transport and Energetics of the Earth and Rocky Planets
DOI: https://doi.org/10.1016/B978-0-12-818430-1.00010-0

267

inventories generated by spin dissipation vary from near zero to significant. Both inventories tend to increase with planetary size, and the power generated by each process has decreased nearly exponentially with time. Additions of heat by large impacts in the early days of the Solar System contributed some heat to shallow zones, in amounts that depend inversely on distance to the Sun. Major conclusions are that planetary interiors are cooler than commonly imagined, and that the inner zones of the largest bodies may be warming even as their outer zones are cooling. Only the near-surface zones have been active, as elucidated below.

Section 10.1 summarizes the scarce available data on the heat flux of extraterrestrial bodies. Section 10.2 focuses on the probable inventories of their internal energy due to spin dissipation and radioactivity. Section 10.3 discusses how surface conditions are controlled, and the effect of molecular volatiles, particularly water, which lowers melting temperatures. Section 10.4 briefly discusses the consequences of these energy inventories and surface conditions on the history and thermal state of the individual rocky bodies. Section 10.5 summarizes.

10.1 Heat flux for extraterrestrial bodies

Actual measurements of extraterrestrial heat flow are available only for the Moon. Limited data will become available for Mars by about 2020.

Measurements of lunar heat flow were made in shallow, 2.5 m-deep holes at the Apollo 15 and 17 landing sites, returning $\Im = 21$ and 16 mW m^{-2}, respectively (Langseth et al., 1976). These values are associated with very low values of thermal conductivity of $\sim 0.01 \text{ W m}^{-1} \text{ K}^{-1}$ for the lunar regolith, which essentially is porous soil. It follows that the temperature gradient just below the surface is very high, but this very shallow value is atypical of lunar crust. Taken at face value, the flux values suggest that the lunar power is about 0.7 TW, corresponding to an average, present-day heat generation of 0.03 µW m^{-3}, or $\sim 10 \text{ pW kg}^{-1}$, presuming quasi-steady state, where the current heat production equals surface heat loss. Power is computed per mass in this chapter, in order to compare rocky bodies of disparate sizes. Approximately steady-state conditions are expected given the small size and antiquity of the Moon and a thermal diffusivity typical of silicates (Chapter 2). As shown below, the deduced values for the specific power of the Moon (power per unit mass, Ψ_{moon}) are nearly 2× higher than expected if the Moon were constituted of ordinary chondritic material, which appears to describe the Earth (Chapter 9).

The paucity of direct measurements necessitates that the thermal condition of extraterrestrial bodies be deduced from indirect data, specifically from their energy generating processes and their probable content of heat-generating radionuclides (Section 10.2).

10.2 Energy inventories

Over geologic time, energy has been supplied to the interiors of the rocky planets from several sources (listed in Section 9.1). The axial spin rates of these objects were formerly much higher than now, and the frictional dissipation of this motion has released heat, especially to the outer layers. Late stage impacts also caused heating, particularly of the shallow layers. In contrast, radioactive decay of K, U, and Th has contributed significant internal heat, which is particularly important at great depths of large bodies, since this heat cannot reach the surface over geologic time. Estimates for the energy contributed by these sources are provided below.

10.2.1 Spin energy arising from planetary accretion from a cold nebula

Thermal models require some understanding of initial conditions. Planetary evolution is anchored in an early state that featured densification of a pre-existing nebula that progressively contracted and spun-up to form the Sun and planets. Subsequent frictional dissipation of the axial spin of the condensed bodies has released heat over time. The following narrative provides a plausible scenario of this complex process.

Spectroscopic data on remote nebulae and galaxies indicate that these objects are cold. Blackbody spectra from stellar nurseries such as the Orion Nebula and Eagle Nebula, and of other gas-rich areas in galaxies, indicate temperatures below 10 K (Bergin and Tafalla, 2007). Interestingly, IR spectra of these objects indicate the presence of the Si-O vibrations associated with solid matter and of CO, CO_2, and H_2O ices (e.g., van Breemen et al., 2011; also see Fig. 9.2).

The collapse of a giant molecular cloud or nebula is a gravitationally-driven process that is governed by Newton's law, momentum conservation, and the progressive loss of energy. The galactic evolution model of Criss and Hofmeister (2018) provides useful information as to how this process operated, because the early evolutionary stages would not depend on the original scale of the cloud. The Virial Theorem, combined with the gravitational self-potential of the cloud, shows that the total energy of an equant cloud can be lowered by vertical collapse into a highly flattened, oblate spheroid whose major radius is about 70% as large as the original, but whose density is at least 15 × greater. The angular spin rate of the object increases during this collapse, but only by a factor of 1.5–2 ×.

The subsequent evolution of the spinning object is more complicated, and involves progressive inward densification and roundup that created the central protoSun. Energy is progressively lowered as the mass "falls" into its own gravitational energy well, yet simultaneous conservation of both momentum and energy during this process is not possible. Energy must be lost by radiation during this densification in view of changes in entropy upon volume reduction, whereas conservation of angular momentum requires that the distal parts of the object move outward as the central zones become denser and faster spinning. Analogous behavior of hurricanes is shown in Fig. 5.1. Movies of evolving hurricanes help one visualize this complex motion, as the outer rain belts clearly move outward as hurricanes make landfall and lose energy (see website list).

Individual planets and their moons formed as secondary, miniature "Solar Systems", within parts of the condensing cloud. The gravitational competition of these subsystems with the much larger central protoSun restricted their size, as discussed by Hofmeister and Criss (2012a). Progressive densification of the contracting nebula produced water ice: details of reactions are given in Chapter 11.

The end result of the collapsed nebula is the highly organized Solar System we see today. Most planetary objects orbit in the same sense, spin in the same sense, and have rotational energy inventories that form the coherent energy trend (Fig 1.12) discovered by Hofmeister and Criss (2012a). The Sun and planets formed simultaneously as cold objects during the complex collapse and spin up of a large cloud. Unlike standard models, this scenario explains the many physical coherencies of Solar System objects, conforms to physical theories of energy, momentum conservation, and gravitation, and is congruent with astronomical observations of remote nebulae.

Much of the initial, axial spin of the planets has been lost via frictional dissipation, but several different mechanisms may be involved. Most heat generated was dissipated in the outer zones of the growing rocky planets. The rate of spin attenuation may have decreased exponentially for some, but different planets have different rate constants, and the process has effectively run to completion for Venus and Mercury. Some details are suggested below.

10.2.1.1 Evidence for axial spin loss

The initial, axial spin of the planets was fast, but is difficult to quantify due to the progressive growth of the planets from local concentrations of dust (sub-nebula) in the larger nebula that was contracting to form the Sun. The current spins of the planets are clearly less than their initial rates because (1) the spin rate of our Sun is very slow compared to young yellow stars in clusters, so it has clearly lost most of its spin, and (2) Mercury, Venus, and our Moon have suffered nearly complete loss of spin, as their remnant axial spins are now coupled with their relatively slow orbital circuits. The mechanism of spin loss for the Sun could differ from these spin-locked, rocky bodies, and the timing likely differs as well. To address the complexities and ambiguities, we first consider the available data.

Differential spin in stars is evident in the long-known latitudinal dependence of the rotation rate of our Sun, and also in the super-rotation of its core (Fossat et al., 2017). Differential rotation is also known in other stars: notably, part of the recent K2 mission was intended to explore differential rotation (e.g., Karoff et al., 2009). Because the interiors of stars rotate differentially, they should also be subject to internal shearing. The giant planets should be similarly affected as they have weak outer layers.

Although solid rock should not behave in this manner, differential rotation is evidenced currently in the Earth by the westward drift of its lithosphere (Chapters 1 and 7). Currently, spin dissipation occurs in the outermost ~ 300 km, and specifically is located between the strong mantle and brittle lithosphere, in the low velocity zone, which is relatively weak. Behavior of the Earth points to viscous dissipation as the basic mechanism for spin down because in the laboratory, viscous dissipation is documented and measured with coaxially rotating cylinders (see websites). The latter experiments are a reasonable parallel for the Earth, which has weak layers between strong ones.

In addition, it is highly unlikely that the terrestrial planets were formed as completely solid rocks. Comets today are mixtures of ices and various silicate minerals (Zolensky et al., 2006) whereas primitive, chondritic meteorites are highly porous, with some carbonaceous types reaching 40% porosity (Consolmagno et al., 2008). Carbonaceous chondrites apparently were subject to reactions between minerals and volatiles. This evidence suggests that the terrestrial planets formed with substantial amounts of ice, forming a matrix that would permit differential spin and its consequent loss. In addition, due to the low density of ice, a planet constituted of a rock-ice admixture would have a lower initial spin rate than would a denser rocky planet. Note that the moons of the outer Solar System are mixtures of ice and rock. Their surfaces are very cold, due to low Solar flux, permitting retention of a large fraction of the volatiles they inherited during accretion. These bodies accreted cold, and have remained cold, as discussed further below.

In short, available data point to shear being the mechanism for spin dissipation in all planetary bodies, consistent with the system-wide trend of Fig. 1.12. Viscous dissipation is a simple and logical explanation for the progressive loss of spin for rotating, self-gravitating fluid bodies. Viscous dissipation is accompanied by heat production, but to gauge its time-dependence, a rate law is needed, which we estimate as follows:

10.2.1.2 A model for frictional spin loss

To probe spin loss, we first consider the simple case of a rotating fluid body with a spherical shape. Stars are useful here, because unlike the planets, large populations are available for study. Both stars and gas giants are largely fluid, based on the average density of our Sun, Jupiter, and Saturn being similar to that of water, and our Sun being in the middle of the main sequence. A Newton fluid is considered here, due to this behavior being common and serving as a starting point for many discussions of fluid behavior (e.g., Tritton, 1977).

Viscous stress in a Newtonian fluid creates a force, $F = A\mu \partial v_x / \partial y$ where A = area, μ = the dynamic viscosity, and flow is along the x direction. The equatorial region dominates, so $y \sim R$, $v_x \sim R\omega$, and A is some function of R^2, where R = body radius, and ω is angular velocity. Using torque = $\mathbf{R} \times \mathbf{F} = I\partial\omega/\partial t$, where $I = JMR^2$ and J is the reduced moment of inertia, and $\partial\omega/\partial R = 0$ for a solid body gives the decline of spin with time (t):

$$\frac{\partial\omega}{\partial t} = -\frac{RF}{JMR^2} \sim -\frac{R\mu A}{JMR^2}\omega \sim -\frac{\mu R}{JM}\omega \sim -f\omega, \tag{10.1}$$

where function f is an inverse time constant that combines several physical properties that depend on size and mass, but does not depend on time or spin rate. Eq. (10.1) predicts that the spin rate of a fluid body will decay exponentially.

Exponential loss of star spin was proposed by Terndrup et al. (2000). Below, we use available data on stellar populations to actually test whether the axial spin rate of astronomical bodies declines exponentially.

10.2.1.3 Exponential spin decay of stars

Many studies have probed the statistics of stars spinning in open clusters based on the variations of their emissions with time, which change periodically due to all stars having spots similar to behavior of our Sun (e.g., Hartman et al., 2010). The populous Pleiades cluster has been measured many times (Fig. 10.1). This very interesting and important cluster has white dwarfs, which are very old, as well as hot and large blue stars, which cannot be old since large stars are short lived (e.g., Zombeck, 2007). Existence of young stars is suggested by the presence of nebulosity in this cluster (see website list). Thus, the Pleiades cluster has stars of many ages, so statistical analysis is appropriate.

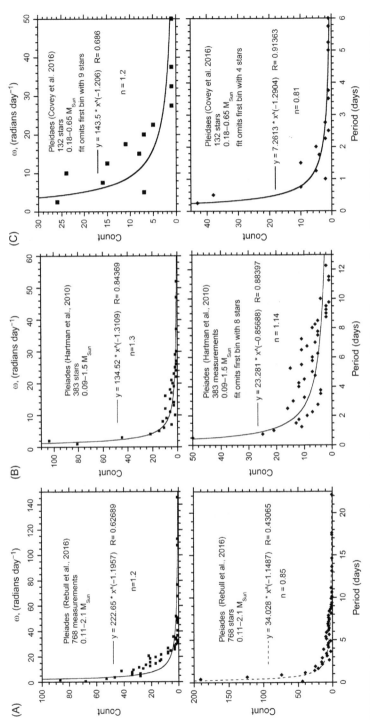

FIGURE 10.1 Histograms of angular velocity (top) and periods (bottom) of small stars in the well-studied Pleiades cluster. (A) and (B) Large datasets on Pleiades open cluster obtained during two different missions and analyzed by different researchers (Hartman et al., 2010, their table 1; Rebull et al., 2016, their table 2). (C) Small, fast spinning stars surveyed by Covey et al. (2016), table 4 therein. For all parts, each histogram bin is shown as a single black point at its top and middle. Black curve = least squares fit with the fit as indicated and *n* from Eq. (10.2). If stated, the depopulated first bin was omitted. These first bins are under-sampled because spins of very slow or very fast stars are difficult to measure.

Spin data are shown as histograms. This commonly used visualization is described by a density function which has a corresponding distribution function that embodies the underlying physics (Apostol, 1969). Paired histograms of some variable and of its reciprocal generally have different shapes because their density functions differ in mathematical form, yet the underlying physical law is the same. However, as discussed by Apostol (1969), similar shapes for such paired histograms indicate that the physical law is exponential in form. Eq. (10.1) predicts that spin rate declines exponentially. To test whether this applies to star spin, we fit actual data to a more general form to determine the power n:

$$\frac{\partial \omega}{\partial t} = - \alpha \omega^n \tag{10.2}$$

where α is a constant whose value depends on n. Power law fits to the top row of histograms in Fig. 10.1 show that n is indeed near unity, i.e., the exponential model holds for spin rate, with some uncertainty. These fits are not perfect because the number of stars in our database is limited and because very fast and very slow spin rates are difficult to measure. For example, data collection must be conducted over an interval that is at least as long as the spin period.

Because period is the inverse of the angular spin rate, the power law fits for period yield n-2 for the rate law (details will be published elsewhere). Power law fits to the bottom row of histograms of Fig. 10.1 are thus also compatible with exponential decay.

Currently popular models known as gyrochronology are based on a "law" derived by Skumanich (1972) which is predicated on the ages of stars and of star clusters derived from assumption-laden models of stellar and cluster evolution being accurate. Our Sun is the only star for which an age is measured, albeit from solids in its vicinity. His exponent of 3 in Eq. (10.2) is clearly not supported by the data on the Pleiades (Fig. 10.1), nor by data from any other stellar population we have investigated.

10.2.1.4 Calculations for stars

To explore initial spin rates, Fig. 10.2 consolidates spin data for stars in open clusters, our Sun, a few young stars, and four surveys of large and/or close stars. All stars are on the main sequence. The techniques involved tracking star spots for the small (dwarf) stars in the open clusters or measuring Doppler broadening effects for the large stars. The initial spin line, from the Virial theorem, describes formation of the star. Stars cannot be observed until they produce light. Ignition may occur after some loss of spin. Parallel lines (isochrones) are drawn through our Sun, which is the only star with a known age albeit one inferred from meteorites, and just past the slowest spinning stars. Finite rotation periods less than 100 days seem to be a characteristic of main sequence stars. Bloated, red giants spin more slowly, consistent with their angular momentum being conserved during expansion.

The number of stars measured is quite large, several thousand, and so Fig. 10.2 should represent their spin distribution and thus their ages. Note that the spin rate of our Sun is much slower than the average for its class. Our Sun appears to be very old compared to other yellow stars, and is clearly not "middle-aged" as popularly believed.

This finding may answer Fermi's paradox as to why sentient life has not been detected elsewhere, when so many stars exist. For complex life to develop requires a planet with a

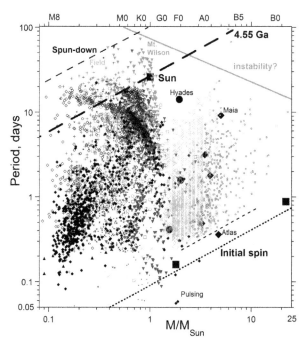

FIGURE 10.2   Period data vs. normalized mass of stars (different symbols) compared to the Virial theorem initial value (dotted line). Cluster data using star spots are shown in black: Diamonds = Pleiades (Rebull et al., 2016) where some stars were noted as having periods affected by pulsations. Large diamonds = large, bright Pleiades stars (see website list). Filled black triangles = IC4665 (Scholz et al., 2009). Open diamonds = M37 (Hartman et al., 2009). X = NGC 2516 (Irwin et al., 2007). Dots = Hyades (Delorme et al., 2011; Douglas et al., 2016). Large dots are the brightest Hyades main sequence stars (see websites). + = Praesepe (Douglas et al., 2017). Large circles = Coma Berenices (Collier Cameron et al., 2009). Small gray dots = M50 (Irwin et al., 2009). Upside down triangles = M35 (Meibom et al., 2009). Open squares = M34 (Meibom et al., 2011). Small circles = NGC 6530 (Henderson and Stassun, 2012). Light gray diamonds = old field type-M (McQuillan et al., 2013). Upside down gray triangles = old Mt. Wilson stars (Baliunas et al., 1996). Large square = the Sun and two young, large stars (Müller et al., 2011; Dufton et al., 2011). For large stars, periods are calculated from Doppler broadening measurements. Dark gray X = A stars (Royer et al., 2002); open squares = B stars (Huang and Gies, 2008 and references therein). The line of instability is suggested by the slowest spinning Mt. Wilson star and by the spin of a 195 M_{sun} Wolf-Rayet star that is rapidly losing its mass. A Wolf-Rayet star with ~25 solar masses anchors the line of instability (see websites). The instability line intersects the formation line at 400 M_{sun} which is just above the largest star masses known (see websites).

certain set of characteristics to be located within a specific distance from a star, and for the star to be stable over a long time. Meeting all criteria simultaneously is highly improbable.

10.2.1.5 Calculations for rocky planets

Based on the above analysis, we conclude that exponential decline of spin with time also holds for rocky planets. Exponential decay governs a wide range of processes, including Newton's law of cooling, and permits explaining very different behaviors with the same mathematics. That is, spin-locked Venus has a much different rate constant than our active Earth.

The amount of energy per kilogram, termed specific energy, that is lost as a spinning spheroid slows down can be estimated as follows. For a homogeneous spheroid of equatorial radius R whose rotational period increases from T_i to T_f, the result is computed as the difference between the rotational kinetic energies (R.E.) at those times:

$$\Delta R.E. \text{ (per kg mass)} = \frac{4\pi^2 R^2}{5}\left(\frac{1}{T_f^2} - \frac{1}{T_i^2}\right)$$ (10.3)

where T_f is the present day rotation period. Note that the change is proportional to R^2. The initial rotational period T_i can be estimated in various ways, but as discussed in Chapters 1 and 9, it was probably about 8 hours for the Earth, and cannot be shorter than its Virial limit of 1.15 hours. If the latter rotation period were used, the energy released would be much larger, but such a high initial spin period is not congruent with other information. The 8 hour rate is similar to previous estimates for the Earth shown in Fig. 1.10, and also is reasonable for Venus and Mercury which have similar densities, but the initial spin period for Mars might have been be about 20% slower due to its lower density. Although the correction is small compared to other uncertainties, the calculations tabulated below use the actual moments of inertia of the individual objects rather than the homogeneous value assumed in Eq. (10.3).

Energies in Table 10.1 were computed for a realistic range of initial rotation rates that spans 4−12 hours. The indicated amount of energy that was released over geologic time is only sufficient to have increased the internal temperatures of Earth and Venus by a few hundred degrees, and is unimportant for the Moon, Mercury, Mars, and Vesta due to their small size. Specifically, in the limit where all energy is internally retained, a megajoule of energy would heat a kilogram of rocky material by about 1000 degrees. The relative ranking of thermal energy per kilogram created by spin dissipation is, from highest to lowest:

$$\text{Earth} \sim \text{Venus} \gg \text{Mars} \sim \text{Mercury} > \text{Moon} \gg \text{Vesta}$$ (10.4)

TABLE 10.1 Estimates of specific energy and average specific power from spin dissipation.

| Object | Current period, hr | Specific energy, MJ kg^{-1} | | | Ψ_{ave}, pW kg^{-1} |
		$T_i = 12$ h	$T_i = 8$ h	$T_i = 4$ h	$T_i = 8$ h
Sun	190	255	570	2300	4052
Mercury	$\sim \infty^a$	0.021	0.048	0.191	0.34
Venus	$\sim \infty^a$	0.128	0.287	1.148	2.03
Earth	24	0.107	0.284	1.243	2.00
Moon	$\sim \infty^a$	0.013	0.028	0.113	0.20
Mars	24.6	0.034	0.089	0.389	0.63
Vesta	5.3	n.a.	n.a.	0.001	

[a]Spin-orbit coupled objects are not freely spinning and so their periods are marked as infinite.

10.2.2 Radionuclide heat generation

The heat generated in the interiors of the rocky planets by the decay of K, U, and Th can be estimated from geochemical models of their bulk compositions, most of which are largely inferred from meteorite data. Abundant compositional data are available for meteorites. In addition, the measured compositions of basalts provide constraints on the K, U, and Th concentrations that the upper mantle provides during melting, while gamma-ray surveys provide reasonably accurate estimates of these concentrations in surface materials. Although the concentrations indicated by basalts and gamma rays are much higher than the K, U and Th concentrations of the bulk planets, due to partitioning during melting, these data nevertheless provide independent means to rank the planets in order of the K, U and Th concentrations in their outer zones, which should be related to bulk values.

The specific power in pW kg^{-1} of rocky materials decreases with time as the radionuclides decay away. Moreover, because each nuclide has its own intrinsic half-life, with that of ^{40}K being much shorter than those of ^{238}U and ^{232}Th, the relative proportion of the total power that is contributed by ^{40}K will also decline with time. The details will vary with any given K, U and Th composition, but can be calculated with Eq. (2.25), which for the present day (0 Ga), reduces to:

$$\Psi_{now} = 0.00351 \times (ppm\ K) + 0.0985 \times (ppb\ U) + 0.0263 \times (ppb\ Th) \qquad (10.5)$$

where Ψ^* is heat generation in pW kg^{-1}, K_{ppm} is the potassium concentration in ppm, and U_{ppb} and Th_{ppb} are the uranium and thorium concentrations of the rock in ppb. Fig. 10.3 shows results for a typical ordinary chondrite.

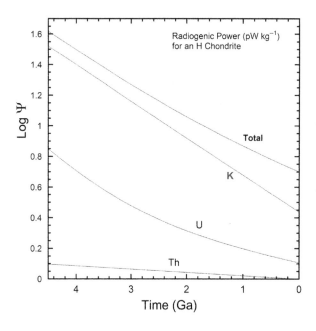

FIGURE 10.3   Heat generation as a function of geologic time for an ordinary H chondrite, assuming a typical composition for K (780 ppm), U (13 ppb) and Th (38 ppb). On this semi-logarithmic plot, the decay trend lines of thorium and potassium are straight because these elements have only a single important radioisotope, while the U trend is curved because both ^{235}U and ^{238}U are important. Over geologic time the total radiogenic power of H chondrites has decreased more than 8-fold, and for this composition, more than $^3/_4$ of the cumulative energy released has been contributed by ^{40}K decay.

Eq. (2.25) can be integrated from 0 to 4.5 Ga to calculate the energy contributed in megajoules per kg of rock over geologic time, providing:

$$\text{Specific Energy, in MJ kg}^{-1} = 0.00222 \times (\text{ppm K}) + 0.0287 \times (\text{ppb U}) + 0.00418 \times (\text{ppb Th})$$

(10.6)

The following subsections will sequentially address the K, U and Th compositions of model bulk planets, chondritic meteorites, basaltic rocks, and surface materials as inferred from gamma ray studies.

10.2.2.1 Model bulk K, U and Th concentrations in the planets

Model bulk compositions of the planets are poorly constrained and estimates of the K, U, and Th concentrations for any given object vary widely. Table 10.2 provides simple averages of the many separate estimates for each planet compiled in the Basaltic Volcanism Study Project (1981); the main advantage of this approach is adopting a single literature source in a field with such divergent opinions. Interestingly, even though the estimated K/U ratios vary widely, the amount of heat generated per kilogram over geologic time is similar for the model compositions of these bodies, and similar to that of chondrites. The latter similarity likely reflects the fact that meteorite compositions greatly or partly underlie most estimates for planetary bulk compositions.

10.2.2.2 K, U and Th concentrations in chondritic meteorites

Table 10.3 provides K, U, and Th concentrations in several types of chondrites (H, L, LL, EH, carbonaceous, and CI). The K, U, and Th contents of the H, L, and LL chondritic meteorites are similar, and these constitute >95% of the chondrites. Thus, the most common types of chondrites will produce comparable amounts of radiogenic heat. At the present time, if the heat produced were immediately lost, the power generated within an Earth constituted of chondritic matter would generate about 30 TW of power, similar to Earth's present day heat flux. This coincidence is of great historical importance as it demolished Kelvin's calculation of Earth's age, which geologists knew was far too young, and it subsequently provided great

TABLE 10.2 K, U and Th in models for bulk planets[a].

Object	K ppm	U ppb	Th ppb	K/U ×1000	Ψ pW kg^{-1}			Specific energy MJ kg^{-1}
					$t = 4.5$ Ga	$t = 3.0$ Ga	$t =$ now	over 4.5 Ga
HED Parent Body	47	15.5	58.5	3	11.86	5.97	3.23	0.79
Mercury	17.3	38.3	143	0.45	25.06	12.88	7.59	1.74
Venus	659	20.8	73.6	32	41.10	18.90	6.30	2.37
Earth	550	23.0	84.0	24	37.93	17.67	6.41	2.23
Moon	79.3	32.0	116	2.5	23.58	11.86	6.48	1.58
Mars	723	20	72	36	43.36	19.86	6.40	2.48

[a]*Average of model compositions from Chapter 4.3 in Basaltic Volcanism Study Project (1981), p. 638–657*

TABLE 10.3   K, U and Th measured in chondritic meteorites, basaltic meteorites, and basalts.

Type/Reference	K ppm	U ppb	Th ppb	K/U ×1000	Ψ pW kg⁻¹			Specific energy MJ kg⁻¹
					$t = 4.5\ Ga$	$t = 3.0\ Ga$	$t = now$	over 4.5 Ga
H[a]	780	13	38	60	41.1	18.4	5.02	2.26
L[a]	920	15	42	61	48.2	21.5	5.81	2.65
LL[a]	880	15	47	59	46.7	20.9	5.80	2.58
EH[a]	840	9.2	30	9	41.4	18.4	4.64	2.25
C (average)	355	15	55				4.17	1.45
CI[a]	550	8	29	69	28.4	12.8	3.48	1.57
Eucrite Juvinas[a]	328	123	297	2.7	86.7	41.5	21.08	5.50
MOR basalt[b]	1330	119	404	11	130.8	62.4	27.01	8.06
Columbia River Basalt[c]	12,000	1080	3710				246	73.1
Mare (Moon)[d]	672	212	910	3.2	167.1	85.57	47.17	11.38
Shergotty(Mars)[a]	1440	105	380	14	127.5	60.7	25.4	7.80

[a]*Lodders and Fegley (1998)*
[b]*Gale et al. (2013)*
[c]*Farmer (2004)*
[d]*Korotev (1998)*

support for chondritic compositional models. However, as shown in Chapters 2 and 9, Earth is much too large for its thermal condition to be in steady state, and most of the heat generated over time still remains inside. Also relevant is the very high K/U ratio of chondrites of nearly 60,000, which ensures that the radiogenic power that they generated long ago was ~8× higher than today, mostly because Earth's age is 3.5× greater than the half-life of ^{40}K.

10.2.2.3 K, U and Th concentrations in basalts and basaltic meteorites

Chemical compositions of basalts and basaltic meteorites provide a very useful way to compare the radiogenic heat generations of planetary bodies, as this common rock type originates at substantial depth, albeit one that is still shallow on a planetary scale. The concentrations in the source region or in the planetary mantles are not known. These should be less than the amounts currently in the basalts, because basalts preferentially uptake K, U, and Th upon melting of their source rocks. However, how much of these radionuclides remain in the source region depends on its size, and it is likely that only the shallow regions were tapped during melting. Nevertheless, the most direct marker of planetary mantles, and one which permits comparison of the rocky bodies, is the composition of heat producing elements in the basalt itself. Moreover, basalt lies on the surface so its radioactive emissions can be remotely sensed.

Many analyses of lunar and terrestrial basalts are available. Of the basaltic achondrites, Shergotty is known to have originated on Mars, and the largest group, the eucrites (e.g.,

Juvinas), has been linked to asteroid Vesta, and has affinities to Mercury (see below). All taken together, of the basaltic rocks from the Solar System, eucrites generate the lowest power, and the shergottites are next. The similarity of the radiogenic power of the shergottites and eucrites is a surprise, considering the huge difference in their K/U ratios. This exemplifies the problems with interpretations based on ratios discussed in Section 9.2.5.3; the low power of shergottites is a consequence of their low Th and U contents.

The heat generation of ocean floor basalts (MORB; Gale et al., 2013) is only slightly higher than that of the basaltic achondrites. However, continental basalts on Earth generate much greater energy, and serve as a better comparison to the extraterrestrial basalts as they better reflect Earth's ancient deep crust, which is known to be greatly enriched in the important radionuclides relative to the mantle. Typical Columbia River basalt is taken as representative of the many continental basalts tabulated in Farmer (2004). The heat generation of average lunar basalts is higher than average MORB, but very much lower than that of continental basalts.

In summary, the present day heat generation per kilogram, and the heat generated over geologic time, is similar for the chondrites and for the model powers of the bulk planets, except for the HED parent body whose estimate is lower. Basalts are geochemically "enriched" so their radiogenic power per kilogram is much higher than those estimates, but they can nevertheless be ranked, from lowest to highest, as follows:

$$Earth's\ continental\ basalts \gg Lunar\ basalts > Shergottites\ (Mars) > Eucrites (Vesta,\ Mercury)$$

(10.7)

The correspondence between this ranking of basalts and that inferred from model bulk compositions of the planets (e.g., Table 10.2) is poor, and calls the latter into question.

10.2.2.4 Gamma-ray spectroscopy

Gamma-ray spectra provide estimates of the Th and K concentrations of surface materials. Uranium is commonly not measured, but if not, its concentration can be estimated reasonably well as $0.27 \times$ the Th concentration. These gamma-emitting radionuclides can be remotely detected only to a depth of 30 cm, so the data mostly reflect the compositions of regolith and soils rather than bedrock. The heat generated per kilogram calculated from these data fall into two groups: Mercury, Vesta, Mars, and the Moon have low indicted power and are treated first, whereas the powers for Venus and Earth are much higher.

Global gamma ray surveys are available for Mercury, asteroid Vesta, the Moon, and Mars (Table 10.4). Peplowski et al. (2012) used results from Mercury's surface to determine average values of 1288 ± 234 ppm K and 155 ± 54 ppb Th, equivalent to a present-day heat generation of 13 pW kg^{-1} if a normal Th/U ratio is presumed. Prettyman et al. (2015) estimated that the surface of asteroid Vesta has average values of 595 ± 35 ppm K and 657 ± 59 ppb Th. The average values tabulated for the Moon from the Lunar Prospector gamma-ray observations are somewhat higher; the values tabulated in Table 10.4 were calculated from the extensive data set of Prettyman (2012). For the surface of Mars, Taylor et al. (2006) estimated average values of 3300 ppm for K and 620 ppb for Th, suggesting that the heat production of average surface materials is about 44 pW kg^{-1}, in agreement with Hahn et al. (2011) who estimated 49 ± 3 pW kg^{-1}. Taylor et al. (2006) also note that their estimated concentrations are higher than those suggested by the SNC meteorites, believed to originate from Mars.

TABLE 10.4   K, U and Th estimates from gamma ray surveys.

Type	K ppm	U ppb	Th[a] ppb	K/U × 1000	$\Psi_{4.5\,Ga}$ pW kg^{-1}	$\Psi_{3\,Ga}$ pW kg^{-1}	$\Psi_{0\,Ga}$ pW kg^{-1}	MJ kg^{-1} 0–4.5 Ga	Reference
Vesta	595	177	657	3.4	137.6	68.9	36.8	9.15	Prettyman et al. (2015): Table 3
Mercury	1288	42	155	31	81.4	37.6	12.7	4.71	Peplowski et al. (2012): Table 4
Venus	11,240	916	2870[a]	12			205.2	63.2	Fegley (2004): Table 7
USA[b]	9000	1700	6500	5.3			370		see text
Moon	735	454	1680	1.6	321.1	162.9	91.9	21.8	Prettyman (2012)
Mars	3300	167	620	20	246.3	115.8	44.3	14.7	Taylor et al. (2006): Table 4

[a]*Th/U assumed to be 3.7 except for Venus.*
[b]*Average values for the conterminous USA were estimated from concentration maps by USGS (2005).*

Because Venus and Earth have dense atmospheres, gamma ray techniques differ and global surveys are not available. However, gamma-ray data are available for the five Venus landing sites (e.g., Surkov et al., 1987) and are summarized by Fegley (2004); each of these sites has higher estimated K, U, and Th concentrations than the average values for Mercury, Vesta, and Mars (Table 10.4). Note that the concentrations at one landing site (Venera 8) included in the average values for Venus are significantly higher than those at the other four sites, suggesting that some enriched lavas were formed during Venus' history.

Global values are likewise not available for Earth, but aerial surveys were made of the USA that are provided as concentration maps (USGS, 2005). The values in Table 10.4 are simple visual estimates of the average surface concentrations we made from these maps for the contiguous 48 states. Though crude, our estimates are reasonable; for example, average soils in Missouri, a typical central state, are 14,300 ppm K, 3800 ppb U (a geometric mean), and 9600 ppb Th (Tidball, 1984), or about 1.5× higher than our visually estimated concentrations for the USA.

In short, the ranking of radiogenic power per kilogram of ancient surface materials suggested by gamma ray data is:

$$Earth \sim Venus \gg Moon > Mars > Vesta > Mercury \qquad (10.8)$$

Before concluding this section, a few remarks are needed regarding gamma ray maps. First, the maps of the USA show that gamma ray intensities are very low for areas covered by water or snow. Vegetation may have a similar but smaller effect, but in any case such effects depress the apparent concentrations. Also, major geologic features such as the Idaho batholith or the prominent belt of Ohio Shale, both representing bedrock with high K, U, and Th concentrations, are not evident on these maps. The gamma ray maps instead provide a very blurred, regionally homogenized picture of the underlying geology. Attempts to make geologic maps from gamma ray surveys are not supported. Second, gamma ray results from Vesta do not reveal that it has "considerable surface compositional diversity," contrary to the assertion of McSween et al. (2013). In our view, remotely-sensed spectral data are routinely over-interpreted, but nevertheless provide useful information averaged over very large scales.

10.2.2.5 Summary

The foregoing sections provide several different measurements and estimates of the K, U, and Th concentrations of planetary materials. Many estimates agree poorly with others. The compositions of the major types of chondritic meteorites are all similar, but do not resemble any other estimates. If the bulk compositions of the rocky planets are approximately chondritic as commonly assumed, much of the potassium must reside in their mantles, as suggested for Earth in Chapter 9. The depth of residence need not be very deep, as suggested by the K-rich kimberlite magmas.

Model estimates of the bulk K, U, and Th concentrations of the planets are poorly constrained and vary widely. Averages of those disparate estimates for any given object do not agree with other data sets.

Gamma-ray spectroscopy provides useful estimates of the Th and K concentrations of average surface materials on very large scales. Although U is not always measured, it can be well estimated from the thorium concentrations, because the Th/U ratios of most rocks are between 3 and 4.

The main internal consistency revealed by the assembled information is that the gamma ray compositions correlate with those of appropriate basalts. In particular, the relative order of planetary rankings deduced from these two independent data sets is similar, as also shown on Fig. 10.4. Thus, the rather low gamma ray values of the Moon's surface conform to the generally low K, U, and Th concentrations of lunar basalts; but even this

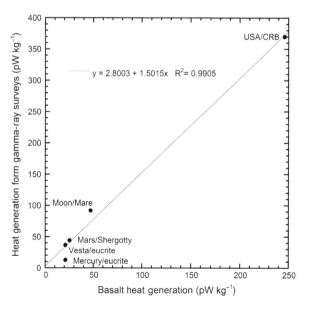

FIGURE 10.4 Estimates of present-day heat generation derived from gamma-ray surveys compared to values indicated by the measured K, U and Th concentrations of appropriate basaltic rocks. The point labeled USA/CRB compares the estimated gamma intensities for the USA with average Columbia River basalt. Note that the gamma intensities of both Vesta and Mercury are compared to eucrites. Mercury has lower gamma intensities than Vesta, probably because what remains of Mercury represents deeper planetary levels than the material that it lost.

gamma intensity may be too high, because widespread ejecta from the Imbrium basin are highly anomalous in Th. Similarly, the even lower low gamma ray intensities of Mars correspond to the low U and Th concentrations of shergotty, a basaltic achondrite believed to have originated on Mars on the basis of many strong geochemical arguments.

Moreover, of the rocky bodies under consideration, the gamma ray intensities are lowest for Vesta and Mercury, in agreement with the very low K, U and Th concentrations of the eucrites. Note that Vesta has long been considered to be the parent body of the eucrites and other members of the HED meteorite family, and that Hofmeister and Criss (2012b) argued that both Vesta and the HED meteorites came from Mercury. Mercury's high density and heavily cratered surface reveal that it was intensely bombarded in the early days of the Solar System, due to its high orbital velocity and its proximity to the Sun, so that practically all of its silicate mantle has been literally blasted away. Even if only 10% of Mercury's present mass of 3.3×10^{23} kg were lost, this is still $10 \times$ the total mass of the asteroid belt. As suggested by Hofmeister and Criss (2012b), all differentiated meteorites must have been produced on bodies that were sufficiently large to foster melting and gravitational segregation of minerals. Whereas Vesta is too small to have a significant gravity field, differentiated meteorites are known to have originated from Mars and the Moon. Mercury likewise provides a suitable geophysical environment for differentiation, and its missing silicate mantle could be represented by both HEDs and Vesta: the gamma ray data support this association (Fig. 10.4).

10.3 Thermal boundary condition of the rocky bodies

The balance of flux between incoming Solar radiation and outgoing blackbody radiation governs the thermal condition of planetary surfaces, thereby establishing the upper boundary condition of their interior temperature structure. This balance is given by

$$L\frac{\pi R^2}{4\pi r^2} = 4\pi R^2 \sigma T_{bb}^4 \tag{10.9}$$

where L is the enormous solar luminosity (3.83×10^{26} W), R is the planetary body radius (which cancels), r is its distance from the Sun, σ is the Stefan-Boltzmann constant, and T_{bb} is the blackbody temperature established by the other known quantities. In effect, the LHS of Eq. (10.9) is the flux received by the diametrical plane of the planet, and the RHS represents the equivalent value of energy returned to space by the blackbody emissions of the planetary surface. This equation is conveniently approximated as:

$$T_{bb} \cong \frac{278}{\sqrt{r_{au}}} \tag{10.10}$$

where r_{au} is the distance to the Sun expressed in astronomical units, with 1 a.u. representing the Earth-Sun distance. Though simple, Eq. (10.10) closely approximates the mean surface temperatures of planetary objects.

Many texts depict Eq. (10.10) where some recent analyses include a term for albedo. Including emissivity or reflectivity of the body in this equation is in error, because temperature is a thermodynamic quantity. Based on the zeroth law of thermodynamics, a black Earth and a white Earth would have the same temperature at 1 a.u., because both are in equilibrium with the Solar flux, and also with each other (zeroth law). Similar average temperatures of the Earth and Moon support our contention, while further illustrating the scant effect of Earth's thin atmosphere. Regarding the physical behavior, electrical insulators have a penetration depth that must be accounted for, in order to provide a correct mathematical analysis of their emission spectra. Kirchhoff's simple relationship (emissivity = absorptivity) needs amending to account for this penetration depth and also for surface reflectivity, as first recognized by McMahon (1950). These requirements have been overlooked in planetary science, see Hofmeister (2014, 2019) for discussion.

In all kinetic models, gas velocities (u) are proportional to the square root of temperature divided by the molecular weight (m). Moreover, any matter whose kinetic energy exceeds the escape velocity intrinsic to any planet can be lost to space. It follows that a simple plot of the square root of blackbody temperature vs. escape velocity is a useful tool for understanding differences between the surface conditions of the planets. In particular, Fig. 10.5 illustrates the opposing effects of high temperature and high escape velocity for different Solar System objects. Objects that are either too small or too hot cannot retain an atmosphere, whereas the huge, cold gas giants retain all gases.

Retention of CO_2 and H_2O molecules are important to planetary evolution, given their ability to transport heat and to promote explosive volcanism. The effect of H_2O is

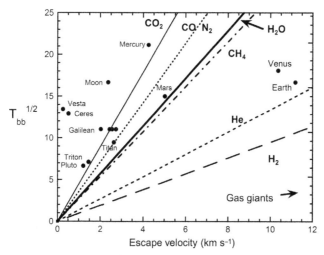

FIGURE 10.5 Plot of the square-root of blackbody temperatures (T_{bb} in Kelvins), estimated with Eq. (10.10), versus the escape velocity of large moons, asteroids, and rocky planets. Both quantities are proportional to the square root of kinetic energy. The indicated gases are lost above and left of the radiating lines, whose positions are correct relative to each other, utilizing that the gas temperature goes as $\frac{1}{2}mu^2$. To anchor the plot, the line for N_2, which is coincident with the line for CO, was placed through Titan, because this body has an N_2 atmosphere. Line placement illustrate how gravitation and temperature have opposing effects on the ability of objects to retain different atmospheric gases, with small or hot objects having lost virtually all gas, and the huge, cold gas giants retaining all. The freezing point of the gas also influences whether it is substantially present in the atmosphere (see below).

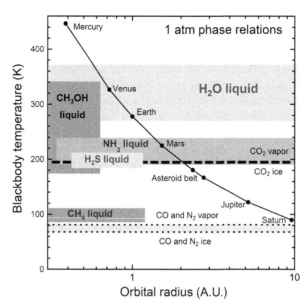

FIGURE 10.6 Comparison of stability fields of important cometary volatiles to the surface temperatures of planets as a function of orbital radius. Gray boxes show the temperature ranges at 1 atm surface pressure of liquid water (which overlaps with HCOOH), ammonia, methanol, methane, H_2S, and CO (which overlaps with that of N_2). The most common ices are represented by boxes that extend across the whole diagram. The liquid field of CO_2 is negligible at 1 atm, but exists at high pressures. In addition, liquid water is an effective solvent, and would dissolve various ices and gases. The transition temperatures are at 1 atm and were obtained from various websites.

particularly important because it can greatly reduce melting temperatures of silicate rocks (e.g., Weinberg and Hasalová, 2014), and also can dissolve and transport chemicals, including the important, heat-generating radionuclides.

Another factor pertains to retention of gas, namely the stability fields of condensed volatiles. If a volatile substance is a liquid or solid phase on a cold surface, then some vapor will be present in the atmosphere, since it is continuously supplied through sublimation, but the amount will be low. Fig. 10.6 compares stability relations of the important ices in comets (Fig. 9.2; Mumma and Charnley, 2011) to blackbody temperatures over most of the Solar System. Venus and Earth are located in the "rust belt" where water would have been an important phase on the surface during accretion, along with methanol and formic acid (HCOOH). The latter exists in comets at the 1% level (Fig. 9.2), and has a stability field very similar to that of H_2O. This diagram should represent atmospheres of the rocky planets and large moons, except for Venus, whose high surface pressure can stabilize supercritical fluid CO_2 (discussed below). Although some moons in the outer solar system are sufficiently large to retain CO_2, they lack CO_2 atmospheres, because temperatures are far below the CO_2 freezing point. Titan has an N_2 atmosphere because this volatile is still a gas at 80 K. Considering Figs. 10.5 and 10.6 together explains why Mars has a dry CO_2 atmosphere. Ammonia could be important to Mars, but is not abundant in comets (Fig. 9.2).

Importantly, cold accretion is required for the tiny, icy moons to have acquired their light molecules as ices. Had light molecules such as H_2O been gases in a hot nebula, neither acquisition not retention would have been possible. Additional implications are discussed below.

10.4 Thermal history of the rocky bodies

Useful inferences of the thermal histories of the planets can be made from the thermal conditions of their surfaces, because these constrain the boundary conditions on their surfaces, and limit their estimated energy inventories. The information assembled above shows that Venus and the Earth have much larger energy inventories per kilogram than the other rocky objects. In particular, the inventories of both spin energy and radiogenic energy for Venus and Earth are far larger than for the others.

Mass also affects processes. Figs. 10.5 and 10.6 show that the size of a rocky body and its surface temperature determine what ices and volatiles can be retained during and after accretion. This is important for several reasons. (1) Because ices are much softer than rocks, their retention or loss during accretion influences dissipation of spin. (2) The interiors of massive bodies are more compressed and their greater surface acceleration also affects the compaction and strength of the outer layers. (3) Volatiles that escape from the interior carry heat with them. The fastest molecules in a gas are the hottest, and light molecules are faster than heavy ones at the same temperature: both are prone to escape. Outgassing of the interior cools a body efficiently. The atmosphere is a separate issue.

Processes are also influenced by the distance to the Sun. (1) Objects close to the Sun lost this spin energy early on. (2) After solar ignition, the Sun controls surface temperatures. (3) As the Sun formed, it drew in material that was not in a stable orbit. Not only does such material accelerate inward, with kinetic energy going as $1/r$, but impactor flux increases as $1/r^2$. Heat provided by the late stage impacts strongly increases towards the Sun, and so Fig. 10.6 represents the relative behavior of the planets before ignition, too. However, the temperature scale needs adjusting. This cannot be constrained because heating during impact is both lost by radiation and by advection of any ices that are volatilized. Both of these effects relate to body mass, and so the effects of mass and distance must be considered simultaneously.

Lastly, compositional gradients are evident in the Solar system (Fig. 1.5). Refractories appear to be concentrated towards the Sun (Hofmeister and Criss, 2012b), which pertains to the distribution of long-lived U and Th.

The brief discussions below note the consequences for the individual objects, and compare these assessments to what is known for those bodies. Statistics on the topography of these objects is added, because this reflects the heterogeneity of their surfaces, which pertains to their internal processes. By comparing the objects, which differ in the main factors of mass and distance, we hope to better understand the process of thermal evolution, and how this creates diversity in the Solar System.

10.4.1 Earth

As much as one MJ kg^{-1} of frictional heat might have been released during spin down of the Earth, more than for the other rocky bodies with the possible exception of Venus. This heat would be sufficient to have increased Earth's internal temperature by about 1000 K. However, most frictional heating probably occurred in Earth's outer layers, which are relatively weak due to their low gravitational compaction. This extra heating of these outer layers by spin down is consistent with differentiation of the Earth above 660 km, as is evident in the rock record. In contrast, temperatures in the lower mantle and core, indicated by phase equilibria, are rather low and consistent with internal heat generation slowly and solely by radioactive decay (Chapter 9).

Not all of this spin-derived heat could have been retained in the Earth. A hot surface radiates as T^4, so losses to space are great for a hot surface. Radiative losses from the surface of the planet were highly important prior to stellar ignition. Notably, these radiative losses would be impeded by a dusty atmosphere. A gradient in temperature across the contracting sub-nebula is expected, and also a gradient in volatiles. Internal incorporation of ices during accretion is expected, yet a large amount of volatiles is not observed currently. In Chapter 9, we argued that ices were lost from the surface as the planet accreted, particularly CO whose retention requires temperatures below 68 K (Fig. 10.6). Figs. 10.5 and 10.6 provide support for the rusting of accreting metal on Earth's surface, which appears to be an early step of differentiation. Oxidation of iron metal (Eqs. 9.3 and 9.4) as the planets formed is needed to explain the linked variation in age and oxidation of Fe-metal in the ordinary chondrites (Chapter 11).

Earth's continental rocks are also much richer in K, U, and Th than any extraterrestrial samples, and the gamma ray intensities of its subaerial surface are largest, supporting that its radionuclide inventory per kilogram is largest. Earth's combined inventory of frictional and radiogenic heat energy is therefore the greatest of the rocky bodies, consistent with Earth being the most active, by far. As discussed elsewhere, Earth is too large for its deeper zones to have attained steady state, so the temperature of its lower mantle might be increasing, and its inner core progressively melting. The dynamics of the Earth-Moon-Sun system also drive plate tectonics and contribute a small amount of additional internal energy (Chapter 5). Much more discussion of the thermal structure and state of the Earth is available in Chapters 2, and 6−9.

10.4.2 Venus

Given that the mass, density, and radius of Venus and Earth are similar, the amount of spin energy that was contributed to these planets was about the same. Whereas Earth is still spinning, the axial spin of Venus is negligible today, and it was likely lost early on, given its proximity to the Sun. Any former moons of Venus were also lost to the Sun, so if Venus ever had any plate dynamics resembling those of Earth, that activity ceased long ago. Proximity to the Sun also provides Venus with a high surface temperature, and more impactors with each having more energy, so the temperature of its outer layers is probably a few hundred degrees hotter than for Earth. The main difference between Venus and Earth suggested by these differences would be that Venus may have been more active than Earth, and heated up more quickly, and had more magmatism in the 1st billion years of its history. This part of the rock record on Earth is nearly obliterated by subsequent activity, but may be preserved on Venus, since its surface is greatly scarred by impacts (Hamilton, 2019 and references therein).

Available gamma ray data suggest that Venus has significant sources of radiogenic power. The same data suggest that some igneous differentiation may have occurred on

Venus, producing enriched magmas. Interestingly, the indicated amounts of K, U, and Th are similar to those of 2–4 Ga old continental shield rocks (Table 9.3). This may be a coincidence, especially insofar as gamma ray data are available only for a few spots, rather than representing an aerial survey. However, extreme differentiation like that on Earth is not demonstrated because the concentrations are far lower than the modern continental crust.

10.4.2.1 Statistical analysis of Venus' topography

Hypsometry provides another measurement that pertains to differentiation. Fig. 10.7 compares inner Solar System bodies by normalizing their topography to their body radius. Titan is included because it is similar in size to the Moon.

Size normalization is essential because topography is related to surface roughness. Comparing elevation directly does not reveal the distribution function that underlies the histogram (e.g., Apostol, 1969). Fig. 10.7A shows that bodies with single peaks have similar half-widths at full maximum. The distribution for Venus differs little from that of Titan,

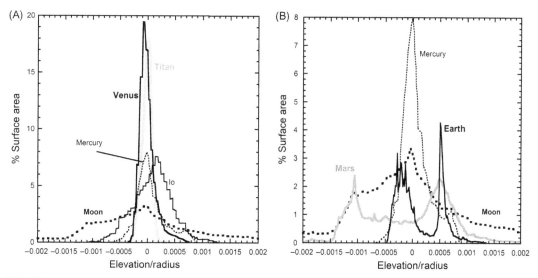

FIGURE 10.7 Hypsograms where elevation is normalized to the planet radius, which are: Moon = 1738 km; Mercury = 2438 km; Titan = 2575 km; Mars = 3396 km; Venus = 6051 km; Earth = 6370 km. (A) Bodies with a single peak, compared to our Moon. Except for the Moon, the widths of all distributions are similar and narrow, indicating the dominance of a single rock type. Data on Titan were reported % per 100 m², which affects the height of the peak somewhat, but changes the profile little, if any. (B) Bodies with complex peaks, compared to Mercury; note that the range of the Y-axis differs from part A. For Earth, the zero reference value is taken as the average radius, rather than as sea level, which is 3.2 km higher (Vérard, 2017). *Data on the Moon are publically available and were were downloaded (see websites). Data on Earth and Mars from Aharonson, O., et al., 2001. Statistics of Mars' topography from the Mars orbiter laser altimeter' slopes, correlations, and physical models; on Venus from Rosenblatt, P., et al., 1994. Comparative hypsometric analysis of Earth and Venus; on Io from Ross, M.N., et al., 1990. Internal structure of Io and the global distribution of its topography. Data on Mercury and Titan from Lorenz, R.D. et al., 2011. Hypsometry of Titan.*

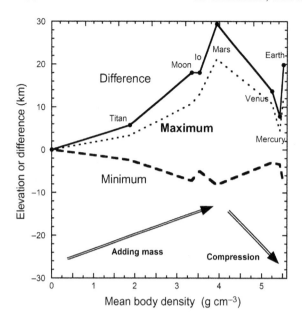

FIGURE 10.8 Dependence of the highest elevation, lowest elevation, and elevation range on mean planet density. Data from websites or White et al. (2014). Although Titan is icy, it is sufficiently cold to be rigid. The Gallilean icy moons are warmer and softer and flow, particularly Europa.

which has a uniform composition, and also resembles that of Mercury, which has crater upon crater. The similarity of these shapes, and that they differ from planets with two or more distinct rock-types (i.e., the Moon, Mars, and Earth) which have broad and/or bimodal distributions, suggests that the surface of Venus has a very uniform chemical composition. Note that Venus, Mercury, and each of Mars' two hypsometric peaks all have similar widths and shapes that resemble the peak representing Earth's oceanic floor. Since topography is related to crustal strength, these surfaces are probably all basaltic. Earth's continents are the oddity. The broad lunar distribution stems from several factors, see below.

The statistics of their topography were also compared to the ratio of elevation divided by the surface acceleration, in m s^{-2}. These are: Titan = 1.354; Moon = 1.624; Mercury = 3.70; Mars = 3.39; Venus = 8.87; Earth = 9.82. In this depiction, Titan's profile is wider than those of the rocky bodies. Because little else differed, we do not show these graphs. Instead, we show how elevations depend on density (Fig. 10.8). A single point has no mass and no topography. As the concentration of mass increases, the topographic relief increases. A maximum is seen for Mars, which is a fairly large body yet one that is not terribly compressed by self-gravitation. Bodies larger than Mars are more compressed and have higher values of g, both of which reduce relief. However, Earth departs from this general trend because its light, continental crust "floats" on its upper mantle. For this reason, the hypsogram peak for Earth's continents (Fig. 10.7) is quite narrow, whereas the peak representing its basalts has a shape much like that of the other planets. Earth and Mars are heterogeneously differentiated, but not Venus, despite its large mass.

10.4.2.2 Heating of Venus

Even with its high surface temperature, Venus would retain water, as it is in the "rust belt" (Figs. 10.5 and 10.6). However, if Venus' temperature had been substantially elevated by impact heating, this could drive the surface temperatures up into the field of water vapor, which would limit differentiation. Based on Fig. 10.7, the surface of Venus is compositionally uniform. Early loss of spin is consistent with limited differentiation of Venus. As noted above, spots on its surface have enhanced concentrations of heat producing elements, but a global gamma ray survey of Venus is not available. The measured spots may be connected with impacts, as on the Moon (see below). Importantly, heat generating elements near the surface contribute nothing towards interior activity.

10.4.2.3 *Atmosphere of Venus and surface temperatures*

The atmosphere of Venus exerts 90 bars pressure on its rocky surface, which is equivalent to the pressure of a ~ 1 km deep ocean on Earth. Effectively, Venus' atmosphere is similar to it having a global ocean about 30% as deep, or about 20% of the volume, of Earth's ocean. Indeed, the phase diagram of CO_2 (Fig. 10.9) shows that Venus' lower atmosphere is a super-critical fluid. Because of this, the thermal surface of Venus is somewhere in its atmosphere, but a thermal surface is not well defined since light of different frequencies has different penetration depths in any medium, due to frequency dependent absorption. The rocky surface of Venus is well-defined, just as Earth's oceans, but the temperature of Venusian rocks are affected by conduction/diffusion of heat through its soupy atmosphere.

Except for size and mass, Venus compares poorly with Earth in many aspects. Its current heat transfer and thermal evolution are disconnected from behavior on the Earth. Venus with its 90 bar atmosphere is a poor analogue to Earth, and its high temperatures do not show that ppm concentrations of CO_2 on Earth cause global warming. The atmosphere of Venus grades from "air" at the top to "fluid" above its rocks. Due to the mass effects and phase stability (Fig. 10.9), there is not an abrupt transition from "air" to "fluid."

Crucially, the supercritical fluid on Venus' surface transfers heat like the liquid state, where thermal conductivity *decreases* as temperature increases. Higher up, where the atmosphere is bona fide gas, its thermal conductivity *increases* as temperature increases

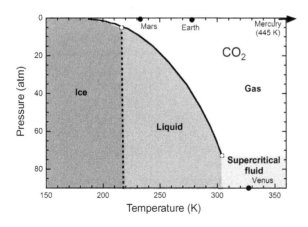

FIGURE 10.9 Phase diagram for CO_2 compared to the surfaces of the terrestrial planets, as represented by their blackbody temperatures. Data from websites and calculations in Section 10.3. Earth is near the triple point of water, and not at the H_2O critical point of 647 K and 218 atm.

(see figures in chapters 5 and 6 of Hofmeister, 2019). Due to this insulating functionality of condensed matter, Venus' temperature is much higher on the rocky surface than its black-body temperature of 327 K. Given the opposite behavior of heat conduction in fluid and gassy states, we propose that global warming on Earth is connected with Solar regulation of the ocean temperature, and the measured correlation of CO_2 with global temperature is due to a warmer ocean releasing more of its dissolved gases. Other gas markers may be useful in understanding ocean-air interactions. We also note that water has a large number of other unusual behaviors (see websites).

10.4.3 Mars

Mars is too small to have had much power generated by spin dissipation. It also lacks significant moons, as Phobos and Deimos are only tiny captured asteroids. Of the rocky planets it is furthest from the Sun, so the blackbody temperature of its surface is coldest, and the energy of its impactors were much lower. It is also small, so its outer layers must be much colder than those of Earth and Venus, and less compacted.

Thus, only radiogenic power is available for Mars. Over geologic time, the radioisotope energy released per kilogram on Mars is probably comparable to that for Venus and Earth, but if Mars is more potassic as believed by many geochemists, much of that power was released early in Mars' history, given the short half life of ^{40}K compared to ^{232}Th and ^{238}U. A sizeable time interval would have been required to heat up Mars' interior from an initially cold state, culminating in a period of volcanism, but the planet is probably cooling off today as its radionuclide inventories diminish and cooling continues. Crater counts over large areas indicate Martian volcanism occurred from about 0.4 to 3.6 Ga ago, with the large shields being the youngest [Basaltic Volcanism Study Project (BVSP), 1981, section 8.7]. This volcanic, midlife episode of Mars' history could have released deep volatiles, particularly carbon dioxide and water, allowing the temporarily formation of an ocean and a significant atmosphere. These volatiles were lost when the volcanic supply dwindled because losses to space have continued from this undersized body. Recent activity has been suggested by crater counts on lava flows, but sampling small areas need not be representative. Also, small areas cannot have many large craters so the distribution may not provide an accurate age (see, e.g., BVSP, 1981).

The hypsogram of Mars is bimodal, like Earth's (Fig. 10.7). However, its two peaks have similar widths. Mars' smooth plains compose the lower elevation peak and its heavily cratered terranes greatly contribute to the higher elevation peak, although the basaltic volcanoes were erupted on top of this crustal type. Given the roughly equal areas of the two terrains (Fig. 10.10), isostatic balance describes the surface of Mars. The bimodality is not due to silica rich differentiation. Likely, the cratered terrains have lower density due to high porosity. This contention is supported by Fig. 10.8. From this viewpoint, the huge volcanoes of Mars represent efficient, advective cooling where the heat source was deep.

Interestingly, the large volcanoes have features indicating eruption under ice (e.g., Hodges and Moore, 1979), which has received much focus subsequently. Possibly, the large volcanoes were produced by volcanic jets that exited the poles, creating a mass

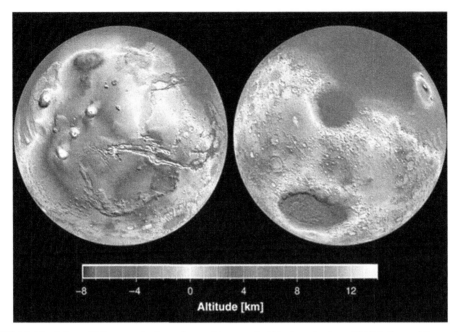

FIGURE 10.10   Map of Mars, made from laser altimetry. Public domain from NASA: https://www.jpl.nasa.gov/spaceimages/details.php?id = PIA02820.

instability and subsequent flopping of Mars' lithosphere. A more stable configuration was attained when the large volcanoes became situated near Mars equator. If this happened, a large amount of spin energy was dissipated, in accord with Fig. 1.12. Nevertheless, even though Mars is still spinning, it is much less active than Earth because of its small size, so it has less spin energy and faster loss of internal heat.

10.4.4 Moon

The Moon is too small to have had much power generated by spin dissipation, and moreover became rotationally locked to the Earth long ago. It is too small to retain volatiles, so it has no flux melting and no continents, but does have anorthosites produced by dry melting. Available data suggest that its inventory of potassium is low, so unlike Mars, the Moon could not have generated significant radiogenic power in its first billion years. The radionuclide inventory of the Moon may nevertheless have been overestimated, because widespread ejecta from the Imbrium basin are highly anomalous in Th.

It follows from these low energy inventories that the Moon has had only weak magmatic activity. The Mare, although prominent to observers due to the contrasting dark color of its basalts, are all about 3 Ga old and associated with major impact basins. It would appear that near-surface melting in these basins was fostered by the augmentation of the Moon's undersized energy inventories by impact energy. As discussed by

Hofmeister and Criss (2012a), the lunar orbit around the Earth gives this body a large cross-section for impacts by material drawn to the Sun, after the planets were largely formed. Regarding the Th anomaly, an impact could have either added material, or dredged material from deeper. The latter seems more likely, since radionuclides are concentrated in melt.

The small Moon is cooling down, and has been in an essentially static condition for ~1.5 billion years. This is reflected in its hypsometric curve, which is quite broad because of the low gravitational acceleration of this small body. In contrast, even though Io has a similar size and gravitational acceleration, its hypsometric curve has a much narrower peak, due to its more homogenous surface. Also, wide-spread melting due to strong gravitational interactions has smoothed the surface.

10.4.5 Mercury

Mercury is too small to have had much power generated by spin dissipation, and moreover became rotationally locked to the Sun long ago. As for Venus, any former moons were lost to the Sun early on. High surface temperature and its relatively small size prohibited the retention of volatiles. Proximity to the Sun has also resulted in the heavy bombardment of Mercury. That is, countless objects with eccentric orbits pass near Mercury because they are gravitationally focused by the Sun. In addition, orbital velocities are very high in the innermost Solar System, so any collisions release great energy per impactor kilogram. For these reasons, not only is Mercury heavily cratered, but most of its mantle was blasted away in the earliest days of the Solar System, leaving little behind other than its remnant, dense iron core.

Because so little silicate material has survived on Mercury, its inventories of radionuclides are small. Indeed, gamma ray survey data indicate that the power of Mercury's surface materials is much lower than that of the other rocky planets, and even less than for the Moon. Mercury is a dead planet, and it has been dead for a very long time. Hofmeister and Criss (2012b) suggested on the basis of geophysical and geochemical arguments that Mercury's missing mantle is probably represented by the HED achondrites, the family of differentiated meteorites that includes the eucrites, and many others. These meteorites represent what has been lost from Mercury, not what remains. The surface is likely very iron rich, which is consistent with low inventory from gamma-ray measurements.

It is unlikely that a wide variety of rocks exist on the surface of Mercury. First, its hypsogram has a single peak like those of Titan, Venus, and Io. The spread in elevation for Mercury is consistent with its strong gravitational acceleration. Second, emission and emissivity data have been improperly analyzed and compared in planetary science due to neglecting effects of reflectivity and surface temperature gradients (Hofmeister, 2014, 2019).

Fig. 10.6 suggests that the iron on Mercury's surface would not have been subject to "rusting" during accretion (Eq. 9.3 and thereafter), if the Sun had ignited prior to complete formation of the planets. Available meteoritic data points to oxidation of the HEDS and other types that appear to be inner Solar System materials (Fig. 1.5). The warmth of planetary surfaces during accretion is attributable to friction during deposition of chondules in the accretionary winds (Chapter 11), and to late arriving impacts.

10.4.6 Vesta

Vesta is much too small to have had significant power generated by spin dissipation. This irregular asteroid is probably a fragment of a much larger planet, as suggested by Hofmeister and Criss (2012b). Mercury is the logical parent body, give its overly large core. Vesta is much too small to have a significant gravity field, so it cannot retain volatiles and is not even round. Contrary to many popular models, it is therefore impossible that igneous differentiation and core formation occurred on Vesta. In contrast, differentiated meteorites are proven to have originated on Mars (the SNC family) and on the Moon, which bodies do have significant gravity fields. So, we find it illogical that no meteorites are conventionally linked to battered Mercury. While we recognize that Vesta could be the recent source of some eucrites, multiple lines of evidence (e.g., abundances in Fig. 1.4) suggest that Vesta and the eucrites probably were once part of Mercury. Gamma ray surveys of Vesta and Mercury show that they have lower intensities than any of the other rocky objects.

10.5 Conclusions

Thermal histories of the rocky planets must be deduced from indirect information, and doing this accurately is as difficult as sketching the Doodang (see opening quote). Nevertheless, physical laws provide powerful constraints, and energy and volatile inventories of different bodies can be estimated and compared. The energy inventories per kilogram of the rocky bodies of the Solar System decrease in the order:

$$\text{Earth} \sim \text{Venus} \gg \text{Moon} > \text{Mars} > \text{Mercury} > \text{Vesta} \tag{10.11}$$

This ranking represents the combination of several separate inventories and estimates, but the ranking determined from any single inventory, considered separately, is very similar. In particular, the ranking of spin inventory is similar to that for the radioisotope inventory, whether the latter is estimated from basalt compositions, gamma ray data, or other arguments. Note that these inventories tend to increase with planetary size, even though the above ranking is made on a per kilogram basis.

Given a chondritic composition, no more than $\sim 2\,\text{MJ kg}^{-1}$ of radiogenic heat has been contributed to any of the rocky bodies over the entire history of the Solar System. This amount of energy, if released all at once, is sufficient to heat up the interiors of large objects by a few thousand degrees, and to cause some melting of localized zones, but of itself is insufficient to sustain a dynamic planet, partly because slow release is accompanied by concomitant loss of heat to space. Venus was probably active in its early history, but available data suggest little activity today. Only the Moon appears to be out of sequence in the above order, but its radionuclide inventory may have been overestimated because widespread ejecta from the Imbrium basin are highly anomalous in Th. Another finding is that the power generated by axial spin has decreased approximately exponentially with time, as has the power generated by radioactivity.

The rocky planets accreted cold, and have remained rather cold. Some magmatism has occurred in their outer layers which have been heated by impacts and frictional spin down, and where radioisotopes are most concentrated. Planetary size and distance from the Sun also matter. Igneous differentiation requires both heat and a gravity field, and is impossible on the cold, tiny asteroids.

Most of the rocky objects are tectonically and magmatically dead, or nearly so. Even the Earth, popularly represented as an active planet, is in truth nearly dead. How many earth-quakes or volcanic eruptions does a typical human being experience in a lifetime? These events are actually quite rare, and considering the problems they cause when they do occur, we are very fortunate that this is so.

References

Aharonson, O., Zuber, M.T., Rothman, D.H., 2001. Statistics of Mars' topography from the Mars orbiter laser altimeter' slopes, correlations, and physical models. J. Geophys. Res. 106, 23723–23735.

Apostol, M., 1969. Calculus: Multi-variable Calculus and Linear Algebra, With Applications to Differential Equations and Probability. Xerox College Publishing, Waltham, MA.

Baliunas, S., Sokoloff, D., Soon, W., 1996. Magnetic field and rotation in lower main-sequence stars: an empirical time-dependent magnetic bode's relation? Astrophys. J. 457, L99–L102.

Basaltic Volcanism Study Project, 1981. Basaltic Volcanism on the Terrestrial Planets. Pergamon Press, Inc. New York. 1286 pp.

Bergin, E.A., Tafalla, M., 2007. Cold dark clouds: the initial conditions for star formation. Ann. Rev. Astron. Astrophys. 45, 339–396.

Collier Cameron, A., et al., 2009. The main-sequence rotation–colour relation in the Coma Berenices open cluster. Mon. Not. R. Astron. Soc. 400, 451–462.

Consolmagno, G.J., Britt, D.T., Macke, R.J., 2008. The significance of meteorite density and porosity. Chemie der Erde––Geochemistry 68, 1–29.

Covey, et al., 2016. Why are rapidly rotating m dwarfs in the Pleiades so (infra)red? New period measurements confirm rotation-dependent color offsets from the cluster sequence. Astrophys. J. 822 (81), . Available from: https://doi.org/10.3847/0004-637X/822/2/81; also available at https://arxiv.org/pdf/1601.07237.pdf.

Criss, R.E., Hofmeister, A.M., 2018. Galactic density and evolution based on the Virial theorem, energy minimization, and conservation of angular momentum. Galaxies 6, 115–135.

Delorme, P., Collier Cameron, A., Hebb, L., Rostron, J., Lister, T.A., Norton, A.J., et al., 2011. Stellar rotation in the Hyades and Praesepe: gyrochronology and braking time-scale. Mon. Not. R. Astron. Soc. 413, 2218–2234.

Douglas, S.T., Agüeros, M.A., Covey, K.R., Cargile, P.A., Barclay, T., Cody, A., et al., 2016. K2 rotation periods for low-mass Hyads and the implications for gyrochronology. Astrophys. J. 822, paper 47. Available from: https://doi.org/10.3847/0004-637X/822/1/47.

Douglas, S.T., Agüeros, M.A., Covey, K.R., Kraus, A., 2017. Poking the beehive from space: K2 rotation periods for Praesepe. Astrophys. J. 842 (83). Available from: https://doi.org/10.3847/1538-4357/aa6e52.

Dufton, P.L., et al., 2011. The VLT-FLAMES tarantula survey: the fastest rotating O-type star and shortest period LMC pulsar—remnants of a supernova disrupted binary? Astrophys. J. [Online] 743, no. L22. Available from: https://doi.org/10.1088/2041-8205/743/1/L22 or http://iopscience.iop.org/0004-637X.

Farmer, G.L., 2004. Continental basaltic rocks. In: first ed. Turekian, K., Holland, H. (Eds.), Treatise on Geochemistry, vol. 3. Elsevier, Amsterdam, pp. 85–122.

Fegley, B., 2004. Venus. Treatise on Geochemistry, V. 1.In: Turekian K. and Holland H., (Eds.), Meteorites, Comets and Planets. Elsevier, Amsterdam, pp. 487–507.

Fossat, E., et al., 2017. Asymptotic g modes: evidence for a rapid rotation of the solar core. Astron. Astrophys. 604 (#A40). Available from: https://doi.org/10.1051/0004-6361/201730460.

Gale, A., Dalton, C.A., Langmuir, C.H., Su, Y., Schilling, J., 2013. The mean composition of ocean ridge basalts. Geochemistry Geophys Geosystems 14 (3), 489–518. Available from: https://doi.org/10.1029/2012GC004334.

Hahn, B.C., McLennan, S.M., Klein, E.C., 2011. Martian surface heat production and crustal heat flow from Mars Odyssey Gamma-Ray spectrometry. Geophys. Res. Lett. 38 (#14), 5. L14203.

Hamilton, W.B., 2019. Toward a myth-free geodynamic history of Earth and its neighbors. Earth Sci. Rev. (in press). Available from: https://www.sciencedirect.com/science/article/pii/S0012825219302636.

Harris, J.C., 1955. The Complete Tales of Uncle Remus. Houghton Mifflin, p. 697.

Hartman, J.D., et al., 2009. Deep MMT transit survey of the open cluster M37. II. Variable stars. Astrophys. J. 675, 1254−1277.

Hartman, J.D., Bakos, G.A., Kovacs, G., Noyes, R.W., 2010. A large sample of photometric rotation periods for FGK Pleiades stars. Mon. Not. R. Astron. Soc. 408, 475−489. Available from: https://doi.org/10.1111/j.1365-2966.2010.17147.x.

Henderson, C.B., Stassun, K.G., 2012. Time-series photometry of stars in and around the Lagoon nebula. I. Rotation periods of 290 low-mass pre-main-sequence stars in NGC 6530. Astrophys. J. 747 (51). Available from: https://doi.org/10.1088/0004-637X/747/1/51 or http://iopscience.iop.org/0004-637X.

Hodges, C.A., Moore, H.J., 1979. The subglacial birth of Olympus Mons and its aureoles. J. Geophys. Res. 84, 8061−8074.

Hofmeister, A.M., 2014. Carryover of sampling errors and other problems in far-infrared to far-ultraviolet spectra to associated applications. Rev. Miner. Geochem. 78, 481−508.

Hofmeister, A.M., 2019. Measurements, Mechanisms, and Models of Heat Transport. Elsevier, Amsterdam (see Ch. 2,5, and 8).

Hofmeister, A.M., Criss, R.E., 2012a. A thermodynamic model for formation of the solar system via 3-dimesional collapse of the dusty nebula. Planet. Space Sci. 62, 111−131.

Hofmeister, A.M., Criss, R.E., 2012b. Origin of HED meteorites from the spalling of Mercury: implications for the formation and composition of the inner planets. In: Hwee-San, L. (Ed.), New Achievements in Geoscience. InTech, Croatia, pp. 153−178. Available from: http://www.intechopen.com/articles/show/title/the-case-for-hed-meteorites-originating-in-deep-spalling-of-mercury-implications-for-composition-and.

Huang, W., Gies, D.R., 2008. Stellar rotation in field and cluster B stars. Astrophys. J. 683, 1045−1051.

Irwin, J., et al., 2007. The monitor project: rotation of low-mass stars in the open cluster NGC 2516. Mon. Not. R. Astr. Soc. 377, 741−755.

Irwin, J., Aigrain, S., Bouvier, J., Hebb, L., Hodgkin, S., Irwin, M., et al., 2009. The monitor project: rotation periods of low-mass stars in M50. Mon. Not. R. Astron. Soc. 392, 1456−1466.

Karoff, C., Metcalfe, T.S., Chaplin, W.J., Elsworth, Y., Kjeldsen, H., Arentoft, T., et al., 2009. Sounding stellar cycles with Kepler − I. Strategy for selecting targets. Mon. Not. R. Astron. Soc. 399, 914−923.

Korotev, R.L., 1998. Concentrations of radioactive elements in lunar materials. J. Geophys. Res. 103, 1691−1701. Available from: https://doi.org/10.1029/97JE03267.

Langseth, M.G., Keihm, S.J. Peters, K., 1976. Revised lunar heat flow values, Proc. Lunar Sci. Conf., 7th, 7, 3143−3171.

Lodders, K., Fegley, B.J., 1998. The Planetary Scientist's Companion. Oxford University Press, Oxford.

Lorenz, R.D., et al., 2011. Hypsometry of Titan. Icarus 211, 699−706.

McMahon, H.O., 1950. Thermal radiation from partially transparent reflecting bodies. J. Opt. Soc. Am. 40, 376−380.

McQuillan, A., Aigrain, S., Mazeh, T., 2013. Stellar rotation periods of the Kepler objects of interest: a dearth of close-in planets around fast rotators. Mon. Not. R. Astron. Soc. 432, 1203−1216.

McSween, H.J., et al., 2013. Dawn; the Vesta−HED connection; and the geologic context for eucrites, diogenites, and howardites. Meteorit. Planet. Sci. 48, 2090−2104.

Meibom, S., Mathieu, R.D., Stassun, K.G., 2009. Stellar rotation in M35: mass−period relations, spin-down rates, and gyrochronology. Astrophys. J. 695, 679−694.

Meibom, S., Mathieu, R.D., Stassun, K.G., Liebesny, P., Saar, S.H., 2011. The color−period diagram and stellar rotational evolution—new rotation period measurements in the open cluster M34. Astrophys. J. [Online]. 733 (2), No. 115. Available from: https://doi.org/10.1088/0004-637X/733/2/115 or http://iopscience.iop.org/0004-637X.

Müller, A., van den Ancker, M.E., Launhardt, R., Pott, J.U., Fedele, D., Henning, Th., 2011. HD 135344B: a young star has reached its rotational limit. Astron. Astrophys. [Online] 530 (A85). Available from: https://doi.org/10.1051/0004-6361/201116732 or http://www.aanda.org/.

Mumma, M.J., Charnley, S.B., 2011. The chemical composition of comets—emerging taxonomies and natal heritage. Annu. Rev. Astron. Astrophys. 49, 471−524.

Peplowski, P.N., Lawrence, D.J., Rhodes, E.A., Sprague, A.L., McCoy, T.J., Denevi, B.W., et al., 2012. Variations in the abundances of potassium and thorium on the surface of Mercury: results from the MESSENGER gamma-ray spectrometer. J. Geophys. Res. 117, E00l04. Available from: https://doi.org/10.1029/2012JE004141.

Prettyman, T.H., 2012. Lunar Prospector GRS Elemental Abundance, LP-L-GRS-5-ELEM-ABUNDANCE-V1.0, NASA Planetary Data System. http://pds-geosciences.wustl.edu/missions/lunarp/grs_elem_abundance.html and http://pds-geosciences.wustl.edu/lunar/lp-l-grs-5-elem-abundance-v1/lp_9001/data/ (Accessed June 2, 2019).

Prettyman, T.H., Yamashita, N., Reedy, R.C., McSween, H.Y., Mittlefehldt, D.W., Hendricks, J.S., et al., 2015. Concentrations of potassium and thorium within Vesta's regolith. Icarus 259, 39–52.

Rebull, L.M., et al., 2016. Rotation in the pleiades with K2. I. Data and first results. Astronom. J. 152 (113), . Available from: https://doi.org/10.3847/0004-6256/152/5/113, also available at https://arxiv.org/pdf/1606.00052.pdf.

Rosenblatt, P., Pinet, P.C., Thouvenot, E., 1994. Comparative hypsometric analysis of Earth and Venus. Geophys. Res. Lett. 21, 465–468.

Ross, M.N., Schubert, G., Spohn, T., Gaskell, R.W., 1990. Internal structure of Io and the global distribution of its topography. Icarus 85, 309–325.

Royer, F., Grenier, S., Baylac, M.O., Gomez, A.E., Zorec, J., 2002. Rotational velocities of A type stars. II. Measurement of v sin i in the northern hemisphere. Astron. Astrophys. 393, 897–911.

Scholz, A., Eislöffel, J., Mundt, R., 2009. Long-term monitoring in IC4665: fast rotation and weak variability in very low mass objects. Astron. Astrophys. 400, 1548–1562.

Skumanich, A., 1972. Time scales for Ca II emission decay, rotational braking, and lithium depletion. Astrophys. J. 171, 565–567.

Surkov, Y.A., Kirnozov, F.F., Glazov, V.N., Dunchenko, A.G., Tatsy, L.P., Sobornov, O.P., 1987. Uranium, thorium, and potassium in the Venusian rocks at the landing sites of Vega 1 and 2. J. Geophys. Res. 92, E537–E540. Available from: https://doi.org/10.1029/JB092iB04p0E537.

Taylor, G.J., et al., 2006. Bulk composition and early differentiation of Mars. J. Geophys. Res. 111, E03S10. Available from: https://doi.org/10.1029/2005JE002645.

Terndrup, D.M., et al., 2000. Rotational velocities of low-mass stars in the Pleiades and Hyades. Astrophys. J. 119, 1303–1316.

Tidball, R.R., 1984. Geochemical Survey of Missouri, USGS Prof. Paper 954-H, 54 p.

Tritton, D.J., 1977. Physical Fluid Dynamics. Van Nostrand Reinhold Co, New York.

USGS 2005. (NURE). https://pubs.usgs.gov/of/2005/1413/.

van Breemen, J.M., Min, M., Chiar, J.E., Waters, L.B.F.M., Kemper, F., Boogert, A.C.A., et al., 2011. The 9.7 and 18 mm silicate absorption profiles towards diffuse and molecular cloud lines-of-sight. Astron. Astrophys. 526, Article A152. Available from: https://doi.org/10.1051/0004-6361/200811142.

Vérard, C., 2017. Statistics of the Earth's topography. Open Access Library J. 4, No. E3398. Available from: https://doi.org/10.4236/oalib.1103398.

Weinberg, R.F., Hasalová, P., 2014. Water-fluxed melting of the continental crust: a review. Lithos 212–215. 158–188.

White, O.L., Schenk, P.M., Nimmo, F., Hoogenboom, T., 2014. A new stereo topographic map of Io: implications for geology from global to local scales. J. Geophys. Res. Planets 119, 1276–1301. Available from: https://doi.org/10.1002/2013JE004591.

Zolensky, M.E., et al., 2006. Mineralogy and petrology of Comet 81P/Wild 2 Nucleus samples. Science 314, 1735–1739.

Zombeck, M.V., 2007. Handbook of Space Astronomy and Astrophysics. Cambridge Univ. Press, Cambridge.

Websites

Planetary fact sheets from NASA (accessed 01.06.19.).
https://nssdc.gsfc.nasa.gov/planetary/factsheet/
Pictures and movies of hurricanes from NOAA. National Oceanic and Atmospheric Administration. (accessed 01.04.18.).
http://www.noaa.gov/

Rotating coaxial cylinders (both accessed 15.06.19.).
https://physics.nyu.edu/pine/hydrodynamic_reversibility.html
https://en.wikipedia.org/wiki/Taylor%E2%80%93Couette_flow
Descriptions of Pleiades cluster (both accessed 15.06.19.).
https://en.wikipedia.org/wiki/Pleiades
https://webda.physics.muni.cz/cgi-bin/ocl_page.cgi?dirname = mel022
Descriptions of massive stars (accessed 15.06.19.).
https://en.wikipedia.org/wiki/R136a2
https://en.wikipedia.org/wiki/List_of_most_massive_stars
https://en.wikipedia.org/wiki/WR_2
Topography of the Moon (accessed 19.06.19.).
http://pds-geosciences.wustl.edu/missions/clementine/gravtopo.html
Water structure and science by M. Chaplin (accessed 01.09.18.).
http://www1.lsbu.ac.uk/water/water_structure_science.html
Phase diagram for CO_2 (accessed 23.06.19.).
https://chem.libretexts.org/Bookshelves/General_Chemistry/Map%3A_General_Chemistry_(Petrucci_et_al.)/
    12%3A_Intermolecular_Forces%3A_Liquids_And_Solids/12.4%3A_Phase_Diagrams

A sedimentary origin for chondritic meteorites

Anne M. Hofmeister and Robert E. Criss

Regarding transport of solids .... "The difficulty has been, and still is, that no one branch of science has attempted to deal with the problem as a whole, or to co-ordinate the vast amount of piecemeal work by students of different outlook in many unrelated fields." **R.A. Bagnold, 1941.**

11.1 Introduction

The rocky planets accreted from dust, ice, and other solid phases in the solar nebula. Since nebulae are cold clouds, rather than hot melts, the protoplanets grew via depositional processes on their surface. We propose that the primitive building blocks of terrestrial planets, chondritic meteorites, are sedimentary deposits. Specifically, these are detrital aggregates that include various combinations of genetically unrelated materials. Some grains formed at high temperature, but some chondrites include low-temperature materials and evidence for fluid infiltration. These characteristics number among many they share with clastic sedimentary rocks: others are detailed below. Importantly, the great abundance of chondrites ($\sim 85\%$ of falls) is consistent with their origin on planetary surfaces; note that sediments cover $>75\%$ of Earth's present surface, and that practically all lunar meteorites are breccias derived from the Moon's surface. Recent processes and conditions are the guide to the past.

Chondritic meteorites are named for their abundant contents of interesting grains called chondrules, which are small (mostly <1 mm), highly rounded grains typically constituted of aggregates of mostly olivine or pyroxene crystals. Current models for the formation of these chondrules involve melting, but this hypothesis does not explain their features satisfactorily, as indicated by table 5 of Rubin (2000). The source of heat causing the melting is not explained (e.g., Nittler and Ciesla, 2016), and existing models are based on many ad hoc assumptions (see Scott and Krot, 2007).

The key characteristic of carbonaceous chondrites is the coexistence of low and high temperature phases, as also occurs in comets, which likewise are considered to be primitive. Although this character of comets was discovered only 13 years ago (Zolensky et al., 2006), the Renazzo carbonaceous chondrite fell almost 200 years ago and immediately attracted scientific study. Remarkably, no model yet advanced has explained this hallmark characteristic of mixed unrelated phases in chondrites, even though this characteristic is common in terrestrial sediments. Additional issues with existing models are mentioned in Section 9.2.2.

The conventional view that chondrules formed by melting stems from their round shapes, combined with the notion of hot accretion, which is based on Kelvin's erroneous model for production of starlight. Importantly, most grains with round shapes in diverse terrestrial rocks were sculpted by abrasive processes, not by melting. Round clasts are observed at the present time in stream beds, and more importantly in dunes, where sand grains are abraded, rounded, and sorted by winds. Sandstones can be aeolian or fluvial, and have formed throughout Earth's history (e.g., Nocita, 1989; Dott, 2003).

Winds produce dunes and other features that are observed on the rocky planets and moons with atmospheres, no matter how low or high the atmospheric pressure (Greeley and Iversen, 1985). Sand and dust storms are devastatingly energetic (Fig. 11.1): the physics of blown sand and dune formation were described in the classic work of Bagnold (1941). Water abrasion is less effective: experiments of Kuenen (1960a,b) showed that wind abrasion reduces grain mass 100−1000 times faster.

The power of wind is connected with several factors, with velocity being most important. Table 11.1 provides some comparisons of damaging winds. Wind speeds are relative

FIGURE 11.1 Picture of an enormous dust storm approaching Stratford, Texas in 1935, the Dust Bowl Era. *Public domain from the George Marsh album of NOAA. http://www.photolib.noaa.gov/htmls/theb1365.htm (accessed 28.06.19.).*

TABLE 11.1 Wind speeds and Earth's tangential spin speeds.

Description	Speed km s^{-1}	Damage or effect	Size[a] of airborne solids mm
Sandstorm	~0.01	Scouring; blinding; car accidents	~3
Tornado (F1)	0.032−0.049	Mobile homes overturned	~40
Tornado (F3)	0.070−0.091	Trains overturned	~400
Tornado (F5)	0.12−0.14	Houses flattened; trees debarked	~400
Commercial jet	0.25	Fast travel; atmosphere impacted	n.a.
Spin at equator today	0.46	Coriolis effects; Hadley cells	n.a.
Spin at formation	1.5−12	See text	n.a.

[a]*These are rough maxima, and do not account for mass of the solids, or the time in the air.*
Data sources are Chapter 10 and the website list.

to the surface of the Earth, which is spinning. Earth's current rotational period of 24 hours is modest compared to its initial rate, but its surface is still moving faster than a jet, and Coriolis forces remain significant today. Just after formation, Earth's equatorial tangential velocity could have been as much as 27× faster, if the Viral value were attained in conservative accretion (Fig. 1.12; Table 11.1). In Chapter 10, we considered that a factor of 3× is more likely, as some dissipation of motion accompanies accretion. In either case, the faster spin of the early Earth would have produced winds and atmospheric disturbances of much greater speed and intensity than today. Perhaps inconceivable tornadoes of Fujita's class F6 (see websites) would have been common.

Not only would wind speeds have been impressive on the newly formed Earth, but the process of accretion is best described as a gas with dust and ices being pulled toward the center of mass. This would generate a cyclone in every sense of the word. Images are shown in Fig. 5.3. Here we note that regions of star formation have much dust and ice (e.g., van Breemen et al., 2011; Section 9.2.2). That matter, including dust, ejected from old stars is recycled into new stars during their formation is long-standing knowledge in astronomy.

Section 11.2 presents a qualitative comparison of chondrules to sand grains, and of chondrites to aeolian sandstones. Section 11.3 quantitatively compares the grain-size distributions of chondrites and aeolian sands. Section 11.4 describes how chondrules formed in the winds of accretion, and then were assembled into chondritic meteorites via deposition, compaction, and finally, ejection by a late impact. Section 11.5 summarizes.

11.2 Characteristics of chondrites shared with aeolian sandstones

Chondritic meteorites and sedimentary rocks are both well studied. Aeolian sandstones (see e.g., Pettijohn et al., 1973) are the sedimentary rock type that is most similar to chondrites, based on their shared characteristics (Table 11.2). Both are detrital. Additional points are that some individual chondrules are aggregates (see below, and figures in Rubin, 2000), and that while sandstones tend to be nearly pure SiO_2, their purity is a function of grain size (Dott, 2003). Many aeolian sandstones contain considerable feldspar, and in fact the classification scheme of Dott is partly based on feldspar content. The enstatite chondrites are nearly monomineralic and are related to the ordinary chondrites (Section 9.2.2). Finer grains in the matrix of chondrites, like those in the pores of sandstones, have a wide range of compositions and origins.

Like most clastic sedimentary rocks (e.g., Davis, 1992), chondrites are composed of phases that are unrelated and predate rock assembly by some span of time. Note that the precursor rocks of even the most ancient terrestrial rock sequences, such as the Isua supercrustals, were greatly modified or disintegrated by erosion, metamorphism and time. Thus, much information on the origins of the components of chondrites has been lost, and multiple origins are to be expected for their constituent grains.

In particular, greatly divergent conditions of origin of constituent phases are obvious for the typical chondrules, the refractory CAIs, the rare pre-solar grains of mainly SiC, and

TABLE 11.2 Petrographic characteristics common to chondritic meteorites and aeolian sandstones.

Grains	Rocks	Fillings in pores
Rounded grains	High porosity	Diverse origins
Angular fragments are common	Different phases in pores	Some low-T phases
Narrow size distribution	Permeable	Fluid infiltration
Typical size $\sim$0.5 mm	Friable	Iron staining
Can be nearly monomineralic[a]	Variable cementation	
Overgrowths common	Dissolution/re-precipitation	
Pitted, dimpled surfaces	Tectonic/impact overprints	
Rounding depends on grain size	Often grain supported	

[a]*Quartz in sandstone; enstatite in EH or EL chondrites; olivine chondrules in many ordinary chondrites.*

the low temperature phyllosilicates and carbonates, which all coexist in the Allende mete-orite. Some other chondrites also contain pre-solar grains that formed in stars that predate our Sun and have huge isotopic distinctions from typical Solar System matter. All the above grain types formed under very different conditions. Furthermore, the hallmark com-ponent of chondrites, specifically the chondrules, are themselves extremely varied in min-eralogy and texture (e.g., Brearly and Jones, 1998; Scott and Krot, 2007), suggesting multiple origins. This diverse assemblage of phases has baffled meteoriticists, as illustrated by the long, descriptive review of chondrites by Brearly and Jones (1998) which offers no unifying themes, and the direct statement by Mittlefehldt et al. (1998) that they do not know how to logically order their chapter on achondrites and metal meteorites. This diver-sity does not support the currently popular assumptions that meteorite components con-densed from hot gas, and that chondrules formed from melt.

The above information points to chondrites being deposits of chondrules (+ /- other phases) on growing planetary surfaces, that were subject to many well-known sedimen-tary processes, and furthermore to impacts that ejected the specimens we have today. But, what are chondrules?

11.2.1 Visual similarity of chondrules and sand grains

Photomicrographs show that sand grains, the main components of aeolian sandstones (Fig. 11.2), are remarkably similar to chondrules, the main component of chondrites (Fig. 11.3). For example, grain supported structures and fractured grains are common in many sandstones (Fig. 11.2A) and also in chondrites (Fig. 11.3B). However the matrix can dominate in some sandstones (Fig. 11.2B) and chondrites (Fig. 11.3C), so a wide distribu-tion of grain sizes and types is present. Rounded grains are also seen on small scales, such as in the pore space between sand grains (Fig. 11.2D) or even inside individual chondrules (Fig. 11.3D).

Grain abutments and fabrics in chondrites are similar to those of sandstones. Pettijohn et al. (1973, p. 91) describe diverse sandstone fabrics that include sutured grains, concavo-convex contacts, point contacts, long-parallel contacts, and also isolated grains in a matrix. The various examples in both Figs. 11.2 and 11.3 show these features.

The fluvial sandstone of Fig. 11.2C has grains that are more angular and a texture like the acapulcoite in Fig. 11.3E, which is not a chondrite, but nonetheless has a texture that is very much like that of the H chondrite in Fig. 11.3A. One simple explanation is that more breakage or shear has occurred for the acapulcoite than the chondrites, i.e., this is similar material but has been metamorphosed after deposition. This finding is supported by the sequence of aeolian sandstones that were subject to tectonic deformation (Fig. 11.2E−G) where shear has packed the grains more tightly and produced more angular shapes while reducing porosity. Note that the aubrite (Fig. 11.3D) looks like a cataclasticized version of an enstatite chondrite (Fig. 11.3C). The pore space was removed, the grains are more sheared and broken, but the compositions of these two meteorite types are essentially the same. With this degree of compaction and shearing, and the presence of many broken chondrules in all chondritic meteorite types, it would be easy to miss a chondrule in a compacted aubrite.

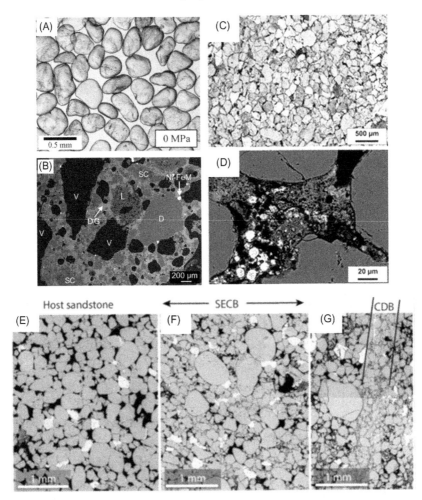

FIGURE 11.2 Images of sandstones on similar scales. (A) Photomicrograph of rounded quartz grains immersed in mineral oil of an aeolian Ordovician sandstone of nearly 100% quartz from St. Peter, Wisconsin (B) Backscattering image of shocked sandstone (labeled SC), showing impact melted glass (labeled DG or D), vesicles (V), lechitelierite (L), and some Ni-Fe from the impactor. (C) Photomicrograph of a fluvial Triassic sandstone from Seeberg, Germany, where (D) is an expanded view using backscattered electron imaging, showing a single pore filled with phyllosilicate and subrounded goethite or limonite particles. Bottom row = SEM images showing tectonic effects on an aeolian Jurassic sandstone, the Aztec in Nevada, which is related to the Navaho sandstones of the Colorado Plateau. All images are from segments within 10 cm of one another. (E) Undeformed sandstone. (F) Sheared and compressed piece, showing lower porosity and a larger fraction of small grains than undeformed host rock. (G) Compressed and cataclastically deformed, showing intense grain-size reduction. Microfractures suggest that grain size reduction is related to grain crushing, but there is also textural evidence for pressure solution. *(A) Figure 4a from Karner, S.L., Chester, F.M., Kronenberg, A.K., Chester, J.S., 2003. Subcritical compaction and yielding of granular quartz sand. Tectonophysics 377, 357– 381; (B) Figure 7b from Fazio, A., D'Orazio, M., Cordier, C., Folco, L., 2016. Target—projectile interaction during impact melting at Kamil Crater, Egypt. Geochim. Cosmochim. Acta 180, 33–50. (C) and (D) from Kowitz A., Schmitt R.T., Reimold W.U. and Hornemann U., The first MEMIN shock recovery experiments at low shock pressure (5–12.5 GPa) with dry, porous sandstone, Meteorit. Planet. Sci. 48, 2013, 99–114. (E) - (G) from Fossen H., Zuluaga L.F., Ballas G., Soliva R. and Rotevatn A., Contractional deformation of porous sandstone: Insights from the Aztec Sandstone, SE Nevada, USA, J. Struct. Geol. 74, 2015, 172–184. All parts reproduced with permissions.*

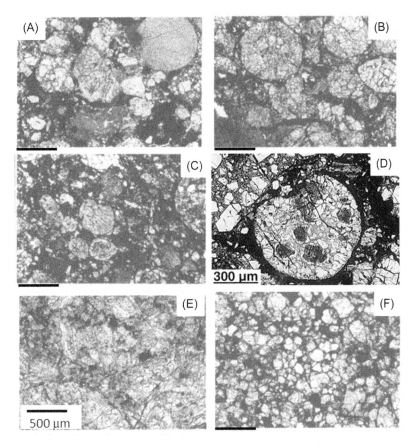

FIGURE 11.3 Photomicrographs of chondritic meteorites and close relatives; black or white scale bars represent 500 μm for all but part D. (A) Ordinary H3 chondrite ALH85121 with chondrules, chondrule fragments, and mineral grains in a dark matrix with limonite staining. (B) Ordinary L chondrite ALH84205 where the dark regions are mostly limonite weathering. (C) Enstatite E4 chondrite ALH84188 with deformed and broken chondrules, and mineral grains in a dark matrix. (D) Chondrule in LL Chainpur containing relict grains of "dusty olivine." The dark color of the grain cores is caused by numerous tiny blebs of low-Ni metallic Fe that had not been oxidized during gas phase reactions. The light-colored overgrowths around the cores consist of low-FeO olivine with the same composition as the brecciated grains in the chondrule. (E) Non-chondritic aubrite ALH84007, consisting of large deformed enstatite crystals, with scattered opaque sulfides. The lower left corner has a very round grain, which is larger than an average chondrule, but not inordinately so. (F) Non-chondritic acapulcoite ALH84190 with olivine, pyroxene, metal and sulfide. Part (D) from Rubin A.E., Petrologic, geochemical and experimental constraints on models of chondrule formation, Earth Sci. Rev. 50, 2000, 3–27, with permissions. All other images from Mason B., MacPherson, G. J., Score, R., Martinez, R., Satterwhite, C., Schwarz, C., and Gooding, J. L., 1992. Descriptions of stony meteorites, In: Marvin, U.B. and MacPherson, G.J. (eds.) Field and Laboratory Investigations of Antarctic Meteorites Collected by United States Expeditions, 1985–1987. Smithsonian Inst. Press, Washington D.C., where the width of each photo is 2.3 mm.

Broken round shapes and dimples are common in these photomicrographs. Overgrowths also exist, but are difficult to see in the quartzites at the scales shown. Dott (2003) provides high magnifications of these effects in aeolian sandstones. Pitted grain surfaces (or frosting) is a hallmark of aeolian deposits and are obvious in the close up of a chondrule in Fig. 11.3D.

The term porphyritic has been applied to chondrites, based on the belief that they crystallized from melt. The term porphyritic is conventionally used to describe igneous rocks that have a bimodal distribution of grain sizes, where the matrix is a very fine grained groundmass that surrounds large crystals called phenocrysts. The latter generally have euhedral faces. Porphyritic textures are visible in hand samples (see websites), which is a consequence of strongly bimodal grain size distributions in the volcanic rocks. Such distributions do not exist in chondrites or in their chondrules, as shown in the images of Fig. 11.3, including the close-up of a chondrule, and many other images in the papers and review articles cited in this chapter. Grain size distributions in chondritic meteorites are generally continuous, and while some structure exists on the distribution curves (see Section 11.3 and Friedrich et al., 2015), these curves are clearly not bimodal. Some chondrules (Fig. 11.3D and other images in Rubin, 2000) appear to be sedimentary rocks in miniature. Pallasites are an example of a bimodal size distribution in a meteorite, but the olivine "eyes" are definitely not chondrules, as they are typically much larger and less round. Pallasite textures could be produced by melting, but not chondritic textures.

Effects of stress are evident in the images of many sandstones and chondrites. Greater shearing and fracturing of chondrites than of sandstones is to be expected, because chondrites were ejected from protoplanetary bodies by impacts, and furthermore experienced another impact when delivered to the full-sized Earth. The pervasive occurrence of impact shock features in meteorites is well known. Impact melting also occurs, which explains the presence of some chondrules that may actually have melted, as well as the occasional existence of glass in interstices. Section 11.3 provides further discussion of impact processes.

11.2.2 Comparison of porosity

A hallmark characteristic of sandstones is high porosity. Ranges reported are 10–40%, where 15% is typical. Porosity decreases as the number of phases increases. For further discussion see Pettijohn et al. (1973) or the website listed, which notes that sandstone porosity is complex.

Porosity in a wide variety of meteorite types has been measured by Consulmagno and colleagues. Average porosities for ordinary chondrites are 7.0% for type H; 7.6% for type L, and 8.2% for type LL from Consolmagno et al. (2008), who note that shock features are associated with decreased porosity. This behavior is consistent with tectonized and shocked sandstones (e.g., Fossen et al., 2015; Kowitz et al., 2013). Thus, the lower porosity of ordinary chondrites than average sandstones is to be expected.

Carbonaceous chondrites have considerably higher porosity, $\sim 35\%$ for type CI. The other types have averages of either $\sim 10\%$ or $\sim 23\%$ (Consolmagno et al., 2008). Values for carbonaceous chondrites are compatible with those of sandstones. Perhaps the higher porosity is connected with their possessing soft phases, which can absorb shock so pore size is reduced, or perhaps the pores represent fluid or ices present in these rocks at the time of impact.

11.2.3 Effect of grain hardness

The main mineral in sandstones, quartz, is fairly hard, with $H = 7$. Feldspar is less hard ($H \sim 6$) and less abundant. The main phases in chondrules and chondrites have similar hardness: olivine with Fe/Mg ~ 0.1 has H = 7, whereas enstatite has H ~ 5, and iron has H ~ 4.5. However, iron is malleable, so the rounding process during abrasion is not relevant. Chemical reactions may also have an effect (Section 11.4). Aluminous chondrules are rare, even though the phases present in this type and in CAIs are hard: corundum has H = 9 and spinel has H = 7.5. Rounding aluminous phases requires other hard phases. That aluminous chondrules are rare is consistent with the overall low abundance of the calcium aluminates in meteorites.

The major mineral phases in these rocks, namely the quartz grains in sandstones and the olivine-rich chondrules in chondrites, have similar hardnesses. Thus, one would expect that sandstones and chondrites would have similar grain size distributions, if chondrites are indeed aeolian deposits.

11.3 Quantification of the grain-size of chondrules and aeolian sand

This section quantitatively compares the grain-size distributions of chondrites and aeolian sands. Importantly, each measurement technique explores a certain range of diameters, which causes the distribution curves to depend in part on the method (Pettijohn et al., 1973, p. 70). The peak of a continuous distribution curve necessarily falls somewhere in the middle of the range investigated by each method since fractions of the total number of grains are presented: this is evident in Fig. 11.4, and is explored in detail in Section 11.3.1.

Most importantly, chondrules are sand-sized grains, and the range of their diameters is typical of terrestrial sands. As seen in Fig. 11.4, and as evident in the cited references and the review of Friedrich et al. (2015), most chondrules range from 0.06 to 2 mm: grains of this size are all defined as "sand" on the Wentworth scale (p. 71 of Pettijohn et al., 1973). The classical method of sieving is based on screens with holes spanning this size range. A few chondrules reach granule and pebble sizes of up to 4 mm, which are also common sizes for clasts in ordinary sandstones. Chondrules as small as 0.002 mm were measured by Simon et al. (2018), a size classified as clay, whereas slightly larger diameters are termed silt by Wentworth. Sandstones likewise incorporate grains of these smaller sizes in their intergranular pores (e.g., Fig. 11.2D).

Although data for average chondrule sizes cannot be precisely compared if determined by different studies using different methods, similar median and mean sizes of ~ 0.3 to ~ 1 mm are observed for chondrules from diverse meteorite types, see the tables in the review of Friedrich et al. (2015). Thus, the mean sizes of chondrules are the same as those of common terrestrial sand-grains.

Based on the range of sizes exhibited by chondrules, and their mean or median sizes, these grains are clearly sand. The famous green sand beaches of Hawaii are constituted of olivine grains, because the parent rock is mafic basalt rather than siliceous continental crust, and because olivine, like quartz, has a hardness of 7 and survives abrasion by wind and waves. This explains why the sand grains in meteorites, i.e., the chondrules, are mostly olivine.

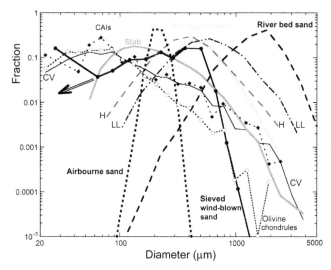

FIGURE 11.4 Grain size distributions, mostly of components in Allende (CV). Bold black lines are sieved sand of three types: solid = wind-blown (data from Bagnold, 1941), where the double arrow indicates the expected trend, had very small sieve sizes been included; dashed = river bed sand; and dotted = sand collected above a dune in a windstorm, when the grains were bouncing off the surface, both from Bagnold and Barndorff-Nielsen, 1980. The remaining curves are measurements of chondrules in images, with the number indicated. Disaggregated = 276 chondrules from Allende (Paque and Cuzzi, 1997). Slab = photomicrograph analysis of 12966 chondrules in the ordinary chondrite NWA5717, which provided a similar distribution to that from a slab of Allende (Simon et al., 2018). H = 261 chondrules of Hammond Downs, by Kuebler et al. (1999). LL = 280 chondrules from Semarkona by Nelson and Rubin (2002). The remainder are SEM-XRD measurements of Simon et al. (2018): CV = 2339 chondrules in Allende; olivine chondrules = 1309 from Allende without pyroxene; CAIs = 180 inclusions of calcium aluminates plus 15 amoeboid olivine aggregates.

11.3.1 Why the distribution curves vary

Grain-sized distributions are well-explored in sedimentology, where the limitations and differences between various techniques have been a focus. All techniques work better for rounder grain assemblages than for elongated ones. Distribution curves present the fraction (or percentage) of grains with a certain size or size range (Fig. 11.4). These are essentially smoothed histograms.

For sand-sized grains, the most accurate method is sieving and weighing the fractions. This method accurately measures all grain-sizes: however, the fraction that passes the smallest sieve contains a very large range of tiny sizes. With the average diameter for the smallest grain fraction not actually being known, the curve artificially upturns for the smallest diameter (Fig. 11.4).

The grain size range that can be investigated by sieving unconsolidated materials includes quite small sizes. Grain sizes determined for disaggregated rocks tend to be larger, probably because fine grained material is lost in the process. Disaggregating Allende provided a peak at larger grain sizes than the other methods used (Fig. 11.4).

Image processing of sandstones accesses small sizes, but distorts the profiles, as shown in Schäfer and Teyssen (1987, their Figure 2). Analyses of grain size distributions from 2-dimensional pictures are affected by the smallest diameter that can be resolved in the image (the cutoff): From Fig. 11.4, downturns at small diameter are controlled by the cutoff (cf., the curve marked CV to the curve marked slab). The profile is also affected, due to the "corpuscle" effect, whereby grain diameters from pictures are underestimated because slicing infrequently cuts through the maximum diameter (e.g., Schäfer and Teyssen, 1987). The corpuscle effect causes apparent grain-sizes to be smaller than true values, where the impact is greater for larger sizes, since fewer of these are present in any given section. Corrections are applied in sedimentology and meteoritic studies, but the combination of various problems discussed above make the corrections inexact, rendering cross-comparisons of results from different methods imprecise.

Given the limitations of 2-dimensional assessments of grain-size distributions, the patterns for the chondrules are indeed similar to those of unconsolidated sand grains (Fig. 11.4). The curves for chondrules and quartz sand differ in detail because individual, wind-blown sand grains are round, unfragmented, and extremely well-sorted. Even for quartz sand, different environments yield different distributions. Grains collected in motion above a dune have very small range of sizes, whereas river bed sand is dominated by larger sizes, partly because the finest grains remained suspended in the water.

Unlike chondrules, aeolian sands are highly uniform in mineralogical composition, and they have not been compacted or fragmented during lithification nor have they been subjected to highly damaging high-speed impacts. Considering these effects, the similarities between the distribution curves of aeolian sands and chondrites are more compelling than their differences, including the grain diameter at the curve peak and the size range. Thus, the distribution curves for ordinary terrestrial sands are similar in shape to those for chondrules, and the average sizes are quite similar. Most chondrules are sand sized on the Wentworth scale. Chondrites are aeolian deposits, produced by nebular and surface winds on rapidly spinning, accreting protoplanets.

11.3.2 Inter-comparisons of meteorite types and components

To best compare grain distributions for different meteorites requires use of the same method. Application of the same technique to different components in Allende (Simon et al., 2018) shows that its inclusions (mostly CAI types) have similar size distributions as its chondrules (Fig. 11.4). CAIs are not round, which means that the grain distribution function for the chondrules from Simon et al. (2018) is largely based on fragments, not on round grains. In contrast, aeolian dune grains and fluvial quartz grains are round. This suggests that grain shape differences cause the size distribution data on meteorites to differ from that on unconsolidated sand grains. As also shown by these authors, the various divisions of chondrules (olivine vs pyroxene) yield similar distribution curves as do CAIs. Thus, not only chondrules, but moreover the inclusions in chondrites, have all been subject to similar degrees of abrasive rounding. The CAIs are less rounded because these are very hard material, and have low abundance. Iron metal grains are also not round, due to their contrasting property of being ductile.

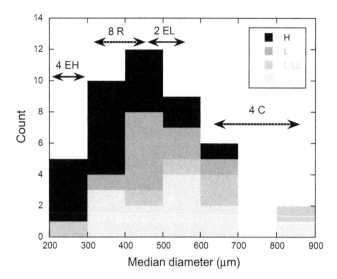

FIGURE 11.5 Dependence of chondrule apparent size on chondrite type. Ordinary chondrites of classes H, L, L/LL and LL are shown as the shaded blocks. Additional meteorite types are shown as double arrows with the number of samples indicated; e.g., 4 C denotes 4 measurements of carbonaceous chondrites. Data from various sources, summarized in tables by Friedrich et al. (2015). Means and medians are usually similar. The rough average diameter for all chondrules studied is ∼0.5 mm.

Friedrich et al. (2015) summarized the data on photomicrograph analysis of chondrules, which explores larger sizes than the SEM measurements of Simon et al. (2018). Fig. 11.4 shows representative distribution curves for H and LL types, which distributions were studied by several different authors using different methodologies. Because the peak position depends in part on the method used, we use histograms to determine whether different meteorite classes actually have different sizes (Fig. 11.5). Meteorites with more iron tend to have smaller chondrules. The greater the abundance of metal grains, the more rounded are the brittle silicate grains, which apparently have been impacted by the durable metal.

11.4 Proposal for formation of chondrules and chondrites

Practically all atoms in existence heavier than helium were formed by nuclear reactions in the interiors of stars. This process has occurred over billions of years and involves stars in huge tracts of the universe. Our Sun is at least a second generation star, and thus the accretion of our Solar System, both Sun and planets, is part of this giant recycling project.

This section offers a proposal for the formation of chondules and chondrites, beginning with the birth of grains in the solar nebula and in stellar outflows, to their receipt on the Earth's surface some 4.5 Ga after the planets formed. Observations from astronomy are combined with thermodynamic assessments and sedimentology principles. Our attempt is to merge results from different fields to better understand meteorites and the thermal condition of the forming planets (see Bagnold's quote). The main contribution of this chapter, and the focus of this section, is to explain how round grains are formed, and how these were assembled into meteorites during cold accretion. However, we cannot avoid discussing the earliest steps when those grains formed. The details are elusive, but ample evidence points to dust existing in the original nebula, and to the conservative collapse of a large, rarified, roundish nebula, which regardless of size produces a denser, flattened, rotating oblate discoid (Criss and Hofmeister, 2018).

Our proposal is summarized in two schematics and in two tables, which list the various processes that the grains in chondritic meteorites may encounter between their formation and much later return to the modern Earth. The tables link the processes to the environments that mineral grains are subject to in various stages of the astronomical recycling project. Some processes occur over several steps. If self-explanatory, or familiar, the processes are simply listed in tables. The subsections are organized in accord with the changing conditions experienced by the material as grains were transformed to rocks and then to meteorites.

11.4.1 Formation of grains in mass loss of previous stars

As stars age, many things can happen, including bloating into red giants or exploding as novae. Material is lost in these processes. Spectacular stellar outflows (Fig. 4.1) are known to contain mineral grains of various compositions. Forsterite is unequivocally identified via comparison of the emission spectra of distant nebulae to laboratory data (Molster et al., 2001). Many other peaks exist, but the assignments to minerals are ambiguous (e.g., Hofmeister et al., 2004). Other aging stars with more carbon produce SiC (Speck et al., 1997). The grains, once formed, drift into the vast reaches of space, to eventually be incorporated in a molecular cloud (Fig. 11.6), and then into another star or star system. Such grains are identified in meteorites by their isotopic signatures, and are termed presolar (reviewed by Nittler and Ciesla, 2016).

Spectral peaks of dust in the stellar outflows suggest grain sizes on the order of microns. Importantly, a featureless blackbody spectrum also exists, which requires large grains, on the order of sub-mm for weakly absorbing material (e.g., Hofmeister et al., 2004). Once the grains are sufficiently large to emit as blackbodies, their mineralogy can no longer be identified remotely, and their size is equivocal. In the dense astronomical clouds, the grains reach mm sizes, as shown by the presence of overtones (Bowey and Hofmeister, 2005).

Currently, it is assumed that grains condensed from hot gas shot out of the formed Sun, after Laplace. This idea has been explored in laboratory experiments. However, laboratory conditions differ grossly from those in space. Experiments are conducted over intervals that are likely more than 12 orders of magnitude shorter than times over which stars or planets form. For favorable kinetics, high temperatures are used in experiments, which is unlike conditions in either stellar outflows (~ 50 K, see figures in Hofmeister et al., 2004), or in molecular clouds, which are ~ 8 K. Importantly, densities used in the laboratory are extremely high compared to those in nebulae, which are more rarified than that of vacuums achieved in the laboratory. In addition, chondrules are constituted of common igneous minerals (e.g., olivine) that crystallize in high temperature magmas on the Earth. The familiarity of igneous processes has promoted the notion of high temperature condensation.

However, other experiments suggest that high temperature is not needed. Grains may form during chemical vapor deposition, electrostatic attraction, and cold welding. Sputter coating is performed in a vacuum. In space, cold welding of metals has caused mechanical problems with satellites (Merstallinger et al., 2009). We suggest that the oxides and silicates may also form in molecular clouds, due to surfaces being very clean, in a process

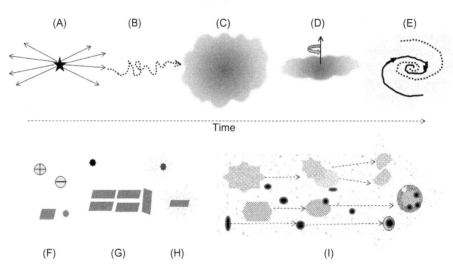

FIGURE 11.6 Processes of grain formation and growth. Top row shows how large-scale features evolve. (A) Previous generations of stars lose mass, which eventually drifts (B) into regions where stars are forming (C), providing nuclei upon which minerals can grow. (D) When a cloud collapses, it spins up and flattens, see Chapter 4. (E) The oblate discoid rounds up, densifies and spins up, eventually forming a star or planet (discussed in Section 11.4.4). Bottom row shows smaller scale events. (F) Charged ions can attract; gravity may also pull grains and atoms together. (G) Processes like chemical vapor deposition can occur, where the atoms gradually grow into a large single crystal or polycrystal. Cold welding is also known to occur in space. (H) Ice can nucleate on crystals, possibly reacting with the material. (I) As the cloud draws to center, it densifies and more interactions can occur. Some encounters are collisions which round, merge, and/or fragment the grains. Chemical reactions with gas or ice can also occur, which reaction rates are enhanced as the collisions produce some heat. These reactions may mostly occur during deposition and burial.

analogous to cold welding. Although the rates of formation would be slow at cold temperatures, probable timescales span many millions of years. Nuclei are also available, in the form of pre-solar grains from ancient stars.

Although the details of formation and growth of meteoritic grains are elusive, the observations point to their origination in dense clouds, which can form stars, and have sand-sized mineral grains. Isotopic data show that a few of these grains are from ancient stars that are very unlike our Sun, whereas the majority of the grains formed in the nebula.

11.4.2 Processing in clouds in space

Not much happens in the dilute interstellar medium (ISM), where grains are meters apart, but gravity is pervasive and gathers material together. Once nebular clouds form, interesting things can happen. Below we focus on ices and how ices may react with metals and various silicates that are present. For discussions of the minerals present, see e.g., Bowey and Adamson, 2002; Aller et al., 2012).

A simple thermodynamic approach is used to explain how the simple speciation of the interstellar medium evolved into the more diverse speciation observed in dense astronomical clouds, which differs little from that of comets (Figs. 9.1 and 9.2; Mumma and

Charnley, 2011). We then again use thermodynamics to describe oxidization of iron metal (also see Section 9.3.1.3), and production of graphite during the formation of the chondrules from dust. The focus is on exothermic reactions, which are highly favorable, even in cold environments. Ices with N and S are not covered, although even though S clearly pertains to FeS in dust, and both N and P are incorporated in iron metal. Similar types of reactions are expected. Notably, Fe-metal is a catalyst in producing ammonia, and Ni and Pt metals catalyze many reactions (see Chapter 9).

11.4.2.1 Ices and ice conversions during densification; graphite production

The ISM contains copious amounts of H_2 gas which exceeds the amount of CO by a factor of 10^4. Several ices (H_2CO = formaldehyde, SO and OCS) are less abundant by another factor of 10^4, and H_3COH (methanol) is even less abundant. Sometimes water ice is present. Many important species (Fe, C, N_2, O_2, S_2) are also likely present, but their IR spectra are weak or nonexistent, rendering them not conducive to detection by their emissions or absorptions. Where ice exists, gas molecules of the same species exist. Hence, rarified nebulae should be considered as a medium of H_2 carrying additional light molecules, their ices, and dust. In dilute nebulae, equilibrium reactions occur during occasional collisions of H_2 (or sometimes H) with rarer types of molecules, and possibly with ice particles, e.g.:

$$H_2 + CO \leftrightarrow H_2CO \tag{11.1}$$

$$H_2 + H_2CO \leftrightarrow H_3COH \tag{11.2}$$

$$H_2 + CO \rightarrow C + H_2O \tag{11.3}$$

Double arrows signify reactions that create an equilibrium speciation of gas molecules. Single arrows signify a reaction which is highly exothermic (Fig. 11.7), such as Eq. (11.3), which produces graphite or diamond, where the latter phase can form as tiny nuclei out of the ordinary stability field.

The ices in dense clouds and comets (Fig 9.2; Mumma and Charnley, 2011) contain much less CO, but much more H_2O, CO_2, and CH_4, than in the ISM. More species exist that have greater numbers of atoms, e.g., HCOOH (formic acid). As a nebula densifies into a cloud, additional exothermic reactions become likely:

$$3H_2 + CO \rightarrow CH_4 + H_2O \tag{11.4}$$

$$CO + H_2O \rightarrow CO_2 + H_2 \tag{11.5}$$

$$2H_2 + CO_2 \rightarrow C + 2H_2O \tag{11.6}$$

Although Eq. (11.4) involves the collision of several molecules, hydrogen is in great abundance and this reaction is strongly exothermic (Fig. 11.7) and thus favored. Thermodynamic calculations show that on balance, densification of nebulae lead to consumption of CO and the production of water, methane, and carbon dioxide ices. Since the hydrogen gas reservoir is effectively boundless, its partial consumption is immaterial. Also, a balance will be produced between the ices and their sublimed gas molecules. Importantly, the drive is to produce solid carbon via Eqs. (11.3) and (11.6).

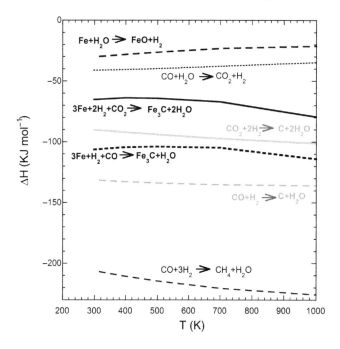

FIGURE 11.7 Exothermic reactions in the nebula, with reactants to products as labeled. Heavy black lines involve oxidation of Fe metal by important ices, in the presence of H_2 gas. Heavy gray lines involve graphitization, which releases water. Light black lines show how carbon dioxide and methane are produced. The latter is highly favored and explains the abundance of water in astronomical clouds. Our calculations were based on the data compiled by Robie and Hemingway (1995).

11.4.2.2 Oxidation of iron in early stages of accretion

Iron metal in ordinary chondrites was subjected to oxidation, as those with more Fe^{2+} in their olivine have more Ni in their metal phase (see tables in Lodders and Fegley, 1998). Apparently, Fe in the metal was lost to the silicates. The following reactions are increasingly exothermic:

$$\{Fe + H_2O\} \rightarrow FeO + H_2 \tag{11.7}$$

$$\{3Fe + CO_2\} + 2H_2 \rightarrow Fe_3C + 2H_2O \tag{11.8}$$

$$\{3Fe + CO\} + H_2 \rightarrow Fe_3C + H_2O \tag{11.9}$$

The curly brackets indicate ice being sequestered on an iron grain. Rust production with pure water is not particularly favorable, but dissolved CO_2 promotes this electrochemical reaction. Rusting is complex, but ubiquitous (e.g., Gradel and Frankenthal, 1990).

It is not necessary that cementite (cohenite) form, because some carbon can dissolve in the iron metal, while the Fe^{2+} goes into solution in the water ice. This process may be more likely. In this case, ice with iron exchanges with Mg in the enstatite. Exchange reactions include:

$$FeO + MgSiO_3(\text{enstatite}) \rightarrow (Mg, Fe)SiO_4(\text{Fe-rich olivine}) \tag{11.10}$$

$$Mg_2SiO_4(\text{forsterite}) + (Mg, Fe)SiO_4 \rightarrow 2(Mg_{1.5}Fe_{0.5}SiO_4)(\text{olivine}) \tag{11.11}$$

Meteoritic evidence, such Ni concentration in juxtaposed iron corresponding with fayalite content, is covered in Chapter 9.

It is not clear at what stage oxidation of iron occurs. A fairly dense medium is required, so oxidation likely occurred while the material was drawn to the center (rounding up). It is likely that oxidation occurred as or after the material was deposited on protoplanetary surfaces. The rate of accretion, rate of heat transfer, and ignition of the Sun are all important to the kinetics. For the inner planets, ignition may be particularly important.

11.4.2.3 Oxygen isotopes

Ordinary, mass dependent fractionation of oxygen isotopes follows a 2:1 line on a three isotope plot (Fig. 1.5). However, a 1:1 line is observed for certain materials in carbonaceous chondrites (CCAM: Clayton and Mayeda, 1999), which is of great interest as this "non mass dependent" behavior is very rare. However, the latter has been observed in Earth's stratosphere and mesosphere, generally involving reactions between gas molecules, in environments where CO and CO_2 are present (Thiemens et al., 1995). Thus, we propose that the two strongly exothermic reactions, Eqs. (11.3) and (11.4) would have been very important in the early solar nebula, both of which removed oxygen from CO to form H_2O. This conversion is evident in the composition of dense astronomical clouds and in comets (Fig. 9.2). Eq. (11.3) produces graphite, which is much more abundant in carbonaceous chondrites than in other meteorites. Such exchange reactions may have produced mass independent fractionations, at least under some circumstances.

11.4.3 Processing in the winds of accretion

Densifying a nebula is essential to planet and star formation (Fig. 11.6D–F). In Chapter 4, we proposed that these bodies first existed as iron protocores. Magnetic forces, friction welding, and cold welding supplement gravitational attraction, and more importantly can hold the iron particles together after a collision. The ductility of iron also promoted initial growth, whereas silicate grains would bounce and/or fracture rather than adhere during grain-to-grain collisions. Once a metal protocore forms, brittle silicate grains would still fracture on impact, but would be held on by gravity. Chapters 4 and 9 discuss formation of early bodies further. The present chapter is concerned with how chondrules and chondrites would form during planetary growth, which also drew in the iron remaining in the nebula with the mineral dust, after the protocore formed.

As the incipient planets drew in nebular matter, strong circling winds are expected (Fig. 11.6E). The winds carry mineral grains toward them, in a stream (Fig. 11.6I), so collisions and abrasions exist, but these are not impacts at high speed because all grains in any given region have the same speed and direction. Heat evolved is minor and rapidly dissipated by radiation. The main process is rounding of grains in the winds of accretion, with some adhesion and growth. Chemical reactions appear to be substantial during this stage (Table 11.3) because oxidized ordinary chondrites are rather old (Table 9.4). This stage can be envisioned as a gathering cyclonic sandstorm of huge proportions. The speed may be quite slow, particularly at first, when a small protocore is the center of gravitational attraction. But, as the protoplanet grows, it pulls more strongly on its surroundings, so the forces will increase. We propose that in the inflows, up to the point of impact, the main process is abrading large mineral grains and aggregating small nebular grains into chondrules.

TABLE 11.3 Processes possibly affecting formation of meteorite components[a].

Grain formation in ejecta	Grain growth in quiet clouds	Tumbling during accretion
Aging stars loose mass[b]	Cold welding	Vertical collapse of nebula
Ions and atoms ejected	Collisions/aggregation	Cyclonic winds during spin up
Cooling of hot gas	$CO-H_2$ gas reactions, etc.	Abrasion/rounding
Nucleation	Icing of grains	Frictional heating/welding
Electrostatic interactions	Catalytic reactions	Coatings (rims) form?
Chemical vapor deposition	Graphitization	Lightening strikes
Drifting/ spreading into space	Metal oxidation	Electrostatic interactions
		Enhanced reactions

[a]*Placement of many processes under the various categories are tentative, as the details of grain growth cannot be ascertained by experiments that reasonably approximate the conditions.*
[b]*Previous generations of stars lose mass in various ways as they age and change. Isotopically distinct material (presolar grains) are identified in meteorites.*

11.4.4 Deposition on surfaces of growing protoplanets

Today, blowing winds move dust across Earth's surface, forming dunes and depositing sediments at distance. Windstorms can be quite vigorous (Fig. 11.1). Wind is a geologic process on other terrestrial planets, e.g., Mars (Greeley and Iversen, 1985). Deposition of chondrules on protoplanetary surfaces would be similar to processes in sandstorms, but much different than those in the quiet nebula. Possibilities are listed in Table 11.4, leftmost column. Deposition involves greater forces than in the collisions in the winds of accretion, but it is not appropriate to describe deposition as an impact, because the wind path is curved and the protoplanet is spinning as well (Fig. 11.8A). However, the surface is a solid, so fragmentation is expected in this stage. Lightening is expected in sandstorms, and occurs today on many planets. Electrostatic interactions should also exist (Greeley, 1979).

It is likely that most of the gas molecules and substantial amounts of ices were lost as the material was laid down, layer upon layer, much like pressing water out of cheese (Fig. 11.8A). But, at the same time, the chondrules, dust particulates, CAIs, metals and other inclusions were warmed, since ballistic radiation was impeded, which enhanced reaction kinetics. Reaction progress depends on the time taken during deposition, radiative transfer, and especially on when the Sun was ignited (Fig. 11.8C). The icy nature of moons beyond the asteroid belt, and the connection of oxidation state of the ordinary chondrites with age (Fig. 9.9; Table 9.4), suggests that the timing of Solar ignition may have affected the chemical composition of the rocky planets. The heat capacity of water is quite high, so heat from the Sun, once ignited, would have warmed the surface, and been retained, therefore enhancing reaction kinetics.

Importantly, as a planet grows, its surface area grows as radius squared, so more material assembles and likely more time elapses during the formation of the same thickness layers at large compared to small radii. Also, just after the protocore formed, the iron

TABLE 11.4 Processes during assembly of meteorite components on protoplanetary surfaces.

Deposition	Burial	Impacts and ejection	Impact upon return
Sandstorms	More compression	High speed collisions	Shock overprinting
Fragmenting collisions	Reduction of porosity	Shock features	Fusion rind added
Abrasion	Fluid infiltration	Compaction	
Degassing	Alteration	Fracturing	
Lightening strikes	Variable cementation	Phase transformations	
Minor compression	Dissolution/precipitation	Melting	
Electrostatic interactions[a]	Rusting[a]	From variable depths	
Warmth from Sun[a]	Diagenesis	Must overcome inflow	

[a]These processes may have occurred during accretion.

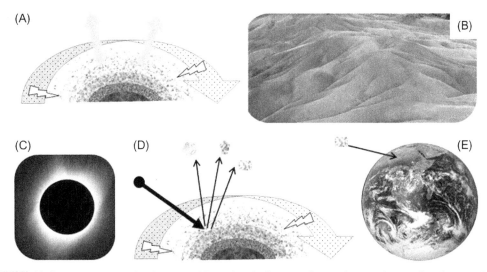

FIGURE 11.8 Processes involved in assembling chondrules into planets, first producing chondrites and then ejecting them. (A) Large scale sedimentary growth. Fuzzy arrows indicate gas and ice being squeezed out by the addition of overlying material. (B) Sand dune, multicolored of volcanic rock from Mauritius. (C) Our Sun has provided high temperatures in the inner Solar System since ignition. (D) Impact ejecting meteoritic materials during accretion. Some material may be melted. The sediments must be fairly consolidated to survive impact. (E) Arrival of a meteorite requires an impact, which may occur billions of years later than formation. *(B) Modified after https://commons.wikimedia.org/wiki/File:Seven_coloured_earths_mauritius.jpg. (C) Image of the solar corona during an eclipse, taken by Aubrey Gemignani from NASA public domain: https://spaceplace.nasa.gov/sun-corona/en/. (E) Blue marble from NASA public domain: https://www.google.com/search?q = nasa + big + blue + marble + image&tbm = isch&source = iu&ictx = 1&fir = vcJR9jkeVnWCXM%253A%252CYi4WR0d7KXkbaM%252C_&vet = 1&usg = AI4_-kSGTq 3AZxsIQm8ssfzsJrgKJeGjnw&sa = X&ved = 2ahUKEwjP34Gy8bfjAhVDA6wKHU7UD1QQ9QEwBXoECAAQDg#imgrc = vcJR9jkeVnWCXM.*

metal in the immediate surroundings would have been depleted. For such reasons, we suggest that the lower mantle was formed from enstatite chondrites, whereas the ordinary chondrites describe the upper mantle (Chapter 9). It is quite possible that enstatite chondrites were deposited on small protoplanets before the Sun had ignited, since these are very old (4563 Ma: Gilmour et al., 2009). The forsterite chondrules which are nearly iron free, are also very old, whereas almost all Fe^{2+} rich chondrules are much younger (4560–4523 Ma), as summarized in Table 9.4.

11.4.5 Processing upon burial

After any given layer of chondrules is deposited in a surficial sandstorm, additional layers may be added on top, particularly at low-lying or submerged sites (Fig. 11.8B). The planet is gradually assembled in this manner. Burial of any given layer of chondrites permits another set of processes to occur, listed in the second column from the left of Table 11.4.

A key step is compaction of the pores between the round chondrules, where the pressurization forces gas and fluid upwards. But as the pores close, some fluid is trapped. At this point, a host of processes known in sedimentology can take place (Table 11.4). Grains are cemented. This may be simply by pressure, or a reaction can be involved. This step produces a rock from the sand sized grains, which we now refer to as a chondrite. All chondrites are variations on a theme. Many have the same components but in different proportions, see above, Chapters 1, 9, 10, and the cited review papers.

Heat is provided by local radioactive decay, friction as the material shifts and compacts, and possible impacts of material heading to the Sun. However, Solar flux eventually becomes the most important heat source for planetary surfaces, and is currently huge compared to radioactive heat emissions. When the Sun ignited is not known, but after this happened, the surface of the Earth was held at ~280 K, like today. This increase in temperature from a few tens of Kelvins to ~280 K would have had an immense effect on the kinetics of reactions, thermodynamic, and transport properties.

Low temperature minerals would form in diagenetic processes: these include carbonates, phyllosilicates, and clays. Rusting is expected. As the material piled on, diagenesis progressed into low grade metamorphism. Sedimentary rocks from this period no longer exist on Earth, as the rock record begins 0.5 Ga after formation (the Acasta gneiss, see e.g., Chapter 9) and is very sparce.

11.4.6 Impacts eject material, which impacts Earth upon return

The rounded grains and many other features observed in chondrites are explained by sedimentary processes that are common on Earth and other rocky bodies today. During formation of the planets, impacts obviously occurred, which is consistent with most of the material in the solar nebula being drawn into the co-accreting Sun. Intercepted impactors would eject material (Fig. 11.8D). Most of this debris would fall back on the planet, but some would be destined for the Sun, and a tiny amount would be shot to the outer Solar System. Hofmeister and Criss (2012) provide the mathematics, and suggest that the asteroid belt is mostly debris left over from accretion. Here we emphasize that the asteroid belt is debris from the surfaces of the growing planets. Some of this debris is fairly large. But only one asteroid, Ceres, is a round, gravitationally accreted body, with an icy rock composition indicative of the outer solar system.

It is highly doubtful that unconsolidated material (e.g., a sand dune: Fig. 11.8B) would be ejected into space. But compacted material could be ejected. Larger impacts might cause melting, particularly if sufficient water is present that melting points are lowered. Thus, some chondrules would be impact melts. Effects of impacts are well-known: Table 11.4 provides a brief summary. Impacts into sandstones have been studied in detail, as this describes Meteor Crater in Arizona (Shoemaker, 1963), where the first natural occurrence of the high pressure coesite phase of silica was discovered (Chao et al., 1960). For reviews of the effect of impacts on sandstones and other rocks, see Dressler and Reimold (2001) and Stöffler et al. (2018).

Undifferentiated meteorites in available collections are quite old (see e.g., Chapter 9). These rocks resided in space for billions of years. A few were ejected more recently from the Moon, Mars, or Mercury, but none of these are chondrites. Most meteorites are chondrites that were originally ejected from the protoEarth and other growing, rocky bodies, destined to "temporally" reside in the asteroid belt.

When intercepted by the Earth or other terrestrial bodies, chondrites suffered yet another impact that exacerbated their fracturing and other shock effects.

11.5 Summary and future work

Chondritic meteorites share many attributes with ordinary terrestrial sandstones, including high porosity, and like them are constituted of mechanical detritus derived from diverse, preexisting source materials. Their remarkable chondrules have the same size as ordinary sand grains, and are highly rounded like aeolian sands. Both high and low temperature phases coexist in some chondrites, as they commonly do in sandstones. Both sandstones and chondrites commonly exhibit grain overgrowths, as well as compositional and textural effects caused by diverse secondary processes. For example, fluid infiltration and recrystallization can produce overgrowths, and can add intergranular matter, including clays and amorphous materials. Tectonism and impacts can cause shearing, fracturing, and many other textures that can even include partial melting. In this view, the great abundance of chondrites (>85% of falls) is fully consistent with sediments covering >75% of Earth's present surface, and with practically all lunar meteorites being breccias derived from the Moon's surface. The specific similarity of chondrites to aeolian deposits is consistent with the high wind velocities expected on the surfaces of accreting, rapidly spinning protoplanets from which they were clearly derived.

As is the case for terrestrial sediments, the origin of individual grains in meteorites is often masked. We have proposed an origin that is consistent with astronomical data on molecular clouds, and with long-known stellar recycling. Processes of grain growth could include electrostatic attraction and cold welding. Dust in the solar nebula predating the accretion of the Sun and planets provided nuclei for growth. Rare presolar grains are recognizable from distinct older stars due to special isotopic ratios. But, the average isotopic signature of the Sun and the rocky bodies in the Solar System was also inherited. This is the norm. Heavy elements in the Sun are compositionally similar to chondritic meteorites, because both accreted from the same dust cloud. Had the hot Sun spit out these grains, as envisioned by Laplace, this non-conservative processes would have provided a gradation in condensation temperature that is contrary to cometary data. Comets are chondrites whose ice has not been largely lost by abrasion, compression, or heating.

Our model explains why chondrites have similar mineral components and compositions but differ more in texture, and addresses virtually all aspects of chondrule and chondrite formation, assuming only that spin is a primary product of accretion. Many of the ideas presented here are testable. Some require new experiments on cold welding under vacuum, and wind abrasion of mineral types found in chondrites. Determining the kinetics of graphitization, rusting, and ion exchange would also be revealing.

References

Aller, M.C., Kulkarni, V.P., York, D.G., Vladilo, G., Welty, D.E., Som, D., 2012. Interstellar silicate dust in the z = 0.89 absorber toward pks 1830-211: crystalline silicates at high redshift? Astrophys. J. 748 (19). Available from: https://doi.org/10.1088/0004-637X/748/1/19.

Bagnold, R.A., 1941. The Physics of Blown Sand and Desert Dunes. William Morrow and Co., New York, New York.

Bagnold, R.A., Barndorff-Nielsen, O.E., 1980. The pattern of natural size distributions. Sedimentology 7, 199–207.

Bowey, J.E., Adamson, A.J., 2002. A mineralogy of extrasolar silicate dust from 10-μm spectra. Mon. Not. R. Astron. Soc. 334, 94–106. Available from: https://doi.org/10.1046/j.1365-8711.2002.05489.x.

Bowey, J.E., Hofmeister, A.M., 2005. Overtones and the 5-8 mm spectra of deeply embedded objects. Monthly Notices of the Royal Astronomical Soc 358, 1383–1393.

Brearly, A.J., Jones, R.H., 1998. Chondritic meteorites. Rev. Miner. 36, 398. Chapter 3.

Chao, E.C.T., Shoemaker, E.M., Madsen, B.M., 1960. First natural occurrence of coesite. Science 132, 220–222.

Clayton, R.N., Mayeda, T.K., 1999. Oxygen isotope studies of carbonaceous chondrites. GCA 63, 2089–2104.

Consolmagno, G.J., Britt, D.T., Macke, R.J., 2008. The significance of meteorite density and porosity. Chemie der Erde: Geochemistry 68, 1–29.

Criss, R.E., Hofmeister, A.M., 2018. Galactic density and evolution based on the Virial Theorem, energy minimization, and conservation of angular momentum. Galaxies 6 (#4), 115–135. Available from: http://www.mdpi.com/2075-4434/6/4/115/pdf.

Davis Jr., R.A., 1992. Depositional Systems: An Introduction to Sedimentology and Stratigraphy, second ed. Prentice-Hall, Inc, Englewood Cliffs, New Jersey.

Dott Jr., R.H., 2003. The importance of eolian abrasion in supermature quartz sandstones and the paradox of weathering on vegetation-free landscapes, J. Geol., 111. pp. 387–405.

Dressler, B.O., Reimold, W.U., 2001. Terrestrial impact melt rocks and glasses. Earth-Sci. Rev. 56, 205–284.

Fazio, A., D'Orazio, M., Cordier, C., Folco, L., 2016. Target–projectile interaction during impact melting at Kamil Crater, Egypt. Geochim. Cosmochim. Acta 180, 33–50.

Fossen, H., Zuluaga, L.F., Ballas, G., Soliva, R., Rotevatn, A., 2015. Contractional deformation of porous sandstone: Insights from the Aztec Sandstone, SE Nevada, USA. J. Struct. Geol. 74, 172–184.

Friedrich, J.M., Weisberg, M.K., Ebel, D.S., Bliltz, A.E., Corbett, B.M., Iotzov, I.V., et al., 2015. Chondrule size and related physical properties: A compilation and evaluation of current data across all meteorite groups. Chem. Erde 75, 419–443.

Gradel, T.E., Frankenthal, R.P., 1990. Corrosion mechanisms for iron and low alloy steels exposed to the atmosphere. J. Electrochem. Soc. 137, 2385–2394.

Gilmour, J.D., Crowther, S.,A., Busfield, A., Holland, G., Whitby, J.A., 2009. An early I-Xe age for CB chondrite chondrule formation, and a re-evaluation of the closure age of Shallowater enstatite. Meteor. Planet. Sci. 44, 573–579.

Greeley, R., 1979. Silt-clay aggregates on Mars. J. Geophys. Res. 84, 6248–6259.

Greeley, R., Iversen, J.D., 1985. Wind as a Geologic Process on Earth, Mars, Venus and Titan. Cambridge University Press, Cambridge U.K.

Hofmeister, A.M., Criss, R.E., 2012. Origin of HED meteorites from the spalling of Mercury: implications for the formation and composition of the inner planets. In: Hwee-San, Lim (Ed.), New Achievements in Geoscience. InTech, Croatia, pp. 153–178.

Hofmeister, A.M., Wopenka, B., Locock, A., 2004. Spectroscopy and structure of hibonite, grossite, and $CaAl_2O_4$: implications for astronomical environments. Geochim. Cosmochim. Acta 68, 4485–4503.

Karner, S.L., Chester, F.M., Kronenberg, A.K., Chester, J.S., 2003. Subcritical compaction and yielding of granular quartz sand. Tectonophysics 377, 357–381.

Kowitz, A., Schmitt, R.T., Reimold, W.U., Hornemann, U., 2013. The first MEMIN shock recovery experiments at low shock pressure (5–12.5 GPa) with dry, porous sandstone. Meteorit. Planet. Sci. 48, 99–114. Available from: https://doi.org/10.1111/maps.12030.

Kuebler, K.E., McSween, H.Y., Carlson, W.D., Hirsch, D., 1999. Sizes and masses of chondrules and metal-troilite grains in ordinary chondrites: possible implications for nebular sorting. Icarus 141, 96–106.

Kuenen, P.H., 1960a. Experimental abrasion. 4. Eolian action. J. Geol. 68, 427–449.

Kuenen, P.H., 1960b. Sand. Sci. Am. (April issue) 202, 95–111.

Lodders, K., Fegley, B.J., 1998. The Planetary Scientist's Companion. Oxford University Press, Oxford.

Mason, B., MacPherson, G.J., Score, R., Martinez, R., Satterwhite, C., Schwarz, C., Gooding, J.L., 1992. Descriptions of stony meteorites. In: Marvin, U.B., MacPherson, G.J. (Eds.), Field and Laboratory Investigations of Antarctic Meteorites Collectedby United States Expeditions, 1985–1987. Smithsonian Inst. Press, Washington D.C.

Merstallinger, A., Sales, M., Semerad, E., Dunn, B.D., 2009. Assessment of Cold Welding Between Separable Contact Surfaces due to Impact and Fretting Under Vacuum. European Space Agency, Leiden, the Netherlands. Available from: http://esmat.esa.int/Publications/Published_papers/STM-279.pdf.

Mittlefehldt, D.W., McCoy, T.J., Goodrich, C.A., Kracher, A., 1998. Non-chondritic meteorites from asteroidal bodies. Rev. Mineral 36, 1529–6466.

Molster, F.J., Lim, T.L., Sylvester, R.J., Waters, L.B.F.M., Barlow, M.J., Beintema, D.A., et al., 2001. The complete ISO spectrum of NGC 6302. Astron. Astrophys. 372, 165–172.

Mumma, M.J., Charnley, S.B., 2011. The chemical composition of comets—Emerging taxonomies and natal heritage. Annu. Rev. Astron. Astrophys. 49, 471–524.

Nelson, V.E., Rubin, A.E., 2002. Size-frequency distributions of chondrules and chondrule fragments in LL3 chondrites: implications for parent-body fragmentationof chondrules. Meteorit. Planet. Sci. 37, 1361–1376.

Nittler, L.R., Ciesla, F., 2016. Astrophysics with extraterrestrial materials. Annu. Rev. Astron. Astrophys. 54, 53–93.

Nocita, B.W., 1989. Sandstone petrology of the Archean Fig Tree Group, Barberton greenstone belt, South Africa: Tectonic Implications. Geology 17, 953–956.

Paque, J.M., Cuzzi, J.N., 1997. Physical characteristics of chondrules and rims, and aerodynamic sorting in the solar nebula. In: 28th Lunar and Planetary Science Conference. Abst. #071.

Pettijohn, F.J., Potter, P.E., Siever, R., 1973. Sand and Sandstone. Springer-Verlag, Heidelberg.

Robie, R.A., and Hemingway, B.S. 1995. Thermodynamic properties of minerals and related substances at 298.15K and 1 bar (105 Pascals) pressure and at higher temperatures, U.S. Geological Survey Bulletin 2131.

Rubin, A.E., 2000. Petrologic, geochemical and experimental constraints on models of chondrule formation. Earth Sci. Rev. 50, 3–27.

Schäfer, A., Teyssen, T., 1987. Size, shape and orientation of grains in sands and sandstones-Image analysis applied to rock thin-sections. Sediment. Geol. 52, 251–271.

Scott E.R.D., Krot A.N., 2007. Chondrites and their components. In: Davis, A.M. (Ed.), Meteorites, Comets and Planets, Chapter 1.07, Treatise on Geochemistry Update 1, Elsevier, online at http://www.sciencedirect.com/science/referenceworks/9780080437514.

Shoemaker, E.M., 1963. Impact mechanics at Meteor Crater, Arizona. In: Middlehurst, B.M., Kuiper, G.P. (Eds.), The Solar System, vol. 4, The Moon, Meteorites and Comets. University of Chicago Press, p. 301.

Simon, J.I., et al., 2018. Particle size distributions in chondritic meteorites: Evidence for pre-planetesimal histories. Earth. Planet. Sci. Lett. 494, 69–82.

Speck, A.K., Barlow, M.J., Skinner, C.J., 1997. The nature of the silicon carbide in carbon star outflows. Mon. Not. R. Astron. Soc. 288, 431–456.

Stöffler, D., Hamann, C., Metzler, K., 2018. Shock metamorphism of planetary silicate rocks and sediments: Proposal for an updated classification system. Meteor. Planet. Sci. 53, 5−49.

Thiemens, M.H., Jackson, T., Zipf, E.C., Erdman, P.W., van Egmond, C., 1995. Carbon dioxide and oxygen isotope anomalies in the mesosphere and stratosphere. Science 270, 969−972.

van Breemen, J.M., Min, M., Chiar, J.E., Waters, L.B.F.M., Kemper, F., Boogert, A.C.A., et al., 2011. The 9.7 and 18 mm silicate absorption profiles towards diffuse and molecular cloud lines-of-sight. Astron. Astrophys. 526, Article A152. Available from: https://doi.org/10.1051/0004-6361/200811142.

Zolensky, M.E., et al., 2006. Mineralogy and petrology of Comet 81P/Wild 2 Nucleus samples. Science 314, 1735−1739.

Websites

Meteoritical data base of The Meteoritical Society.
https://www.lpi.usra.edu/meteor/about.php.
R.A. Bagnold.
https://en.wikipedia.org/wiki/Ralph_Alger_Bagnold.
Planetary fact sheets from NASA (accessed 01.06.19.).
https://nssdc.gsfc.nasa.gov/planetary/factsheet/.
Fujita scale for tornado damage.
http://www.tornadoproject.com/cellar/fscale.htm.
Images of porphyritic rocks.
http://www.pitt.edu/~cejones/GeoImages/2IgneousRocks/IgneousTextures/4PorphyriticFineGrained.html.
https://www.sandatlas.org/porphyry/.
Sand and dust storms.
https://public.wmo.int/en/our-mandate/focus-areas/environment/SDS.
Sandstone porosity.
https://wiki.aapg.org/Porosity#Sandstone_pore_systems.

12

Conclusions and future work

"My observations have convinced me that some men, reasoning preposterously, first establish some conclusion in their minds which, either because of its being their own or because of their having received it from some person who has their entire confidence, impresses them so deeply that one finds it impossible ever to get it out of their heads." *(Galileo Galilei, 1632)*.

"It is always well to keep this in mind in order not to infer from the facts more than can rightly be inferred from them: The various idealistic and realistic interpretations are metaphysical hypotheses which, as long as they are recognized as such, are scientifically completely justified. They may become dangerous, however, if they are presented as dogmas or as alleged necessities of thought. Science must consider thoroughly all admissible hypotheses in order to obtain a complete picture of all possible modes of explanation." *(Hermann Helmholtz, 1878)*.

12.1 Overview

This book summarizes and evaluates data on the Earth and other large bodies in the inner Solar System, and takes a critical look at assumptions and models on the thermal state and evolution of rocky planets. Our recent studies in heat transfer, solid mechanics, and gravitation of oblate bodies have pointed out that several popular notions about fundamental matters are flawed: some of these are connected with initial conditions, and some with the current state and workings of the Earth and planets. This book uncovers additional misconceptions, and tries to improve current understanding of heat transfer in large bodies, by offering new hypotheses that are based on available data, some of which are quite recent. Our goal is to stimulate thinking and discovery.

Our approach follows the scientific method (Fig. 12.1), which recognizes that observations are the basis of scientific progress. Data motivate the formulation of hypotheses, which lead to proposals of experiments to test the ideas. Critical thinking is key, because this is used to both analyze and interpret data, and to make predictions that can be tested by new experiments. The more complicated a phenomenon is, the more trips around the circle of scientific progress are needed. A body of data with multiple confirmations of hypothetical concepts leads to a theory, which is commonly supported by a mathematical analysis. Quantitative confirmation is desirable, but not always possible. Science is based on *concepts*, not numbers. However, leaps are often made with insufficient information: an example is the 1st law of thermodynamics, which is actually a postulate. Direct proof that energy is conserved, if heat is accounted for, does not exist because temperature, not heat, is the quantity being measured.

Clairvoyance being rare is an understatement. As time passes, experiments once considered impossible become realities, and the results can be surprising. New, unforeseen observations can require rethinking theory, and a return to the circle of scientific progress. The alternative path of rationalizing, ignoring, or denying obvious problems disengages

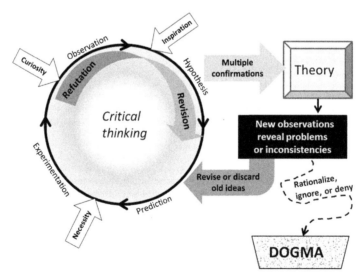

FIGURE 12.1   Schematic of the scientific method in theory and practice. White arrows = possible motivations for pursuing science. Light gray circle with paths indicated by various arrows = the cycle of scientific progress where observations are explained by a hypothesis (or idea), which makes a prediction, that then can be tested via experiments and further observations. If the original idea is refuted, a new or revised hypothesis is needed and the steps are repeated. White circle = critical thinking, which is central to science. Through sufficient and extensive confirmation, a hypothesis is expanded into a theory (upper right). However, theories are seldom perfect, and subsequent observations (black box) can reveal flaws, requiring a return to the cycle of scientific progress (dark gray arrow). Curly dashed arrow = the dead-end path to dogma, which involves human shortcomings.

the scientific community from the scientific method, and converts a theory into dogma, where ideas are accepted and believed, rather than being critically evaluated. Dogma encourages subjective interpretation of data, which process is geared to support the status quo. When consensus is reached, scientific progress in that area stops. Dogma precludes scientific advance and denigrates originality and creativity, creating ideal conditions for its own preservation. Dogma can only be overturned by overwhelming evidence that eventually results in a paradigm shifts and scientific revolution. Dogma can affect fundamental issues in science (e.g., the currently popular stance that matter which does not interact with light not only exists, but is abundant) or can be parochial (e.g., Vesta is differentiated).

Dogma is tied to reputation, as Galileo recognized (see quote). When reputations are huge, dogma mutates, rather than being eliminated One example is Kelvin's discounted explanation for the production of starlight, which currently survives in models of core formation, and in the concept that stars need to be hot before they ignite. Another example is Kelvin's age of the Earth, which was refuted by radioactivity, yet then morphed into the dogma that radioactivity controls Earth's thermal state.

Peripheral research problems can be affected by a faulty theory. The problem is magnified for phenomena associated with remote or inaccessible conditions, such as Earth's deep interior, or its early history, for which observation and proof are elusive. For this reason, many hypotheses about the conditions and workings of Earth's interior are suspect, especially where heat transfer is concerned, since historical errors exist in thermodynamics and the kinetic theory of gas, which affect models of heat transfer and interpretation of measurements (Hofmeister, 2019). Several questionable assumptions underlie research into the thermal state and evolution of Earth and other rocky bodies (Fig. 12.2, top row) and have led to many popular ideas (second row) which upon inspection are not supported by fact (middle row). When errors in basic assumptions are not discussed, new ideas are not just impeded, they are treated with hostility. The fourth row of Fig. 12.2 lists key ideas presented in the present book, reflecting our attempt to better explain the observations. These ideas led to several new postulates for behavior of the Earth (bottom row). Section 12.2 summarizes these new findings, and places them in the context of previous models. Section 12.3 covers future work, and Section 12.4 returns to the scientific method in theory and practice, and the need to promote critical thinking.

12.2 New and old ideas on rocky planets

Below, each chapter is summarized, focusing on the key points and important findings. The first 5 chapters, composing Part I, provide the tools and information needed to improve our understanding of the thermal state and evolution of the Earth and sister rocky bodies. In view of Helmholtz's quote, various scenarios are proposed in Part I, but some of these are shown to be unlikely in Chapters 6−8 (Part II), which use thermal models to probe different layers in the Earth. Chapters 9−11 (Part III) address thermal evolution of the Earth and other rocky bodies, ending with chondrules, which are the tiniest and oldest round bodies in the Solar System.

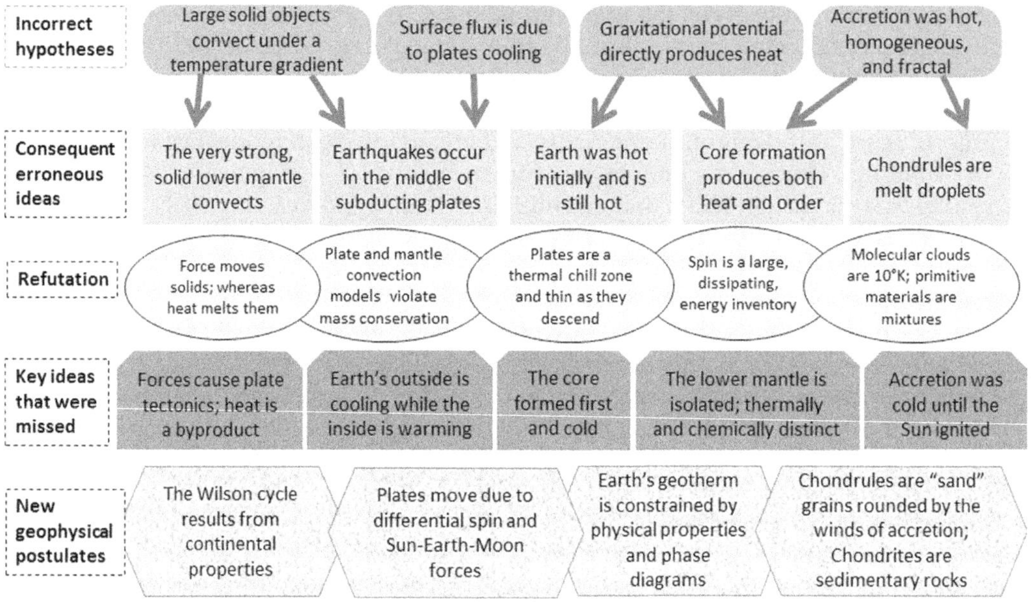

FIGURE 12.2 Schematic illustrating how certain assumptions concerning heat and heat transfer involving the Earth and terrestrial planets have led to popular, but unsupportable concepts. The left column summarizes each row. From top to bottom: Rounded rectangles = underlying assumptions, some of which are ~100 years old. Dark gray arrows connect these to recent ideas, which are also faulty. Ovals = counter evidence discussed in this book. Barn-shaped boxes = ideas developed recently and in this book by recognizing that key information was missing. Hexagons = new models, presented in this book.

Although the focus of this book is on thermal physics, it is impossible to ignore the chemical compositions and chemical evolution of planets, due to these being tied through phase boundaries. Since the main result of substantially heating solids is the production of buoyant melts, gravity figures prominently in evolution. Although early studies in physics adopted Newton's emphasis on forces and momentum, energy has been a primary focus of physical science for quite some time. The directional nature of forces and momentum makes these more difficult to understand than energy, which is a scalar, but information is lost in the latter approach. Another impediment is the conventional omission of gravity from classical thermodynamic studies. But discussion of gravity for the round planets is essential to provide reasonably accurate thermal models. The book therefore covers much more than heat transfer in order to probe the evolution of rocky bodies.

- **Chapter 1** summarizes diverse observations pertaining to heat flow inside the Earth. Setting boundary and initial conditions requires information on temperatures, flux, various material properties, heat sources, internal layering, and lateral differences. We show how the presence and location of molten layers constrain both temperatures and flux, and discuss implications for Earth's thermal structure and evolution. Problems in assuming adiabatic gradients and internal homogeneity are discussed. Energy sources in addition to radionuclide decay are covered: we show that dissipation of spin via friction during differential rotation is significant, yet overlooked. Abundant evidence reviewed here shows that spin exerts crucial control on plate tectonics, for which a new mechanism is presented later in the book.

- **Chapter 2** reviews important results of Fourier's theory of heat conduction, and discusses their applications to planets and the associated limitations. New solutions are derived for radiative cooling, heat generation, and cooling time scales. The relevance of these results to rocky objects that range from individual grains to planets is demonstrated.

- **Chapter 3** evaluates mechanisms for heat transfer in large bodies. Planetary interiors are layered, self-compressed spheroids, and moreover are open systems with internally graded gravitational acceleration, high temperatures, and high pressures. Advection of fluids coexisting with solids is very important in planets, due to fluid buoyancy, mobility, low viscosity, and their ability to carry heat, latent heat, and heat-producing elements, and is covered in detail.

  Microscopic mechanisms are also highly relevant. The conventional treatment of heat transport in solids as being analogous to gases is in error due for several reasons (Hofmeister, 2019), and has led to problematic extrapolation of properties. Importantly, mantle convection models are also based on gas behavior, since the Rayleigh's formulation assumes that heat and mass diffusivities are equal to the kinematic viscosity, after the kinetic theory of gas. Instead, the viscosity of solids arises in Newtonian drag, which retards rather than promotes motion. Another implicit assumption in applying the Ra number to Earth is that matter will flow under any applied stress, which is inapplicable to rocks (Hofmeister and Criss, 2018). Chapter 3 also discusses problems with mass conservation in mantle convection models, and summarizes the new, radiative diffusion model for heat transport in condensed matter (Hofmeister, 2019).

- **Chapter 4** discusses formation of the Solar System and the initial conditions of planets. Self-gravitation is essential to the thermodynamic behavior of large bodies yet is neglected in classical thermodynamic analyses. Accretion predominantly converts gravitational potential energy into orbital and axial spin energies, not into primordial heat. This deduction is supported by planetary rotation being regular and close to Virial Theorem predictions (Hofmeister and Criss, 2016). Any heat produced during accretion was quickly lost to the surroundings to meet 2nd law requirements. Late stage impacts added heat to the outer layers of planets but did not raise interior temperatures. Subsequent dissipation of spin gradually released heat.

  Chapter 4 proposes that accretion began with formation of iron cores by magnetic attraction, friction welding, and cold welding. These protocores probably grew by gravitational sorting of subsequently accreted, heterogeneous material. This sorting sent radionuclides upward, cooling the Earth, as did outgassing, consistent with energy minimization and entropy maximization.

- **Chapter 5** focuses on fully accreted planets, and evaluates heat produced during the gravity driven planetary scale motions which constitute plate tectonics. These motions are important because planetary heat sources are insufficient to drive whole mantle convection, a hypothetical process with several other well-known problems that are summarized previously and here. Differences between radial, axial, and lateral motions associated with gravitational forces explain the different tectonic and volcanic behaviors of Earth and other planets. Dissipation of axial spin is covered, emphasizing the role of friction. Torques produced by externally derived gravitational forces explain each of lunar drift and plate tectonics. Polar jets of magma may be relevant to Mars. We also consider whether addition of heat and matter by late-arriving impacts can explain the high temperatures in the Hadean/Archean, since hot accretion is incompatible with the immense energy reservoir of spin.

- **Chapter 6** shows that the thermal profile of the continental lithosphere is controlled by the fixed surface temperature, the fixed temperature of its partially melted base, the thermal conductivity of its rocks, and the distribution of its heat-generating radionuclides. Length scales are sufficiently small that steady state calculations provide useful guides. New analytical solutions are provided that treat both thermal conductivity and internal heat generation as variables. Calculations that incorporate geologic and experimental constraints yield vertical temperature profiles that resemble those observed for continents. Basal flux primarily reflects the depth to melting, yet decreases as the heat production of overlying crust increases, contrary to standard interpretations. These findings help explain the magmatism of continental interiors, and lead us to suggest that the Wilson cycle is a direct result of the high heat generation and low thermal conductivity of continental lithosphere.

- **Chapter 7** shows that plate models of the oceanic lithosphere, which purportedly describe their cooling, violate conservation of mass and do not solve Fourier's 2-dimensional equation. Steady-state thermal conditions are a better description of Earth's lithosphere over the last 0.3 Ga. We use surface flux measurements and thermal conductivity for a layered lithosphere to calculate geotherms. Thermal boundary thicknesses of ~ 69 km agree with various seismic determinations. From petrologic evidence, the low velocity zone is buffered by the peridotite solidus where water content decreases downwards to nothing by 320 km. We show that most subducting slabs thermally equilibrate above 320 km, which entails thinning, but the few slabs that are thick at the surface, steep, and fast reach 660 km. Inferred times for upper mantle recycling are consistent with ophiolite ages and modest chemical inhomogeneity. Our findings show that layers above and below 660 km are independent systems, thermally and mechanically, and are consistent with the deductions of Chapter 5, which offered a new mechanism for plate tectonics. It follows that these layers are also chemically distinct, since accretion was heterogeneous, as shown in Part III.

- **Chapter 8** confirms that thermal transport in the lower mantle and core today are independent of that in the outer layers, on the basis of Earth's large size and the low thermal diffusivity of oxide minerals, and the even lower thermal diffusivity of silicate minerals. This situation permits construction of thermal profiles of the deep Earth: namely, the two-phase metallic core follows the solidus of iron alloy, and the top of the lower mantle lies below the dry peridotite solidus. Thermal conductivity of the core is irrelevant. We use the most recent data, which have shown that temperatures of pure iron melting grossly overestimate melting of alloy material that describes the core. Our estimates are likely high, due to effects of Ni and P not being yet investigated experimentally, and overemphasis on lithophiles, which are scarcely present in core analogues, meteoritic alloys.

Without samples, mantle thermal conductivity is estimated in Chapter 8 considering data and models for dense solids, leading to "hot silicate" and "cold oxide" geotherms. Temperatures are reasonably constrained, except for the height and position of the local maximum in the lower mantle. Meteoritic values for radionuclide contents of the lower mantle are consistent with present-day temperatures and progressive melting of the core.

- **Chapter 9** evaluates the thermal history of Earth's outer layers (above 660 km) separately from thermal evolution of the interior, due to large cooling time-scales making these independent. Internal radioactive heating warms the lower mantle, which heats and melts the core. Thermal isolation permits estimating a lower mantle chemical composition. Evolution of the outer layers involves external and internal gravitational forces. Briefly, Earth began with nucleation of an iron core, followed by cold accretion of the lower mantle from an iron-depleted sub-nebula. The outer layers were assembled from chondrites and ices, which oxidized as accretion progressed, as quantified by studies of meteorite age which correlate with oxidation of Fe in these samples. Spin dissipation and high radioactivity, coupled with melts and volatiles being buoyant, caused warming and differentiation above 660 km and provided high temperatures and thin crusts in the Archean. Impact heating is shallow. Cooling proceeded slowly until plate tectonics developed due to force imbalances in the Sun-Earth-Moon system, which adds some heat while promoting cooling and mass recycling above 660 km.

- **Chapter 10** shows that the thermal histories of the rocky bodies are controlled by their initial inventories of energy, and by the subsequent production and loss of heat, all of which are largely governed by object mass. Available data constrain these inventories and the processes relevant to the large rocky bodies of the inner Solar System, which began with cold accretion. Inventories per unit mass of heat generated over geologic time by radionuclides vary threefold among these bodies, whereas the inventory generated by spin dissipation varies from near zero for spin-locked bodies to significant. Both correlate with body mass, and the power generated by both has decreased exponentially with time. Distance from the Sun controls heat supplied to the surface by impacts and radiation, which controls volatile speciation and affects thermo-chemical evolution. Deep planetary interiors are cooler than commonly imagined, and some may be warming even as their outer zones are cooling. We argue that the tectonic and magmatic activity of the rocky bodies is weak and shallow, and has declined with time.

- **Chapter 11** addresses a long-standing puzzle as to the origin of chondrules. Although melting is currently proposed to explain their shape, many inconsistencies remain, and no known heat source can explain how this primitive, early material could have undergone ubiquitous melting. We show instead that chondritic meteorites share many characteristics with clastic sedimentary rocks, being detrital aggregates that can include both high- and low-temperature materials. Their distinctive chondrules are rounded, well-sorted, sand-sized grains whose quantitative size distribution indicates aeolian processing. High-speed winds are consistent with the rapid spin rates and huge Coriolis forces of accreting planets. Most grains originally formed in the nebula, possibly by chemical vapor deposition in a vacuum, along with cold welding. Upon deposition on the surface of an accreting planet, sedimentary assemblies of chondrules plus dusty matrix were variously subjected to compaction, fluid infiltration, recrystallization, overgrowth development, cementation, and other processes that produced textural and petrographic variety. Impacts ejected some consolidated material into space and provided shock structures and some melting. A sedimentary origin of chondrites is consistent with their petrographic character, their large relative abundance on planetary surfaces, and the assembly of the rocky planets from the diverse types of condensed matter known to exist in nebulae and molecular clouds.

The timing of Solar ignition is associated in this book with the oxidation state and age of chondritic meteorites, which are correlated as discussed in Chapter 9. Solar ignition elevated the temperature of the inner Solar System by several hundred Kelvins. In contrast to the terrestrial planets and our Moon, bodies which formed in the asteroid belt (Ceres) and beyond have surfaces too cold to have outgassed the ices acquired during accretion. Because these bodies are too small to have acquired and retained gas, their current composition proves that accretion was cold. The importance of solar ignition to planetary development also follows.

The proposal for the origin of chondrites in Chapter 11 is consistent with the findings in the previous chapters of a cold beginning, histories of planetary surfaces and corresponding interiors differing, physical properties of materials being important to planetary behavior, and with gravitation playing a key role over much of geologic history. We have found flaws in many equations and concepts that are widely used in Earth science, and shown that new measurements, coupled with reanalysis of both theory and data, can provide sensible answers to many long-standing, hotly debated issues.

12.3 Future work

Many of the ideas presented in the book are testable. A short list follows. Other avenues are possible.

Regarding the beginnings (Chapters 4 and 9) and chondrule formation (Chapter 11), grain size of olivine beach sands has not been investigated. Wind tunnels can provide controlled environments, and could quantify the effects of malleable iron metal or hard CAI grains being present. Reactions of enstatite chondrites with gas at and below room temperature need to be investigated. Experiments need to be conducted under rarified conditions (laboratory vacuum), and the effect of sunlight should be probed.

Additional information on the planets would improve thermal models (Chapter 10). Gamma ray data from Venus are spotty, and their average intensity for Earth's surface has not been computed. Laboratory measurements of material properties are needed below 298 K for the Moon and Mars. Specifically, laser flash measurements of thermal diffusivity are needed, which are accurate because these remove spurious ballistic transport gains while avoiding thermal contact losses. Data on surface materials is important to ascertain the shallow thermal gradient and model its evolution. All previous models consider only progressive cooling, yet Earth probably experienced a phase of near-surface warming upon formation (Chapter 9). Measurements of heat transfer in chondrites with volatile phases in the pores are needed to address early processes.

Quantifying the nature of Earth's interior (Chapter 8) is challenging, since no samples originating below ~ 300 km are available. Knowledge of the thermal state of the core can be advanced by phase equilibria experiments that focus on the solidi, not liquidi, and include Ni, P, and N in the iron alloy, in addition to S and C which have previously been explored. Experiments probing the amount of solids that a melt can hold, yet still not conduct shear waves, are needed. Regarding the lower mantle, mineralogical models of seismic velocities need to be constructed with appropriate temperature gradients. Adiabats explored previously, as summarized in Chapter 1, are not relevant, since heat is flowing irreversibly, and because the formulae used represent isentropes. Mineralogic models of seismic profiles may be able to establish whether a temperature peak exists in Earth's thermal profile.

Of paramount importance to advancing geophysics is considering alternate mechanisms for plate tectonics than mantle convection, which is a failed hypothesis for reasons discussed in this book and previously (e.g., Foulger, 2010). We proposed one such mechanism (Chapter 5), and showed that viewing plate thickness as constant has led to misinterpreting seismic data on deep earthquakes (Chapter 7). This incorrect interpretation prevented recognition that plates thin and re-equilibrate in the upper mantle and/or transition zone. Thermal models of subducting slabs that use modern thermal transport data are needed to address their rate of thinning. Existing models all assume constant thickness, which is impossible for a chill zone.

Our new formulation for cooling of continents (Chapter 6) can be applied locally. Existing calculations and their representations in review papers, which include ballistic radiative transfer, are invalid. The distribution of radionuclides needs to be reconsidered, as well as its time-evolution via magmatism.

Rayleigh's critical number (Chapter 3) should be tested against solids more viscous than syrup, which is currently the most viscous substance measured. This test would reveal whether our recasting his formula to account for different behavior of the three transport properties provides the correct critical number. Rayleigh's formulation assumes that temperature and depth are independent, and that physical properties are constants: these stipulations do not apply to Earth. Hence, additional tests are needed to quantify the effect of temperature on physical properties.

The analytical and numerical solutions to Fourier's equations (Chapters 2 and 6) provided here are sufficient to address many problems in geophysics and planetary science. If these could be extended to address multiple heat transfer mechanisms, temperature and depth variations, and time dependence, then a wider range of problems could be investigated.

12.4 New and old problems in science: are remedies possible?

Many roadblocks to the advancement of science are rooted in human nature (see Galileo's quote). It is far easier and less risky to defer to authority, than to think things through. We have been indoctrinated in education to not challenge the status quo. Yet, inculcation of students with current paradigms is counter productive in the long run (see Kuhn, 1970; Hamilton, 2002). The need to master the literature in a subject area is obvious, but making a contribution to the natural sciences also requires training in laboratory work, field observations, analytical mathematics, and most importantly, critical thinking. Hands on experience is essential.

Human beings confuse cause with effect, commonly jump to conclusions, and are overly influenced by group think. Circular reasoning and consensus provide support for dogma. For example, as stated by Bercovici (2015): "The plates are cooling thermal boundary layers that are both driven by and become slabs, which are in themselves convective downwellings; the plates are thus convection." His statement is equivalent to: *The Chicken is the Egg: problem solved!* To the contrary, a process is not the same as the matter it affects. Regarding the specific error in his statement, plates do not bend over to form the circular convective flow patterns seen in laboratory experiments. Circular reasoning is detected through critical analysis.

Another source of confusion is of numbers with facts, and holding the printed word as truth. For example, people believe and obey signs. Numerology is a significant problem in modern science, due to the proliferation of computers and software (Truesdell, 1984). Computer output can be very precise, but need not resemble anything real. Unlike engineering, verification or validation of model solutions for most problems in Earth and planetary science is not possible. No instrument can measure temperature or convective patterns deep in planetary interiors, nor turn back time. Evaluating accuracy in numerical studies of complex natural phenomena requires comparison to analytical solutions, and critical evaluation of the implicit and explicit assumptions. Even when a computer model supports a popular idea (e.g., "the generalized potential" allegedly proves that Vesta is differentiated), its mathematical underpinnings still need to be investigated. Exposure of flaws such as in "the generalized potential" (Hofmeister et al., 2018) is the first step towards improvement. The bottom line is that quantitative understanding of the real world is not garnered through computer mock-ups of postulates, but is instead based on measurements of natural phenomena and experimentation (Fig. 12.1).

Humans are also social animals, and so agreement lends confidence to the correctness of one's beliefs or behavior. Through this trait, mediocrity in science is fostered, where discoveries that support the status quo or the views of some eminent scientist receive kudos. History shows that most important new ideas are received with hostility, and even if published through dedication and perseverance, tend to be ignored. Because obtaining funding requires approval by peers, groupthink is promoted and critical thinking is impeded. Is the goal of science to seek the truth and make discoveries, or as stated by Jones (1999) is science "merely a game for distributing prestige and grants among competing players."

Only recently have some societies and governments advocated diversity, and adopted the viewpoint that religions and cultures other than those of the majority have validity. This change comes down to recognizing that different people think in different ways. Although multiple working hypotheses are touted in science (see Helmholtz's quote) this practice is not followed. Hamilton's (2002) discussion of such problems in geoscience holds today.

The information explosion supports dogma. So much "stuff" exists on the web and in print that a person can selectively choose to believe whatever best matches their own point of view, see Fig. 12.1. The information explosion seems connected with an increased number of papers being summarily rejected, where negative "reviews" don't even mention the new data or findings that were supposedly evaluated. Anonymous review is a particular problem for any research area with dogmatic underpinnings. Changes in current practices are overdue, as evidenced by the recent practice of presenting the evaluations of associate editors as anonymous.

Important scientific advances are not predictable. History has repeatedly shown that paradigms fall and their supporters are forgotten. This book has shown that many key hypotheses in Earth science have serious flaws. We have provided alternative hypotheses that address long-standing problems in the geologic sciences and related fields (Fig. 12.2). We hope that the reader will be stimulated to think critically, and beyond what is proposed here.

References

Bercovici, D., 2015. Mantle dynamics: an introduction and overview. In: second ed. Schubert, G. (Ed.), Treatise on Geophysics, vol. 1. Elsevier, Amsterdam, The Netherlands, pp. 1−22. (head ed.).

Foulger, G.R., 2010. Plates vs Plumes: A Geological Controversy. Wiley-Blackwell, Chichester, UK.

Galilei, G., 1632. Dialogue Concerning the Two Chief World Systems (Originally published in Italian by Fiorenza, Per Gio: Batista Landini) English translation by Stilman Drake in 1953 is available in English at: <http://www.webexhibits.org/calendars/year-text-Galileo.html> (accessed 19.07.19.).

Hamilton, W.B., 2002. The closed upper-mantle circulation of plate tectonics. In: Stein (eds), Plate Boundary Zones, Geodynamic Series 30, American Geophysical Union, Washington D.C., p. 359−410.

Helmholtz, H., 1878. *The Facts of Perception.* (Address in German given during the anniversary celebrations of the Friedrich Wilhelm University in Berlin, in 1878; reprinted in Vorträge und Reden, vol. II, pp. 215−247, 387−406) Available in English at: <https://www.marxists.org/reference/subject/philosophy/works/ge/helmholt.htm> (accessed 20.07.19.).

Hofmeister, A.M., 2019. Measurements, Mechanisms, and Models of Heat Transport. Elsevier, Amsterdam.

Hofmeister, A.M., Criss, R.E., 2016. Spatial and symmetry constraints as the basis of the Virial theorem and astrophysical implications. Can. J. Phys. 94, 380−388.

Hofmeister, A.M., Criss, E.M., 2018. How properties that distinguish solids from fluids and constraints of spherical geometry suppress lower mantle convection. J. Earth Sci. 29, 1−20. Available from: https://doi.org/10.1007/s12583-017-0819-4.

Hofmeister, A.M., Criss, R.E., Criss, E.M., 2018. Verified solutions for the gravitational attraction to an oblate spheroid: implications for planet mass and satellite orbits. Planet. Space Sci. 152, 68−81. Available from: https://doi.org/10.1016/j.pss.2018.01.005.

Jones, D. 1999. Citing to infinity. Nature 402, 600.

Kuhn, T.S., 1970. The Nature of Scientific Revolutions. Univ. Chicago Press, Chicago, IL.

Truesdell, C., 1984. An Idiots's Fugitive Essays on Science. Springer-Verlag, New York, NY.

Appendix

Conventions, abbreviations, and variables used

Conventions used

d	differential or total derivative
∂	partial derivative
δ	very small change; delta function
Δ	large change
e	mathematical function; the number 2.718281828...
π	geometrical constant; the number 3.141592654...
$\zeta(3)$	Riemann Zeta function of 3 (Apéry's constant), 1.202056903159594...
f, f^*	some unspecified function, when not a subscript (F in Chapter 5)
i	root of -1, when not a subscript
f, i	when used as subscripts, refer to the initial and final states
g	gravity, when used as a subscript
j, k, n	used in summations, refers to an integer
$\rightarrow$	as an overbar, indicates vectors with magnitude
$\wedge$	as an overbar, indicates direction only
0	initial or reference value (commonly subscripted)
∞	limit of infinity
$^\circ$	degrees, superscripted

Abbreviations

d	dimensions (not italicized)
LHS	left hand side (of an equation)
RHS	right hand side (of an equation)
EM	electromagnetic
IR	infrared
UV	ultraviolet
vis	visible
rfrt	refracted
ln	natural logarithm

log	common logarithm
exp	exponential function
erf	error function
erfc	complementary error function
tan	tangent, tangential
LFA	laser-flash analysis
SEM	scanning electron microscopy
Gr	dimensionless Grashof number
Ra	dimensionless Rayleigh number
Le	dimensionless Lewis number
Sc	dimensionless Schmidt number
V.T.	Virial Theorem of Clausius
RE, R.E.	rotational kinetic energy
KE, K.E.	kinetic energy
LE, L.E.	linear kinetic energy
TE	total energy
MOR	mid-ocean ridge
LVZ	low velocity zone
UM	upper mantle
TZ	transition zone
LM	lower mantle
CMB	core mantle boundary
D″	region above the CMB
CAI	calcium aluminate inclusions
I, III	metallic meteorite types
CV, CB	carbonaceous meteorite types
LL, L, H	ordinary chondrite meteorite types
K, U, Th	heat producing radioactive elements

Physical constants and their values

c	speed of light$=2.99792458 \times 10^8$ m s^{-1}
h	Planck's constant$=6.62609923 \times 10^{-34}$ J s
k_B	Boltzmann's constant$=1.380658 \times 10^{-23}$ J atom^{-1} K^{-1}
N_a	Avogadro's number$=6.0221 \times 10^{23}$ atoms mol^{-1}
R_{gc}	gas constant$=N_a k_B = 8.314$ J mol^{-1} K^{-1}
σ_{SB}	Stephan-Boltzmann constant$=5.670 \times 10^{-8}$ W m^{-2} K^{-4}
G	gravitational constant$=6.67 \times 10^{-11}$ m^3 kg^{-1} s^{-2}

Useful conversion factors

1 atm$=101325$ J m^{-3}

hc/k_B=1.43877 (for use with wavenumbers)

Fitting constants

a, b, g, h	fitting parameters in Chapter 6
F, G, H	fitting parameters for thermal diffusivity data
b	fitting constant in many equations
A, b, H, k	fitting parameters for thermal conductivity data

Rs   residual in fitting
χ^2 chi squared, indicates goodness of fits

Subscripts

wat   water
mtl   mantle
sfc   surface
lng   length
vol   volume
self   self-potential

Variables

$a\ddagger$	scale length
A	absorption coefficient
Å	area
b	an unknown constant; fitting parameter; attenuation
B	bulk modulus
C	heat capacity (per mole and per volume are apparent from units in the equations)
c_m	specific heat* (heat capacity per mass)
d	grain size or depth in Chapter 7
D	thermal diffusivity, sometimes D_{heat} for clarity
D_∞	thermal diffusivity, long length limit
D_m, D_{mass}	mass diffusion coefficient
$\mathfrak{J}$	flux in energy (Joules per time per area)
F	Helmholtz free energy; also fitting parameter for $D_{heat}(T)$
F_g	Force of some type, generally subscripted
g	gravitational acceleration, when not subscripted
G	Gibbs function (free enthalpy); also fitting parameter for $D_{heat}(T)$
G_S	Shear modulus
G_{sfc}	Gradients in temperature (Chapter 6)
h	layer height (Chapter 4)
H	enthalpy; also fitting parameter for $D_{heat}(T)$
I	intensity or moment of inertia (Chapters 5 and 10)
J	reduced moment of inertia
K	thermal conductivity (used by Fourier)
l	path length
L	length or angular momentum in Chapter 4
M, m	mass (sometimes per mole)
N	number of atoms, molecules, or moles
p_x	momentum; subscripts indicate the direction
P	pressure
Q	applied external heat or heat loss (Chapter 7 appendix)
R	body radius
r	radius in cylindrical coordinates
s	radius in spherical coordinates
S	entropy
t	time
T	temperature*
T_i, T_f	period in Chapter 10
U_g	gravitational potential

u	speed or velocity; dummy variable in integrals
V	volume
x,y,z	Cartesian, orthogonal directions
X,Y,Z	directions in Chapter 7
Z	number of formula units
α	thermal expansivity
β	compressibility
ε	coefficient of friction
γ_{th}	Grüneisen parameter
γ_i	mode Grüneisen parameter
η	dynamic viscosity (sometimes called absolute viscosity)
κ	surface conductance in Chapter 7 appendix
λ	wavelength or lumped parameter in Chapter 7 appendix
Λ	mean free path
ν	frequency
θ, ϕ	angle variables or dummy variables
Π	power
ρ	density
σ_{xy}	stress, with directions
τ	time constant or lifetime or relaxation time (note subscripts)
υ	kinematic viscosity
ϖ	lumped parameter
ω	circular frequency
ψ	internal heat generation per volume (Chapters 2, 6, 8, and 9)
ψ^*	internal heat generation per mass (Chapter 2)
Ψ	specific power, internal heat generation per mass (Chapter 10)

Index

Note: Page numbers followed by "*f*" and "*t*" refer to figures and tables, respectively.

Printed in the United States
By Bookmasters